The Theory of Anisotropic Elastic Plates

Mathematics and Its Applications

Volume 476

The Theory of Anisotropic Elastic Plates

by

Tamaz S. Vashakmadze

Vekua Institute of Applied Mathematics (VIAM),
Faculty of Applied Mathematics and Computer Science
of Javakhishvili Tbilisi State University,
Georgia

SPRINGER-SCIENCE+BUSINESS MEDIA, B.V.

A C.I.P. Catalogue record for this book is available from the Library of Congress.

ISBN 978-90-481-5215-5 ISBN 978-94-017-3479-0 (eBook)
DOI 10.1007/978-94-017-3479-0

Printed on acid-free paper

This monograph is dedicated to the memory of

Solomon Michlin, Ilya Vekua and my parents

Anna and Sergy

Contents

Foreword

The main purpose of this work is construction of the mathematical theory of elastic plates and shells, by means of which the investigation of basic boundary value problems of the spatial theory of elasticity in the case of cylindrical domains reduces to the study of two-dimensional boundary value problems (BVP) of comparatively simple structure.

In this respect in sections 2–5 after the introductory material, methods of reduction, known in the literature as usually being based on simplifying hypotheses, are studied. Here, in contradiction to classical methods, the problems, connected with construction of refined theories of anisotropic nonhomogeneous plates with variable thickness without the assumption of any physical and geometrical restrictions, are investigated. The comparative analysis of such reduction methods was carried out, and, in particular, in section 5, the following fact was established: the error transition, occuring with substitution of a two-dimensional model for the initial problem on the class of assumed solutions is restricted from below.

Further, in section 6, Vekua's method of reduction, containing regular process of study of three-dimensional problem, is investigated. In this direction, the problems, connected with solvability, convergence of processes, and construction of effective algorithms of approximate solutions are studied.

The investigation of these different methods of reduction and application of some necessary representations, obtained in sections 2–3, made it possible to construct in section 7 the theory of elastic plates with the following values:

a) simplicity and physical evidence, based on hypotheses and the possibility of investigating the problems of error estimation and convergence of the process, characterizing methods, containing regular processes, are combined ibid;

b) the differential operator, corresponding to this theory, is factorized in such a way that the search for an approximate solution of the initial three-dimensional problem is carried out by means of a parallel procedure, i.e., by inversion of simple uniform operators, being the subject of investigation in two-dimensional theory of elasticity.

In section 8 the application of some results from sections 3, 4, 6, 7 is represented in some (more general) problems of the theory of elasticity: 1) three-dimensional problems for unbounded cylindrical bodies , from the point of view of the application of projective methods from section 6, are studied with substantiation of Saint-Venant principle; 2) problems for domains of quasi-cylindrical form (corresponding to shell bodies) are considered ibid., and the methods of calculation, based on Schwarz alternative technique, are proposed; 3) algorithms for problems in the case of thick orthotropic plates are constructed and realized when a horizontal section D appears to be single or bi-connected boundary domain ; 4) the methods set forth are extended to dynamical problems of thermoelasticity; 5) two-dimensional models for multi-layer plates are suggested; 6) the possibility of the application of methods, known in literature as the theory of cracks and problems of magnetoelasticity for plates, constructed in sections 2–4 is discussed.

In section 9 we consider the application of some results of Chapter I when piezoelectric and electric conductive thermodynamic elastic plates are anisotropic and nonhomogeneous.

In section 10 research designs and a list of some unsolved mathematical problems of non-linear solid mechanics are given. Here also refined theories are constructed for some linear BVPs of thermoelasticity and methods are developed for solving three-point operator equations arised and in particular for Vekua system when Ω_h are nonhomogeneuos with variable thickness of plates or shallow shells.

In the section 11 there are briefly described results for developing some parts of section 6 in the case of non-shallow nonhomogeneous isotropic elastic shells.

In chapter III methods are considered for the approximate solution of two-dimentional boundary value problems for the system of differential equations, which also arise as a necessary step for solving the BVPs for chapter I and II.

In section 12 methods are considered for solving two-dimensional BVPs for strongly-elliptical differential systems.

In sections 13–16 a method is considered for solving a one-dimensional BVP for second order nonlinear ordinary differential equations with Newtonian boundary conditions .

The section 17 has applicable character and here are given numerical results and correspondently programs of design some concrete one and two-dimentional BVP by methods, developting in sections 12–16.

The material of this investigation is based,in particular, on the monographs *Package of applied programs of designing spatial constructions [1982]*, *Some problems of mathematical theory of anisotropic elastic plates [1986]*, as well as on some of the author's articles published between 1964 to 1996.

The material of this book was reported systematically in the Seminar of VIAM from 1968 till 1997, on VI, VII, VIII, X, XI Symposiums of ISIMM and at the conferences Boundary integral equations, problems, programms, numerical realizations (Puschino–Moscow 1984–1989). Some part of this monograph contains this special course of lectures for students of Tbilisi State University and in particular in September 1997 the author read the cycle of lectures on *Tbilisi International Centre of Mathematics and Informatics(TICMI)*.

The author wishes to express his thanks to all participants of these meetings especially to Sergei Ambartsumyan, Grigori Barenblatt, Oleg Belotsercovski, Aleksander Chekin, David Gordeziani, George Jaiani, Allan Jeffrey, Aleksander Khvoles,Karl Kirchgässner, Ekkehart Kröner, Michail Lazarev, Thomas Leviňski, Leo Magnaradze, George Manjavidze, Hamlet Meladze, Sitiro Minagava, Lev Nikitin, Petr Perlin, Vladimir Petviashvili, Jose Rodrigues, Wolfgang Schneider, Hans Troger, Natali Vadchva, Ram Vadchva, Iosif Vorovich, Wolfgang Wendland, Leonidas Xantis, Vakhtang Zgenti, Franz Ziegler, Henrick Zorski.

The author takes this opportunity to express his thanks to a group of mathematicians from the Vekua Institute of Applied Mathematics, especially to Mary Bitsadze, Tsitsino Gabeskiria, Iurii Morozov, Aliki Muradova, Eka Shavlakadze, Tamar Siboshvili,Diana and Tamriko Vashakmadze for their careful revision of the manuscript and valuable contributions to the improvement of this work.

The author wishes to thank the ISF (Foundation of George Soros), Grant KZB 200, and the Rector of Tbilisi State University, Roin Metreveli, for their cardinal support.

Lastly, it is a pleasure to express my deep gratitude to Ivo Babušhka, Philippe Ciarlet, Michiel Hazewinkel, Anatolii Fomenko, Tengiz Meunargia.

Tamaz Vashakmadze,
March, Octouber 1997.

Notations

Below we introduce the list of basic notations which will be used afterwards.

The repetition of an the index denotes summation; small Latin and Greek indices assume the values of 1, 2, 3 and 1, 2 accordingly, unless otherwise stipulated.

In a reference to a subsection of a section the first number denotes the number of the section, the second one denotes the subsection.

$x = (x_1, x_2, x_3)$ or (x, y, z) are rectangular Cartesian coordinates

$x = (r, \varphi, z)$ are cylindrical coordinates

$\Omega_h(x) = D(x_1, x_2) \times]h^-, h^+[$, $(h^\pm = h^\pm(x_1, x_2))$ is a three-dimensional cylindrical domain

$D(x_1, x_2)$ is a two-dimensional open connected domain

$\bar{\Omega}_h$ is a closure, $\partial\Omega_h$ is the boundary of the domain Ω_h

$S^\pm = D(x_1, x_2) \times \{h^\pm\}$ are surfaces

$S = \partial D \times [h^-, h^+]$ is a lateral surface of Ω_h

$2h = h^+ - h^-$ is thickness

$\hbar = \frac{1}{2}(h^- + h^+)$ is an middle surface of plate Ω_h

$h^- = h_1$; $h^+ = h_2$

$u = (u_1, u_2, u_3)^T$ is a displacement vector (in a Cartesian system)

$\sigma = \{\sigma_{ij}\} = (\sigma_{11}, \sigma_{22}, \sigma_{33}, \sigma_{32}, \sigma_{31}, \sigma_{21})^T$ is a symmetric stress tensor

$\varepsilon = \{\varepsilon_{ij}\}$ is a tensor finite and $e = \{e_{ij}\}$ is a tensor of small deformations

E is modulus of elasticity

ν is Poisson's ratio

λ, μ are Lame coefficients

$E = \mu(3\lambda + 2\mu)/(\lambda + \mu)$, $\nu = \lambda/2(\lambda + \nu)$

$\lambda = E\nu/(1 - 2\nu)(1 + \nu)$, $\nu = E/2(1 + \nu)$

$D = 2Eh^3/3(1 - \nu^2)$ is cylindrical rigidity with bending

$\lambda^* = 2\lambda\mu(\lambda + 2\mu)^{-1}$

$\nabla^2 = \partial_{11} + \partial_{22} + s_{33} = \partial_{ii}$ is the three-dimensional Laplacian operator

$\Delta = \partial_{\alpha\alpha}$ is the two-dimensional Laplacian

$\dfrac{\partial}{\partial x_i} = \partial_i =_{,i}$ is a derivative by x_i, $\dfrac{\partial}{\partial t} = \partial_t$ is the derivative with respect to time

δ_{ij} is Kronecker's symbol

A and B are symmetric matrices of 6×6

M_α are bending moments

$M_{\alpha\beta}$ are twisting moments

$Q_{\alpha 3}$ are generalized shearing forces

$T_{\alpha\beta}$ are surfaces efforts

u^*_α are components of a rotation of normals

u^*_3 is Reissner's averaged deflection

$\overline{u}_\alpha$ are horizontal averaged components of u

$R = R(x^1, x^2, x^3)$is a radiusvector at a point $x = (x^1, x^2, x^3)$

$R_i = \partial_i R$ is a local basis

$R^i R_j = \delta^i_j = \begin{cases} 1, i = j, \\ 0, i \neq j, \end{cases}$ is a biorthogonal basis of R_j

$q_{ij} = q_{ji} = R_i R_j$ are covariant components of metric tensors of a space
$q^{ij} = q^{ji} = R^i R^j$ are contravariant components of metric tensors of a space
$q_i^j = q_j^i = R_i R^j = \delta_i^j$ are mixed type components of metric tensors of a space
$ds^2 = dR^2 = g_{ij} dx^i dx^j$ is a square of an element of length
$g = det\{q_{ij}\} > 0$
σ^{ij},σ_{ij} are contravariant and covariant components of stress tensors
e^{ij},e_{ij} are contravariant and covariant components of deformation tensors
u^i,u_i are contravariant and covariant components of displacement vectors
∇^i,∇_i are contravariant and covariant derivatives respect x^i or x_i
$\sqrt{\frac{g}{a}} = 1 - 2Hx^3 + K(x^3)^2 \equiv \Theta$
k_α are main curvature lines
$H = \frac{1}{2}(b_1^1 + b_2^2) = \frac{1}{2}(k_1 + k_2)$ is a middle curvature line
$K = b_1^1 b_2^2 - b_1^2 b_2^1 = k_1 k_2$ is Gaussian curvature line of a middle-surface S_h of a shell Ω_h
S is lateral surfaces of Ω_h
$r(x^1, x^2)$ and $n(x^1, x^2)$ are a radiusvector and an unit normal to $S \ni (x^1, x^2)$
x^1,x^2 are Gaussian parameters of S_h
$a_{\alpha\beta}$,$b_{\alpha\beta}$ are coefficients of first and second square forms of S_h respectively:
$a_{ij} = a_{ji} = r_i r_j$, $a_i^j = r_i r^j$, $r_3 = r^3 = n$
$b_{ij} = b_{ji} = -r_i n_j = -n_i r_j$, $b_i^j = -r^j n_i$, $n_\alpha \equiv \partial_\alpha n$
$a = a_{11} a_{22} - a_{12}^2 > 0$
BVP– boundary value problem, PDEs–partial differential equations.

L'esprit humain éprouve moins (des
difficultés) à se parter en avant,
qu'à se replier sur lui-même.
Laplace.

Introduction

1. The Basic Equations and Boundary Value Problems in the Theory of Elasticity of Anisotropic Bodies

1.1. The Basic Equations

The information necessary for the theory of elasticity in the case of anisotropic nonhomogeneous bodies is given below; it is based on the papers by Ciarlet [1988]; Knops , Payne [1971]; Lekhnitski [1977]; Love [1959]; Muskhelishvili [1966]; Novožilov [1951]; Parton, Perlin [1981] .

The equlibrium equation of an elastic body has the form:

$$\partial_j(\sigma_{ij} + \sigma_{kj} u_{i,k}) = f_i, \quad x \in \Omega_h = D(x_1, x_2) \times]h^-(x_1, x_2), h^+(x_1, x_2)[. \tag{1.1}$$

Boundary conditions:

$$l^{\pm}[u] = \sigma_{i3} + \sigma_{j3} u_{i,j} - d_{ij} u_j = g_i^{\pm}, \quad x \in S^{\pm} = D \times \{h^{\pm}\}, \tag{1.2}$$

$$l(\partial_1, \partial_2, \partial_3)[u] = g, \qquad x \in S = \partial D \times]h^-, h^+[. \tag{1.3}$$

In connection with the domain's Ω_h structure, the condition on surfaces $S^{\pm}$ is singled out. Besides, without concretizing the type of boundary linear operator l, condition (1.3) is considered to be one of the classic boundary conditions , $d = \{d_{ij}\} = d^T \geq 0$ is a matrix with constant coefficients.

The connection between tensor ε and vector u lies in:

$$\varepsilon_{ij} = \frac{1}{2}[u_{i,j} + u_{j,i} + A_{ij}], \quad A_{ij} = u_{k,i} u_{k,j}, \quad x \in \bar{\Omega}_h. \tag{1.4}$$

The generalized Hooke's law is:

$$\sigma = B\varepsilon, \quad \varepsilon = A\sigma, \quad x \in \bar{\Omega}_h. \tag{1.5}$$

It is assumed, that in any point of body $\bar{\Omega}_h$, even if one plane of elastic symmetry spreads, which is parallel to the coordinative plane Oxy, that is to say, matrices A and B depend on thirteen indepedent elastic constants at most.

Hooke's law in an isotropic case is:

$$\sigma_{ij} = \lambda \varepsilon_{kk}[u]\delta_{ij} + 2\mu\varepsilon_{ij}[u], \quad \varepsilon_{ij} = \frac{1}{2\mu}\sigma_{ij} - \frac{\lambda\delta_{ij}}{2\mu(3\lambda + 2\mu)}\sigma_{kk}. \tag{1.6}$$

The equilibrium equation in terms of the displacemant vector in the case of a small deformation (or, when the values A_{ij} and $\sigma_{ij} u_{j,k}$ are lacking) takes the form:

$$\mu \nabla^2 u + (\lambda + \mu) \operatorname{grad} \operatorname{div} u = f, \quad x \in \Omega_h. \tag{1.7}$$

The equilibrium equation in terms of displacement vector in anisotropic linear case is:

$$L_{ij}(\partial_1, \partial_2, \partial_3)u_j = f_i, \quad x \in \Omega_h. \tag{1.8}$$

Here

$$L_{\alpha\alpha} = b_{\alpha\alpha}\partial_\alpha^2 + 2b_{\alpha 6}\partial_{12} + b_{66}\partial_{3-\alpha}^2 + b_{6-\alpha,6-\alpha}\partial_3^2,$$

$$L_{12} = L_{21} = b_{16}\partial_{11} + (b_{12} + b_{66})\partial_{12} + b_{26}\partial_{22} + b_{45}\partial_3^2,$$

$$L_{\alpha 3} = L_{3\alpha} = [(b_{36} + b_{45})\partial_{3-\alpha} + (b_{\alpha 3} + b_{6-\alpha,6-\alpha})\partial_\alpha]\partial_3,$$

$$L_{33} = b_{55}\partial_{11} + 2b_{45}\partial_{12} + b_{44}\partial_{22} + b_{33}\partial_3^2.$$

Relations (1.1) – (1.5) define the basic boundary value problems of statics of elastic body; thus, the stress–strain state Ω_h should be determined when the body is exposed to volume forces f, and exterior forces g and $g^\pm$ applied to its surface. These problems will be the subject of our investigation below.

1.2. The classification of transition methods, literature review

Further, after Vorovich [1965], we shall give the classifications of methods of transition from three-dimensional problems of the theory of elasticity (1.1) – (1.5) to two-dimensional differential problems and introduce the notion of error transition.

All methods of solution of three-dimensional problems of the theory of elasticity can be conventionally divided into direct methods immediately in three-dimensional formulation (methods of the potential theory, integral transformations, series), which were developed mainly for domains with sufficiently smooth boundaries, or of canonical shape (sphere, semisphere, space, space layer etc., see Kupradze, Gegelia, Basheleishvili, Burchuladze [1976]; Lourie [1955]; Novački [1970]) and into methods of reduction, reducing the initial problems to the problems of low dimension. *Methods of reduction are more often applied to the case of elastic bodies with small thickness and with surfaces containing angular (conic) points and lines (edges).*

Methods of reduction, in their turn, can be divided into two main groups after the classification, accepted in literature.

A) All methods based on physical and geometrical assumptions can be combined within this group. By means of these methods a three-dimensional problem is substituted by a two-dimensional one and the last one appears to be the adequate for the initial one. *We shall term two-dimensional models of this group finite models, or refined theories in the wide sense, including the classic Kirchhoff–Love theory together with the problem of the definition of a generalized plane stressed state, the system of nonlinear differential equations of von Karman, dynamic models of Rayleigh–Lamb's type, the refined theories of plate bending in proper sense, etc.*

B) We refer the methods containing the regular process of investigation of three-dimensional problems by means of the sequence of two-dimensional ones to this group. We shall term two-dimensional models of this group reduced (cut) systems in the case where from the countable set of equations we limit ourselwes to the finite number.

The problem of transition (from multi-dimensional problems of the theory of elasticity in the case of bodies with small thickness to the problems of lower dimension) has about two hundred years of history; its investigation was encouraged by **Chlandni's** experiments, referring to nodal figures of vibrating plates. The earlier period of development of the theory of plates and shells with detailed analysis and the richest bibliography is given in the well known monographs by Love [1959]; Muskhelishvili [1968]; Timoshenko [1957], Todhunter, Pearson [1886,v.I,1893,v.II]. The detailed interpretation of initial tendencies, connected with the construction of two-dimensional differential problems for the bodies with small thickness, with a rather comprehensive (in our opinion) bibliography is given in survey articles by Babuška, Li [1992]; Bourquin, Ciarlet [1989]; Dudchenko, Lourier, Obraztsov [1984]; Galinsh [1967]; Grigoliuk, Selezov [1973]; Kohn, Vogelius [1984]; Vorovich [1965]; Yeh Kai-yuan, Rimrott [1993]. The results of the last few years, obtained in the theory of plates and shells, are reflected in monographs by the following authors: Ambartsumian [1967]; Antman [1995]; Berdichevski [1973]; Ciarlet [1990,1997]; Ciarlet, Rabier [1980]; Destyunder, Salaun [1996]; Donnel [1976]; Goldenveizer [1976]; Lukašiewicz [1979]; Nečas, Hlaváček [1981]; Panč [1975]; Vekua [1985]; Washizu [1982].

Some key problems of the theory of elasticity in the case of cylindrical domains of finite length are considered in this work. Nevertheless, in subsection 8.1 the problems of construction of corresponding two-dimensional models in the case of semi-infinite cylindrical bodies are examined as the supplement of Vekua's method. In this respect it is necessary to stress the following. As is known, the reduction of spatial problems in the case of semi-infinite cylindrical rods to the plane problems of the theory of elasticity is connected with the principle introduced by Saint-Venant [1855]. The grounds of this principle were started with the works of Lax [1957]. The first substantial results were obtained by Toupin [1965]. The rigorous mathematical statement and the proof of Saint-Venant's principle were shown in this paper. Subsequently, many of FSU and foreign scientists (e. g. see Oleinik, Iosifian, Shamaev [1990]; Parton, Perlin [1981] ch. III, §2, Trabucho, Viaño [1988]) dealt with this problem.

Been following Vorovich [1965], let us *introduce the notion of error transition, appearing in the process of reduction of three-dimensional problems of the theory of elasticity (1.1) - (1.5) to two-dimensional differential problems.*

As has already been stressed, in the case of semi-infinite cylindrical domains the advanced method of energy inequalities gives us the opportunity of substantiation of transition of three-dimensional problems to two-dimensional ones, obtaining the values for error transition.

On the other hand, in case of elastic plates (shells) the situation is quite different, to be more exact as it was remarked in the paper by Vorovich [1965]: "the transition from three-dimensional problems to two-dimensional problems makes up the main content of the theory of plates and shells — practically, the most important section in the theory of elasticity. Still, from the *mathematical point of view this problem is almost unstudied.* That is why it is quite expedient, now, to draw the attention not only of mathematicians, but, "elasticians", as well, to this important and interesting problem", or " It would be even more important to have bounds on the errors of the solutions of the approximate equations, as compared with the solutions of the three-dimensional equations. *It would seem that this question is at present still a completely open problem*" (Koiter [1971]).

We should add that engineering-physical assumption in the nature of phenomena, occuring in elastic bodies of small thickness, and the elaboration of the great formal mathematical apparatus, in this respect, were the achievements of science and technique in theoretical and, particularly, in applied studies; nevertheless, a number of mathematical problems, formulated in papers and related directly to the creation of a rigorous substantiated theory of plates and shells, seem to be still unsolved. These problems are:

I. The construction of finite models of plates without simplifying hypotheses as the methods of group A;

II. The investigation of problems of convergence, the estimation of error transition and effective solvability of two-dimensional models of group B, containing deterministic process.

Chapter I. Refined Theories

2. The Method of the Construction of Refined Theories without Simplifying Hypotheses

A number of authors (e.g., see Ambartsumian [1967], p.19; Khoroshun [1978], p.3) consider the difficulty of the estimation of error transition to be the main fault in the method of hypotheses and it, evidently, accounts for the lack of a rigorous substantiated rule by means of which the refined theories would have followed from three-dimensional problems of the theory of elasticity, as was the case in Kirchhoff's classic theory (e.g., see Goldenveiser and Koloss [1965]; Friedrichs [1950]; Koiter [1971]; Morgenstern [1959]; Prager, Singe [1947]; Shoychet, see Morozov [1978], ch.III and others).

Some advance has been outlined in the construction of refined theories without simplifying hypotheses for the latter years.

Dodnikov and Nazarov [1982] deduced equations corresponding to Reissner's refined theory by means of asymptotic method according to E. and F. Cosserat's equations of three-dimensional couple–stress theory of elasticity. At the same time the additional algebraic relations between elastic constants should be fulfilled. The analogous concept is developed in Palmov's work [1983].

The special place within the range of these investigations is taken by works of Ciarlet [1997]; Destyunder, Salaun [1996] and their collaborators. They constructed two-dimensional , corresponding to different theories of plates on the basis of the asymptotic method. Although these results are temporary, if the substantiation of the corresponding two-dimensional models is connected, they stimulate obtaining a transition estimation of asymptotic character. At the same time we should note that the problem of construction of finite models without simplifying hypotheses of physical and geometrical character based on restrictions, admitting the rigorous mathematical interpretation in terms of the theory of functions, was stressed for the first time in these papers. The important conclusions which I want to underline regarding these works are the following:

(i) *These results contribute only to asymptotic type of estimation of transition error.*

(ii) *Each of the methods of construction corresponds to a certain concrete mathematical model, which means there is no mathematical formal rule common for construction of any of the refined theories.*

(iii) *In these works the value characterizing the transition approximation of the initial spatial problem by the mathematical model is represented as a remainder term of the asymptotic series. There is no appropriate effective estimation of the transition error.*

To a great extent, in our view, Berdichevski [1973] advanced the investigation of this problem. He deduced the solvable equations with respect to bending, corresponding to the models of Kirchhoff, Reissner, Donnel [1976]; evidently, irrespective of Kennard [1953], he used the idea of asymptotic equivalence of different refined theories and distribution of displacement vector with respect to the plate thickness. The dynamic models of Rayleigh–Lamb type and the questions of asymptotic exactness are studied in this work Berdichevski [1973], too. In subsection 3.2 we shall return to the discussion of this case again.

In the works of Vashakmadze [1981a,1981b,1985,1988] the method of construction of refined theories of group A and new analogous models were elaborated without any simplifying hypotheses; all this was based on special relations called *exact nonlocal representations* in the main characteristics of a stress–strain state of elastic plates, originated from three-dimensional problems of the theory of elasticity (1.1) – (1.5). At the same time, for *the error vector, characterizing the approximation of initial three-dimensional problems by two-dimensional ones, the analytical expressions were found.* In the linear case it is the sum of remainder terms of the quadrature formulae of Simpson, trapezoids, or Gauss.

The following subsections of this section are devoted to the statement of these results.

2.1. Isotropic case

For clearness let us consider first the homogeneous isotropic medium and use the equations (1.1) – (1.3), (1.5), (1.6). In addition, within this paragraph we consider $h = -h^- = h^+ = \text{const}$, d to be a zero matrix.

Let us introduce the averaged values (with respect to plate thickness) values , connected with components of the displacement vector and stress tensor. Analogously to the linear case we shall call this procedure the Reissner averaging [1944]:

$$M_\alpha = \int_{-h}^{h} t(\sigma_{\alpha\alpha} + \sigma_{k\alpha}u_{\alpha,k})dt, \tag{2.1}$$

$$M_{\alpha\beta} = \int_{-h}^{h} t(\sigma_{\alpha\beta} + \sigma_{k\alpha}u_{\beta,k})dt, \tag{2.2}$$

$$Q_{ij} = \int_{-h}^{h} (\sigma_{ij} + \sigma_{kj}u_{i,k})dt, \tag{2.3}$$

$$u_\alpha^* = \frac{3}{2h^3}\int_{-h}^{h} tu_\alpha dt, \tag{2.4}$$

$$u_3^* = \frac{3}{4h^3}\int_{-h}^{h} (h^2 - t^2)u_3 dt, \tag{2.5}$$

$$\psi_\alpha = \frac{1}{2}\int_{-h}^{h}(h^2-t^2)\sigma_{\alpha 3}dt. \tag{2.6}$$

Here and below the integration occurs over the variable $x_3 = t$ if the inverse procedure is not taken into account.

Now let us use formulae (1.5), (1.6). If we integrate these expressions at $i=\alpha$, $j=3$ from 0 to t, further multiply the resulting relation by t, and again integrate from $-h$ to h, then for u_α^* we obtain what follows below.

Thus for rotation of normals we have:

$$u_\alpha^* = -u_{3,\alpha}^* + 2(1-\nu)^{-1}D^{-1}\psi_\alpha - B_\alpha, \tag{2.7}$$

$$B_\alpha = \frac{3}{4h^3}\int_{-h}^{h}(h^2-t^2)A_{\alpha 3}dt.$$

The generalized moments M_α and $M_{\alpha\beta}$ are related to the functions u_3^*, ψ_α and σ_{33} by

$$M_\alpha = -D(\partial_\alpha^2 + \nu\partial_{3-\alpha}^2)u_3^* + \frac{2}{1-\nu}(\psi_{\alpha,\alpha} + \nu\psi_{3-\alpha,3-\alpha})$$
$$+\frac{\nu}{1-\nu}\int_{-h}^{h} t\sigma_{33}dt + M_\alpha^{NL}, \tag{2.8}$$

$$M_{\alpha\beta} = -(1-\nu)Du_{3,\alpha\beta}^* + \psi_{\alpha,\beta} + \psi_{\beta,\alpha} + M_{\alpha\beta}^{NL}, \tag{2.9}$$

where

$$M_\alpha^{NL} = \int_{-h}^{h} t\sigma_{\alpha k}u_{\alpha,k}dt + D(A_\alpha^* + \nu A_{3-\alpha}^* - B_{\alpha,\alpha} - \nu B_{3-\alpha,3-\alpha}),$$

$$M_{\alpha\beta}^{NL} = \int_{-h}^{h} t\sigma_{\alpha k}u_{\beta,k}dt + \frac{1-\nu}{2}D(2A_{\alpha\beta}^* - B_{\alpha,\beta} - B_{\beta,\alpha}),$$

$$A_{\alpha\beta}^* = \frac{3}{4h^3}\int_{-h}^{h} tA_{\alpha\beta}dt, \qquad A_{\alpha\alpha}^* = A_\alpha^*.$$

Proof. From (1.6) for ε_{33} there follows:

$$\varepsilon_{33} = (\lambda + 2\mu)^{-1}(\sigma_{33} - \lambda\varepsilon_{\alpha\alpha}).$$

Then for M_α we have:

$$M_\alpha = \int_{-h}^{h} t(\lambda\varepsilon_{kk} + 2\mu\varepsilon_{\alpha,\alpha} + \sigma_{k\alpha}u_{\alpha,k})dt$$

$$= \int_{-h}^{h} t[(\lambda + 2\mu)\varepsilon_{\alpha\alpha} + \lambda\varepsilon_{3-\alpha 3-\alpha}$$

$$+\lambda(\lambda + 2\mu)^{-1}(\sigma_{33} - \lambda\varepsilon_{\beta\beta}) + \sigma_{k\alpha}u_{\alpha,k}]dt$$

$$= \int_{-h}^{h} t\Bigg[(\lambda^* + 2\mu)u_{\alpha,\alpha} + \lambda^* u_{3-\alpha,3-\alpha}$$

$$+\frac{1}{2}((\lambda^* + 2\mu)A_\alpha + \lambda^* A_{3-\alpha}) + \lambda(\lambda + 2\mu)^{-1}\sigma_{33}\Bigg]dt.$$

As

$$u_\alpha(x_1, x_2, t) = u_\alpha(x_1, x_2, 0) - \int_0^t u_{3,\alpha}dt + \frac{1}{\mu}\int_0^t (\sigma_{\alpha 3} - \mu A_{\alpha 3})dt, \qquad (a)$$

for M_α now we have the expression:

$$M_\alpha = -\int_{-h}^{h} t\Bigg[\int_0^t \Big((\lambda^*\Delta u_3 + 2\mu\partial_\alpha^2 u_3) - \mu^{-1}\Big((\lambda^* + 2\mu)\sigma_{\alpha 3,\alpha} + \lambda^*\sigma_{33-\alpha,3-\alpha}\Big)$$

$$+(\lambda^* + 2\mu)A_{\alpha 3,\alpha} + \lambda^* A_{33-\alpha,3-\alpha}\Big)dt - \frac{1}{2}(\lambda^* + 2\mu)A_\alpha$$

$$-\frac{1}{2}\lambda^* A_{3-\alpha} + \lambda(\lambda + 2\mu)^{-1}\sigma_{33}\Bigg]dt$$

$$+\int_{-h}^{h} t\sigma_{k\alpha}u_{\alpha,k}dt.$$

If we use the identity

$$\int_{-h}^{h} tdt\int_0^t f(t)dt = \frac{1}{2}\int_{-h}^{h}(h^2 - t^2)f(t)dt, \quad \forall f \in L(-h,h),$$

for M_α these immediately follows:

$$M_\alpha = -\frac{1}{2}\int_{-h}^{h}(h^2-t^2)\left[\lambda^*\Delta u_3 + 2\mu\partial_\alpha^2 u_3 - \mu^{-1}\left((\lambda^*+2\mu)\sigma_{\alpha 3,\alpha} + \lambda^*\sigma_{33-\alpha,3-\alpha}\right)\right.$$

$$\left. +(\lambda^*+2\mu)A_{\alpha 3,\alpha} + \lambda^* A_{33-\alpha,3-\alpha}\right]dt$$

$$+\frac{1}{2}\int_{-h}^{h} t\left((\lambda^*+2\mu)A_\alpha + \lambda^* A_{3-\alpha}\right)dt + (\lambda+2\mu)^{-1}\lambda\int_{-h}^{h} t\sigma_{33}dt$$

$$+\int_{-h}^{h} t\sigma_{k\alpha}u_{\alpha,k}dt.$$

If we change λ and μ by ν and E and use (2.5), (2.6), follows (2.8) from the last expression

By analogy, if we use the representation for $M_{\alpha\beta}$:

$$M_{\alpha\beta} = \mu\int_{-h}^{h} t(u_{\alpha,\beta} + u_{\beta,\alpha} + A_{\alpha\beta})dt + \int_{-h}^{h} t\sigma_{k\alpha}u_{\beta,k}dt.$$

and (a), we obtain (2.9).

The generalized intersecting forces $Q_{\alpha 3}(x_1,x_2)$ *and deflection* $u_3^*(x_1,x_2)$ *in case* $(x_1,x_2)\in D$ *satisfy the equations:*

$$Q_{\alpha 3} = -D\Delta u^*_{3,\alpha} + \Delta\psi_\alpha + \frac{1+\nu}{1-\nu}\psi_{\beta,\alpha\beta} + \frac{\nu}{1-\nu}\int_{-h}^{h} t\sigma_{33,\alpha}dt$$

$$+h(g_\alpha^+ + g_\alpha^-) - \int_{-h}^{h} tf_\alpha dt + Q_{\alpha 3}^{NL}, \tag{2.10}$$

where

$$Q_{\alpha 3}^{NL} = \int_{-h}^{h} t\partial_\beta(\sigma_{\alpha k}u_{\beta,k})dt$$

$$+D\left[A^*_{\alpha,\alpha} + \nu A^*_{3-\alpha,\alpha} + \frac{1-\nu}{2}(2A^*_{\alpha 3-\alpha,3-\alpha} - \Delta B_\alpha) - \frac{1+\nu}{2}B_{\beta,\alpha\beta}\right].$$

$$D\Delta^2 u_3^* - \frac{2}{1-\nu}\Delta\psi_{\alpha,\alpha} - \frac{\nu}{1-\nu}\int_{-h}^{h} t\Delta\sigma_{33}dt - \int_{-h}^{h} t\partial_{\alpha\beta}(\sigma_{\alpha k}u_{\beta,k})dt$$

$$-D[(\partial_{11}+\nu\partial_{22})A_1^* + 2(1-\nu)\partial_{12}A_{12}^* + (\nu\partial_{11}+\partial_{22})A_2^* - \Delta B_{\alpha,\alpha}]$$

$$+\int_{-h}^{h}\partial_\alpha(\sigma_{k3}u_{\alpha,k} - \sigma_{\alpha k}u_{3,k})dt$$

$$= h(g^+_{\alpha,\alpha} + g^-_{\alpha,\alpha}) + g^+_3 - g^-_3 - \int_{-h}^{h} t f_{\alpha,\alpha} dt - \int_{-h}^{h} f_3 dt. \tag{2.11}$$

Proof. From equations (1.1), when $i = \alpha$ we have:

$$Q_{\alpha 3} = M_{\beta\alpha,\beta} + h(g^+_\alpha - g^-_\alpha) - \int_{-h}^{h} t f_\alpha dt. \tag{a}$$

Taking into account (2.8) and (2.9) here, it immediately follows that the relations (2.10) are true.

If we use equations

$$M_{\alpha\beta,\alpha\beta} = Q_{\alpha 3,\alpha} + \int_{-h}^{h} t f_{\alpha,\alpha} dt - h(g^+_{\alpha,\alpha} + g^-_{\alpha,\alpha})$$

$$Q_{3\alpha,\alpha} = \int_{-h}^{h} f_3 dt - g^+_3 + g^-_3$$

$$M_{\alpha\beta,\alpha\beta} = Q_{\alpha 3,\alpha} - Q_{3\alpha,\alpha} + \int_{-h}^{h} t f_{\alpha,\alpha} dt - h(g^+_{\alpha,\alpha} + 8 g^-_{\alpha,\alpha}) - g^+_3 + g^-_3 + \int_{-h}^{h} f_3 dt,$$

which follow from (1.1) and expressions (2.8) and (2.9), the truth of representation (2.11) will become evident.

(2.11), as we can see below, corresponds to the first equation in von Karman's nonlinear differential system (e.g., see Ciarlet [1994]).

Now let us start with the construction of the second basic relation, corresponding to the other equation of this system.

Let us assume that

$$\sigma_{\alpha\beta} = (-1)^{\alpha+\beta}(\partial_{3-\alpha}\partial_{3-\beta} F(x_1, x_2, x_3) + \frac{1}{2h} P_{\alpha\beta}(x_1, x_2)). \tag{2.12}$$

Then equations

$$Q_{\alpha\beta,\beta} = \int_{-h}^{h} f_\alpha dt - (g^+_\alpha - g^-_\alpha),$$

will be identically satisfied if functions $P_{\alpha\beta}$ are chosen as follows:

$$P_{\alpha\alpha,\alpha} - P_{\alpha 3-\alpha,3-\alpha} = \Phi_\alpha[u], \tag{2.13}$$

$$\Phi_\alpha[u] = \int_{-h}^{h} [f_\alpha - \partial_\beta(\sigma_{\beta k} u_{\alpha,k})] dt - (g^+_\alpha - g^-_\alpha).$$

Then, let us consider the integral

$$I[\varepsilon] = \frac{1}{2h}\int_{-h}^{h}(\partial_{11}\varepsilon_{22} - 2\partial_{12}\varepsilon_{12} + \partial_{22}\varepsilon_{11})dt,$$

and use formula (1.6).
Then by virtue of (2.12) and (2.13) we shall have:

$$I[\varepsilon]=\frac{1}{2Eh}\int_{-h}^{h}(\partial_{11}\sigma_{22}-\nu_{11}\partial_{11}\sigma_{11}-2(1+\nu)\partial_{12}\sigma_{12}+\partial_{22}\sigma_{11}-\nu\partial_{22}\sigma_{22}-\nu\Delta\sigma_{33})dt$$

$$= \frac{1}{2Eh}\left[\int_{-h}^{h}(\Delta^2 F - \nu\Delta\sigma_{33})dt + (\Delta P_{\alpha\alpha} - (1+\nu)\Phi_{\alpha,\alpha}[u])\right],$$

or

$$\Delta^2 F^* = EI[\varepsilon] + \frac{\nu}{2h}\int_{-h}^{h}\Delta\sigma_{33}dt + \frac{1}{2h}[(1+\nu)\Phi_{\alpha,\alpha}[u] - \Delta P_{\alpha\alpha}],$$

where

$$F^* = \frac{1}{2h}\int_{-h}^{h} F dt.$$

On the other hand, by virtue of formula (2.4) we have

$$I[\varepsilon] = \frac{1}{2h}\int_{-h}^{h}\left[\partial_{11}(u_{2,2} + \frac{1}{2}A_{22}) - \partial_{12}(u_{1,2} + u_{2,1} + A_{12}) + \partial_{22}(u_{1,1} + \frac{1}{2}A_{11})\right]dt$$

$$= \frac{1}{4h}\int_{-h}^{h}(\partial_{11}A_{22} - 2\partial_{12}A_{12} + \partial_{22}A_{11})dt = \frac{1}{2}I[A].$$

Taking into account the last expression, the desired second basic relation will have the form:

$$\Delta^2 F^* = \frac{E}{2}I[A] + \frac{\nu}{2h}\int_{-h}^{h}\Delta\sigma_{33}dt + \frac{1}{2h}[(1+\nu)\Phi_{\alpha,\alpha}[u] - \Delta P_{\alpha,\alpha}], \qquad (2.14_1)$$

or by (2.13)

$$\Delta^2 F^* = \frac{E}{2} I[A] + \frac{\nu}{2h} \int_{-h}^{h} \Delta\sigma_{33} dt$$

$$+\frac{1+\nu}{2h}\Big(\int_{-h}^{h} f_{\alpha,\alpha} dt - (g^+_{\alpha,\alpha} - -g^-_{\alpha,\alpha})\Big)$$

$$-\frac{1+\nu}{2h}\int_{-h}^{h} \partial_{\alpha\beta}(\sigma_{\beta k} u_{\alpha,k}) dt - \frac{1}{2h}\Delta P_{\alpha\alpha}. \qquad (2.14_2)$$

We shall call formulae (2.7) – (2.14) nonlinear *exact nonlocal representations* (e.n.r.) of basic characteristics, introduced by means of (2.1) – (2.6). These formulae make it possible to construct two-dimensional models in case of elastic plates without any simplifying hypotheses.

For this purpose for values ψ_α and $\int_{-h}^{h} t\sigma_{33} dt$ appearing in (2.7) – (2.11) let us use the parametric representation and quadrature formula of Simpson, trapezoid, or Gauss (with one node).

Evidently, we have:

$$\int_{-h}^{h} t\sigma_{33} dt = (1-\gamma)\int_{-h}^{h} t\sigma_{33} dt + \gamma \int_{-h}^{h} t\sigma_{33} dt$$

$$= \frac{1+2\gamma}{3} h^2 [\sigma_{33}(x_1, x_2, h) - \sigma_{33}(x_1, x_2, -h)]$$

$$+(1-\gamma)\rho_{Sm}[t\sigma_{33}] + \gamma\rho_{tr}[t\sigma_{33}]$$

$$= \frac{1+2\gamma}{3} h^2 [\sigma_{33}(x_1, x_2, h)$$

$$+\sigma_{j3}(x_1, x_2, h) u_{3,j}(x_1, x_2, h) - \sigma_{33}(x_1, x_2, -h)$$

$$-\sigma_{j3}(x_1, x_2, -h) u_{3,j}(x_1, x_2, -h)]$$

$$-\int_{-h}^{h} t\sigma_{j3}(x_1, x_2, t) u_{3,j}(x_1, x_2, t) dt$$

$$+((1-\gamma)\rho_{Sm} + \gamma\rho_{tr})[t(\sigma_{33} + \sigma_{j3} u_{3,j})]$$

or

$$\int_{-h}^{h} t\sigma_{33}dt = \frac{(1+2\gamma)}{3}h^2(g_3^+ - g_3^-) + r_1[t\sigma_{33};\gamma], \tag{2.15}$$

$$r_1[t\sigma_{33};\gamma] = ((1-\gamma)\rho_{Sm} + \gamma\rho_{tr})[t(\sigma_{33} + \sigma_{j3}u_{3,j})] - \int_{-h}^{h} t\sigma_{j3}u_{3,j}dt$$

$$= ((1-\gamma)\rho_{Sm} + \gamma\rho_{tr})[t\sigma_{33}]$$

$$-\frac{1+2\gamma}{3}h^2\left[\sigma_{j3}u_{3,j}|_{x_3=h} - \sigma_{j3}u_{3,j}|_{x_3=-h}\right].$$

Here ρ denotes the remainder term of quadrature formula, γ is an arbitrary parameter .

Similarly

$$\psi_\alpha = \frac{h^2(1+2\gamma)}{3}Q_{\alpha 3} + r_2\left[t\int_0^t \sigma_{\alpha 3}dt;\gamma\right]. \tag{2.16}$$

$$r_2\left[t\int_0^t \sigma_{\alpha 3}dt;\gamma\right] = ((1-\gamma)\rho_{Sm} + \gamma\rho_{tr})\left[t\int_0^t \sigma_{\alpha 3}dt\right] - \frac{(1+2\gamma)h^2}{3}\int_{-h}^{h}\sigma_{j3}u_{\alpha,j}dt.$$

Remark 2.1. By using results of theory of quadrature formulae (e.g., see Sard [1949] or Nikolski [1979]), in our case remainder terms take the following form.

Now we consider the case when $\rho = \rho_{Sm}$ is the remainder term of Simpson's rule:

$$\rho_{Sm}[tf(t)] = \int_{-h}^{h} F_{r,Sm}[t;-h,h]\partial_t^r(tf(t))dt, \quad r \le 4.$$

Here

$$F_{r,Sm}[t;-h,h] = \frac{1}{(r-1)!}\left[\frac{(h-t)^r}{r} - \frac{2h}{6}\left(K_r(-h-t) + 4K_r(-t) + K_r(h-t)\right)\right],$$

$$K_r(u) = \begin{cases} u^{r-1}, & u \ge 0, \\ 0, & u < 0. \end{cases}$$

As $F_{r,Sm}[h;-h,h] = 0$, we obtain immediately the unimprovable in Sard's sense estimates of type:

$$|\rho_{Sm}[tf(t)]| = \left|\int_{-h}^{h}(tF_{r,Sm}[t;-h,h] + rF_{r+1,Sm}[t;-h,h])\partial_t^r f(t)dt\right|$$

$$\leq \ (2h)^{r+2} c_{pr} ||\partial^r f||_{L_q(-h,h)},$$

where

$$c_{pr} = \left\{ \int_0^1 \left| \left(t - \frac{1}{2}\right) F_{r,Sm}[t; -0, 1] + r F_{r+1,Sm}[t; 0, 1] \right|^p dt \right\}^{1/p},$$

$$||\partial^r f||_{L_q(-h,h)} = \left\{ \int_{-h}^{h} |\partial^r f|^q dt \right\}^{1/q}.$$

Other cases, when we used the formulae of the trapezoid or Gauss' (with one node), of course, are investigated similarly and the corresponding results will be stated in section 5.

Remark 2.2. For convenience of writing down in remainder terms r let us denote:

$$t\sigma_{33} = \tau_3, \quad t(\sigma_{33} + \sigma_{j3} u_{3,j}) = \bar{\tau}_3, \quad t \int_0^t \sigma_{\alpha 3} dt = \tau_\alpha, \quad t \int_0^t \sigma_{ij} u_{j,k} dt = \overline{\sigma_{ij} u_{j,k}}.$$

Now we return to e.n.r. (2.7) – (2.14) and change their form.

Following the classical scheme (compare e.g. with Reissner [1944]; Panč [1975]; Timoshenko, Voinovski–Krieger [1959]), the functions $u^*_\alpha, M^*_\alpha, M_{\alpha\beta}$ are presented by deflection u^*_3, shearing forces $Q_{\alpha 3}$, data of initial problems (1.1) – (1.5)— known functions f, g as well as by remainder terms.

Components of rotation of normals satisfy the expressions:

$$u^*_\alpha = -u^*_{3,\alpha} + \frac{2h^2(1+2\gamma)}{3D(1-\nu)} Q_{\alpha 3} + R_\alpha[u^*_\alpha; \gamma]; \quad R_\alpha = \frac{2}{D(1-\nu)} r_2[\tau_\alpha; \gamma] - B_\alpha. \quad (2.17)$$

Proof. These e.n.r. follow from (2.7) if we take into account (2.16).

For the bending moments M_α we have:

$$\begin{aligned} M_\alpha = & - D(\partial^2_\alpha + \nu \partial^2_{3-\alpha}) u^*_3 \\ & + \frac{2}{3} h^2 (1+2\gamma)(Q_{\alpha 3,\alpha} + \nu Q_{3-\alpha 3,3-\alpha} - \frac{\nu}{2(1-\nu)} (g_3^+ - g_3^-)) \\ & + \frac{\nu}{1-\nu} \left[\int_{-h}^{h} ((h^2 - t^2)(f_3 - \underline{\partial_\beta(\sigma_{k\beta} u_{3,k})}) - 2t\sigma_{k3} u_{3,k}) dt - r_1[\tau_3; \gamma] \right] \\ & + 2 r_{2,\alpha}[\tau_\alpha; \gamma] + M^{NL}_\alpha, \end{aligned}$$

Let us consider (2.8). The member depending on ψ will be rewritten in the form

$$\frac{2}{1-\nu}(\psi_{\alpha,\alpha} + \nu \psi_{3-\alpha,3-\alpha}) = 2\psi_{\alpha,\alpha} + \frac{2\nu}{1-\nu}(\psi_{\alpha,\alpha} + \psi_{3-\alpha,3-\alpha}),$$

and (2.3), (2.6) and equation (1.1) with $i = 3$ will be used. Then by virtue of formulae (2.15), (2.16) this expression follows.

The summand (underlined) following f_3, evidently, can be represented in the form (compare with (2.16)):

$$\int_{-h}^{h} t dt \int_{0}^{t} \partial_\beta(\sigma_{k\beta} u_{3,k}) dt = \frac{h^2(1+2\gamma)}{3} \int_{-h}^{h} \underline{\sigma_{\beta\delta} u_{3,\beta\delta} dt}$$

$$+\frac{1}{2}\int_{-h}^{h}(h^2-t^2)(\sigma_{\beta\delta,\beta}u_{3,\delta}+\partial_\beta(\sigma_{3\beta}u_{3,3}))dt$$

$$+r_2[\overline{\sigma_{\beta\delta}u_{3,\beta\delta}};\gamma].$$

In its turn, the first (underlined) summand from the last formula based on (2.12) can be represented in the form:

$$\int_{-h}^{h} \sigma_{\beta\delta} u_{3,\beta\delta} dt = 2h(\partial_{11}u_3^*\partial_{22}F^* - 2\partial_{12}u_3^*\partial_{12}F^* + \partial_{22}u_3^*\partial_{11}F^*)$$

$$+u^*_{3,11}P_{11} - 2u^*_{3,12}P_{12} + u^*_{3,22}P_{22}$$

$$+\int_{-h}^{h}(u_{3,\beta\delta}-u^*_{3,\beta\delta})\sigma_{\beta\delta}dt.$$

Thus, consideration of these formulae in the expression for M_α gives:

$$\begin{aligned} M_\alpha =& - D(\partial_\alpha^2 + \nu\partial_{3-\alpha}^2)u_3^* \\ &+ \frac{h^2(1+2\gamma)}{3}\Big[2(Q_{\alpha3,\alpha} + \nu Q_{3-\alpha3,3-\alpha}) \\ &\qquad - \frac{\nu}{1-\nu}(g_3^+ - g_3^- + 4hL[u_3^*, F^*])\Big] \\ &+ \frac{\nu}{1-\nu}\int_{-h}^{h}(h^2-t^2)f_3dt + R_{2+\alpha}[M_\alpha;\gamma], \end{aligned} \tag{2.18}$$

where

$$L[u,v] = \partial_{11}u\partial_{22}v - 2\partial_{12}u\partial_{12}v + \partial_{22}u\partial_{11}v,$$

$$R_{2+\alpha}[M_\alpha;\gamma]$$

$$= M_\alpha^{NL} - \frac{\nu}{1-\nu}\Bigg\{ \int_{-h}^{h} [2t\sigma_{k3}u_{3,k} + (h^2-t^2)(\sigma_{\beta\delta,\beta}u_{3,\delta} + \partial_\beta(\sigma_{3\beta}u_{3,3}))]dt$$

$$+\frac{2}{3}h^2(1+2\gamma)\left[(-1)^{\beta+\delta}u_{3,\beta\delta}P_{\beta\delta} + \int_{-h}^{h}\sigma_{\beta\delta}\partial_{\beta\delta}(u_3-u_3^*)dt\right]$$

$$+r_1[\tau_3;\gamma] + 2r_2[\sigma_{\beta\delta}u_{3,\beta\delta};\gamma]\Bigg\}$$

$$+2r_{2,\alpha}[\tau_\alpha;\gamma].$$

Similarly, owing to (2.9) *we shall have for the moments* $M_{\alpha\beta}$:

$$M_{\alpha\beta} = -(1-\nu)Du_{3,\alpha\beta} + \frac{1}{3}(1+2\gamma)h^2(Q_{\alpha3,\beta} + Q_{\beta3,\alpha}) + R_5[M_{\alpha\beta};\gamma], \quad (2.19)$$

$$R_5[M_{\alpha\beta};\gamma] = M_{\alpha\beta}^{NL} + r_2[\tau_{\alpha,\beta} + \tau_{\beta,\alpha};\gamma].$$

Shearing forces $Q_{\alpha3}$ *satisfy the following equations:*

$$Q_{\alpha3} - \frac{1+2\gamma}{3}h^2\Delta Q_{\alpha3}$$

$$= -D\Delta u_{3,\alpha}^* + \frac{h^2(1+2\gamma)}{3(1-\nu)}\partial_\alpha(g_3^+ - g_3^- + +2h(1+\nu)L[u_3^*,F])$$

$$+h(g_\alpha^+ + g_\alpha^-)$$

$$-\int_{-h}^{h} tf_\alpha dt + \frac{1+\nu}{2(1-\nu)}\int_{-h}^{h}(h^2-t^2)f_{3,\alpha}dt + R_{5+\alpha}[Q_{\alpha3};\gamma], \quad (2.20)$$

where

$$R_{5+\alpha}[Q_{\alpha3};\gamma]$$

$$= Q_{\alpha3}^{NL} + r_2[\Delta\tau_\alpha;\gamma]$$

$$-\frac{1+\nu}{1-\nu}\partial_\alpha\Bigg[\frac{1}{1+\nu}r_1[\tau_3;\gamma] + r_{2,\alpha}[\overline{\sigma_{\beta\delta}u_{3,\beta\delta}};\gamma]$$

$$+\frac{1}{2}\int_{-h}^{h}(h^2-t^2)(\sigma_{\beta\delta,\beta}u_{3,\delta} + \partial_\beta(\sigma_{3\beta}u_{3,3}))dt + \int_{-h}^{h} t\sigma_{k3}u_{3,k}dt$$

$$+\frac{h^2(1+2\gamma)}{3}\Big((-1)^{\beta+\delta}u_{3,\beta\delta}^*P_{\beta\delta} + \int_{-h}^{h}\sigma_{\beta\delta}\partial_{\beta\delta}(u_3-u_3^*)dt\Big)\Bigg]$$

Proof. Correctness of (2.20) follows if we apply the abovegiven procedure to the member $\psi_{\beta,\alpha\beta}$ from (2.10), which contains also an underlined term with an integrand $\partial_\beta(\sigma_{k\beta}u_{3,k})$.

The basic equation which corresponds to the first equation with respect to deflection of von Karman system has the following form:

$$D\Delta^2 u_3^* = \left(1 - \frac{h^2(1+2\gamma)(2-\nu)}{3(1-\nu)}\Delta\right)(g_3^+ - g_3^-)$$
$$+2h\left(1 - \frac{2h^2(1+2\gamma)}{3(1-\nu)}\Delta\right)L[u_3^*, F_*] + h(g_{\alpha,\alpha}^+ + g_{\alpha,\alpha}^-)$$
$$-\int_{-h}^{h}\left(tf_{\alpha,\alpha} - \left(1 - \frac{1}{1-\nu}\Delta(h^2-t^2)\right)f_3\right)dt + R_8[u_3^*;\gamma], \quad (2.21)$$

where

$$R_8[u_3^*;\gamma] = D\left((\partial_{11}+\nu\partial_{22})A_1^* + 2(1-\nu)\partial_{12}A_{12}^* + (\nu\partial_{11}+\partial_{22})A_2^* - \Delta B_{\alpha,\alpha}\right)$$
$$+\int_{-h}^{h}[t\partial_{\alpha\beta}(\sigma_{\alpha k}u_{\beta,k}) - \partial_\alpha(\sigma_{k3}u_{\alpha,k})]dt - \frac{2}{1-\nu}\Delta\int_{-h}^{h}t\sigma_{k3}u_{3,k}dt$$
$$+\int_{-h}^{h}\left(1 - \frac{1}{1-\nu}\Delta(h^2-t^2)\right)(\sigma_{\alpha\beta,\alpha}u_{3,\beta} + \partial_\alpha(\sigma_{\alpha 3}u_{3.3}))dt$$
$$+\left(1 - \frac{2h^2(1+2\gamma)}{3(1-\nu)}\Delta\right)\left[(-1)^{\alpha+\beta}u_{3,\alpha\beta}P_{\alpha\beta} + \int_{-h}^{h}\sigma_{\alpha\beta}\partial_{\alpha\beta}(u_3 - u_3^*)dt\right]$$
$$-\frac{1}{1-\nu}\Delta\left(r_1[\tau_3;\gamma] + 2r_2[\overline{\sigma_{\alpha\beta}u_{3,\alpha\beta}};\gamma]\right);$$

Proof. Such a representations follows from (2.11), if we use for $\Delta\psi_{\alpha,\alpha}$ the above-given procedure the same we did for $M_{\alpha\beta}$ and $Q_{\alpha 3}$, when we constructed e.n.r. (2.17) - (2.20).

Remark 2.3. Here we noted that the presence of introduced by A. Föppl and von Karman bilinear differential form $L[u_3^*, F^*]$ necessitated on account of *non-symmetricity of functions* $Q_{\alpha 3}$ *and* $Q_{3\alpha}$.

Let us return to the second basic relation (2.14). By virtue of (1.4):

$$\partial_{11}A_{22} - 2\partial_{12}A_{12} + \partial_{22}A_{11} = -L[u_3, u_3] - L[u_\alpha, u_\alpha].$$

Therefore

$$\frac{E}{2}I[A] = -\frac{E}{2}L[u_3^*, u_3^*] - \frac{E}{4h}\int_{-h}^{h}(L[u_3, u_3] - L[u_3^*, u_3^*] + L[u_\alpha, u_\alpha])dt.$$

Besides, we also have (see i.e. Nikolski [1979]):

$$\frac{\nu}{2h}\int_{-h}^{h}\Delta\sigma_{33}dt = \frac{\nu}{2}\Delta(g_3^+ + g_3^-) - \frac{\nu}{2h}\Delta\left[\int_{-h}^{h}(\sigma_{k3}u_{3,k})dt - \rho_{tr}[\sigma_{33} + \sigma_{k3}u_{3,k}]\right].$$

Thus, the second basic relation will take the form:

$$\Delta^2 F^* = -\frac{E}{2} L[u_3^*, u_3^*] + \frac{\nu}{2}\Delta(g_3^+ + g_3^-) + R_9[F^*], \tag{2.22$_1$}$$

or

$$\begin{aligned}\Delta^2 F^* =& -\frac{E}{2} L[u_3^*, u_3^*] + \frac{\nu}{2}\Delta(g_3^+ + g_3^-) \\ &+ \frac{1+\nu}{2h}\left(\int_{-h}^{h} f_{\alpha,\alpha} dt - (g^+_{\alpha,\alpha} - g^-_{\alpha,\alpha})\right) + R_9'[F^*],\end{aligned} \tag{2.22$_2$}$$

where

$$\begin{aligned}R_9[F^*] =& \frac{\nu}{2h}\Delta\rho_{tr}[\sigma_{33} + \sigma_{k3}u_{3,k}] - \frac{E}{4h}\int_{-h}^{h}(L[u_3; u_3] - L[u_3^*; u_3^*] \\ &+ L[u_\alpha, u_\alpha])dt + \frac{1}{2h}[(1+\nu)\Phi_{\alpha,\alpha}[u] - \Delta P_{\alpha\alpha}] - \frac{\nu}{2h}\int_{-h}^{h}\Delta(\sigma_{k3}u_{3,k})dt,\end{aligned}$$

$$R_9'[F] = R_9[F^*] - \frac{1+\nu}{2h}\left(\int_{-h}^{h} f_{\alpha,\alpha} dt - (g^+_{\alpha,\alpha} - g^-_{\alpha,\alpha})\right)$$

Proof. (2.22$_1$) follows from (2.14$_1$), (2.22$_2$) corresponds to (2.14$_2$).

The system (2.17) - (2.22) is an integro-differential relation, originated from the initial system (1.1) - (1.6) without any mechanical and geometrical hypotheses. These e.n.r. (2.17) - (2.22) make it possible to construct different two-dimensional models, corresponding also to known refined theories.

Before starting this procedure we should pay attention to some summands appearing in (2.21) - (2.22). In addition, let us extend the above-given procedure to an anisotropic case (1.5)

Remark 2.4. The relation (2.21) contains the summands of the form:

$$R_8'[u_3^*; \gamma] = \frac{4h^3(1+2\gamma)}{3(1-\nu)}\Delta L[u_3^*, F^*]. \tag{2.23}$$

Remark 2.5. By virtue of

$$\int_{-h}^{h} \sigma_{\alpha 3}u_{3,\alpha} dt = u^*_{3,\alpha}Q_{\alpha 3} + r_3[\sigma_{\alpha 3}u_{3,\alpha}], \tag{2.24}$$

$$r_3[\sigma_{\alpha 3}u_{3,\alpha}] = \int_{-h}^{h}(\sigma_{\alpha 3}(u_{3,\alpha} - u^*_{3,\alpha}) - u^*_{3,\alpha}u_{\alpha,k}\sigma_{k3})dt,$$

the remainder term of formula (2.22) can be represented as follows:

$$R_9 = R_9'' - \frac{\nu}{2h}\Delta[(u^*_{3,\alpha}Q_{\alpha 3}) + r_3[\sigma_{\alpha 3}u_{3,\alpha}]].$$

2.2. Anisotropic case

The above given procedure on construction of exact nonlocal representations (2.17) - (2.22) is valid with respect to anisotropic case, when the generalized Hooke's law is represented in the form of (1.5).

First, let us construct the analogue of the relation (2.22).

Let us define the value ε_{33} from the equations (1.5); substituting it into other equations, we obtain:

$$\begin{aligned}
\sigma_{\alpha\alpha} &= c_{1\alpha}\varepsilon_{11} + c_{2\alpha}\varepsilon_{22} + 2c_{\alpha 6}\varepsilon_{12} + b_{33}^{-1}b_{3\alpha}\sigma_{33},\\
\sigma_{12} &= 2c_{16}\varepsilon_{11} + 2c_{26}\varepsilon_{22} + 2c_{66}\varepsilon_{12} + b_{33}^{-1}b_{36}\sigma_{33},
\end{aligned} \tag{2.25}$$

where

$$c_{\alpha\alpha} = b_{\alpha\alpha} - b_{33}^{-1}b_{\alpha 3}^2, \qquad c_{12} = c_{21} = b_{12} - b_{13}b_{23}b_{33}^{-1},$$
$$2c_{\alpha 6} = b_{\alpha 6} - b_{\alpha 3}b_{36}b_{33}^{-1}, \quad 2c_{66} = b_{66} - b_{36}^2 b_{33}^{-1}.$$

Hence, there obviously follows:

$$\begin{aligned}
\varepsilon_{\alpha\alpha} &= a_{1\alpha}\sigma_{11} + a_{2\alpha}\sigma_{22} + a_{\alpha 6}\sigma_{12} - l_{\alpha\alpha}\sigma_{33},\\
\varepsilon_{12} &= a_{16}\sigma_{11} + a_{26}\sigma_{22} + a_{66}\sigma_{12} - l_{12}\sigma_{33},
\end{aligned}$$

where numbers a and l are expressed by means of b and c are numbers in formulae corresponding to (2.8) - (2.11) (Ambartsumian [1967], p.28).

Now let us calculate $I[\varepsilon]$. By virtue of(2.12) we shall have:

$$\begin{aligned}
I[\varepsilon] &= \frac{1}{2h}\int_{-h}^{h}(\partial_{11}\varepsilon_{22} - 2\partial_{12}\varepsilon_{12} + \partial_{22}\varepsilon_{11})dt\\
&= L_4^1(\partial_1, \partial_2; a)F^*(x_1, x_2) + \frac{1}{2h}(L_{11}'P_{11} - 2L_{12}'P_{12} + L_{22}'P_{22})\\
&\quad -\frac{1}{2h}\int_{-h}^{h}(-1)^{\alpha+\beta}l_{\alpha\beta}\partial_{3-\alpha}\partial_{3-\beta}\sigma_{33}dt,
\end{aligned}$$

where

$$\begin{aligned}
L_4^1(\partial_1, \partial_2; a) &= a_{22}\partial_1^4 - 3a_{26}\partial_1^3\partial_2 + 2(a_{12} + a_{66})\partial_1^2\partial_2^2 - 3a_{16}\partial_1\partial_2^3 + a_{11}\partial_2^4,\\
L_{\alpha\alpha}' &= a_{\alpha\alpha}\partial_{3-\alpha}^2 - 2a_{\alpha 6}\partial_{12} + a_{1\alpha}\partial_\alpha^2;\\
2L_{12}' &= a_{16}\partial_2^2 - 2a_{66}\partial_{12} + a_{26}\partial_1^2 = 2L_{21}'
\end{aligned}$$

On the other hand, by virtue of (1.5), as before we have:

$$I[\varepsilon] = \frac{1}{2} I[A].$$

Therefore, for the stress function $F^*(x_1, x_2)$ in the anisotropic case the representation of the following form is correct:

$$L_4^1(\partial_1, \partial_2; a)F^* = -\frac{E}{2} L^*[u_3^*, u_3^*] + (-1)^{\alpha+\beta} l_{\alpha\beta} \partial_{3-\alpha} \partial_{3-\beta} (g_3^+ + g_3^-) + R_9^A[F^*], \quad (2.26)$$

where

$$R_9^A[F^*] = \frac{1}{2h}(-1)^{\alpha+\beta} l_{\alpha\beta} \partial_{3-\alpha} \partial_{3-\beta} \left[p_{tr}[\sigma_{33} + \sigma_{k3} u_{3,k}] - \int_{-h}^{h} \sigma_{k3} u_{3,k} dt \right]$$

$$-\frac{E}{4h} \int_{-h}^{h} (L[u_3, u_3] - L[u_3^*, u_3^*] + L[u_\alpha, u_\alpha])\, dt - \frac{1}{2h}(-1)^{\alpha+\beta} L'_{\alpha\beta} P_{\alpha\beta}.$$

Now, let us derive formulae, similar to the equations (2.7) – (2.11). First, let us use the fourth and the fifth equations of Hooke's law and define $\varepsilon_{\alpha 3}$. Then, evidently, we shall obtain

$$u_\alpha^* = -u_{3,\alpha}^* + \psi_\alpha^A - \frac{3}{2h^3} B_\alpha, \quad (2.27)$$

where

$$\psi_\alpha^A = \frac{3}{h^3 \delta} \int_{-h}^{h} t dt \int_0^t (b_{\alpha+3,\alpha+3} \sigma_{\alpha 3} - b_{45} \sigma_{33-\alpha}) dt$$

$$= \frac{3}{h^3 \delta} \int_{-h}^{h} t \Phi_\alpha dt,$$

$$\delta = b_{44} b_{55} - b_{45}^2.$$

On the basis of formulae (2.27) and using the relation (2.25) for generalized moments we shall have:

$$M_\alpha = -\frac{2h^3}{3} \Big[(c_{\alpha\alpha} \partial_\alpha^2 + 2c_{\alpha 6} \partial_{12} + c_{\alpha 3-\alpha} \partial_{3-\alpha}^2) u_3^*$$

$$- ((c_{1\alpha} \partial_1 + c_{\alpha 6} \partial_2) \psi_1^A + (c_{\alpha 6} \partial_1 + c_{\alpha 2} \partial_2) \psi_2^A) \Big]$$

$$+ b_{33}^{-1} b_{\alpha 3} \int_{-h}^{h} t \sigma_{33} dt + M_\alpha^{NL}, \quad (2.28)$$

where

$$M_\alpha^{NL} = \frac{1}{2}\int_{-h}^{h}\Big[t(2\sigma_{k\alpha}u_{\alpha,k} + c_{1\alpha}A_{11} + 2c_{\alpha 6}A_{12} + c_{\alpha 2}A_{22})$$
$$-(h^2-t^2)((c_{1\alpha}\partial_1+c_{\alpha 6}\partial_2)A_{13}+(c_{\alpha 6}\partial_1+c_{\alpha 2}\partial_2)A_{23})\Big]dt,$$

and

$$M_{\alpha\beta} = -\frac{2h^3}{3}\Big[(c_{16}\partial_{11} + 2c_{66}\partial_{12} + c_{26}\partial_{22})u_3^*$$
$$-\Big((c_{16}\partial_1 + c_{66}\partial_2)\psi_1^A + (c_{66}\partial_1 + c_{26}\partial_2)\psi_2^A\Big)\Big] + M_{\alpha\beta}^{NL}, \quad (2.29)$$

where

$$M_{\alpha\beta}^{NL} = \int_{-h}^{h} t(\sigma_{k\alpha}u_{\beta,k} + c_{16}A_{11} + c_{66}A_{12} + c_{26}A_{22})dt$$
$$-\frac{1}{2}\int_{-h}^{h}(h^2 - t^2)\left[(c_{16}\partial_1 + c_{66}\partial_2)A_{13} + (c_{66}\partial_1 + c_{26}\partial_2)A_{23}\right]dt.$$

Further, multiplying the primary two equations from the system (1.1) by x_3 and integrating within the range of $-h$ to h and taking into account equalities (2.28) and (2.29) for generalized intersecting forces $Q_{\alpha 3}$ we shall obtain:

$$Q_{\alpha 3} = -\frac{2h^3}{3}L_{\alpha 3}(\partial_1, \partial_2; c)u_3^* + \frac{2h^3}{3}L_{\alpha\beta}(\partial_1, \partial_2; c)\psi_\beta^A + +h(g_\alpha^+ + g_\alpha^-)$$
$$+ b_{33}^{-1}\int_{-h}^{h} t\left[(b_{\alpha 3}\partial_\alpha + b_{36}\partial_{3-\alpha})\sigma_{33} - b_{33}f_\alpha\right]dt + M_{\beta\alpha,\beta}^{NL}, \quad (2.30)$$

where

$$L_{\alpha 3}(\partial_1, \partial_2; c) = c_{\alpha\alpha}\partial_\alpha^3 + 3c_{\alpha 6}\partial_\alpha^2\partial_{3-\alpha} + (c_{12} + 2c_{66})\partial_\alpha\partial_{3-\alpha}^2 + c_{3-\alpha 6}\partial_{3-\alpha}^2,$$

$$L_{\alpha\alpha}(\partial_1, \partial_2; c) = c_{\alpha\alpha}\partial_\alpha^2 + 2c_{\alpha 6}\partial_\alpha\partial_{3-\alpha} + c_{66}\partial_{3-\alpha}^2,$$

$$L_{21}(\partial_1, \partial_2; c) = L_{12}(\partial_1, \partial_2; c) = c_{16}\partial_1^2 + (c_{16} + c_{66})\partial_{12} + c_{26}\partial_2^2.$$

Finally, the equation corresponding to the first basic relation (2.11) for the anisotropic case, if based on (2.30), will take the form:

$$-\frac{2h^3}{3}L_4^0(\partial_1,\partial_2;c)u_3^* - \frac{2h^3}{3}L_{\alpha 3}\psi_\alpha^A - L_{33}(\partial_1,\partial_2;b)\int_{-h}^{h} t\sigma_{33}dt - \partial_{\alpha\beta}M_{\alpha\beta}^{NL}$$

$$+\int_{-h}^{h}\partial_\alpha(\sigma_{k3}u_{\alpha,k} - \sigma_{k\alpha}u_{3,k})dt \tag{2.31}$$

$$= h(g_{\alpha,\alpha}^+ + g_{\alpha,\alpha}^-) + g_3^+ - g_3^- - \int_{-h}^{h}(tf_{\alpha,\alpha} + f_3)dt,$$

where

$$L_4^0(\partial_1,\partial_2;c) = c_{11}\partial_1^4 + 4c_{16}\partial_1^3\partial_2 + 2(c_{12}+2c_{66})\partial_1^2\partial_2^2 + 4c_{26}\partial_1\partial_2^3 + c_{22}\partial_2,$$
$$L_{33}(\partial_1,\partial_2;b) = b_{33}^{-1}(b_{13}\partial_1^2 + 2b_{36}\partial_{12} + b_{23}\partial_2^2).$$

While comparing formulae (2.7) – (2.11) and (2.27) – (2.31) we clarify that the relation (2.15) is suitable for the anisotropic case. A somewhat different situation holds for summands, depending on ψ_α^A. In this case we shall define ψ_α^A beforehand from formula (2.27):

$$\frac{2h^3}{3}\psi_\alpha^A = \frac{2}{3\delta}h^2(1+2\gamma)\left[b_{3+\alpha,3+\alpha}Q_{\alpha 3} - b_{45}Q_{3-\alpha 3}\right] + r_2^A[\psi_\alpha^A;\gamma], \tag{2.32}$$

where

$$r_2^A[\psi_\alpha^A;\gamma] = \frac{2}{\delta}r_1[t\Phi_\alpha;\gamma] - \frac{2h^2}{3\delta}(1+2\gamma)\int_{-h}^{h}\sigma_{k3}(b_{3+\alpha,3+\alpha}u_{\alpha,k} - b_{45}u_{3-\alpha,k})dt.$$

Then the account of (2.15) and (2.32) in formulae (2.27) – (2.31) immediately gives us the representation for values u_α^*, M_α and $M_{\alpha\beta}$ by means of generalized intersecting forses $Q_{\alpha 3}$, deflection u_3^*, normal pressure $g_3^+ - g_3^-$ and remainder terms R^A:

$$u_\alpha^* = -u_{3,\alpha}^* + \frac{1+2\gamma}{3h\delta}\left[b_{3+\alpha,3+\alpha}Q_{\alpha 3} - b_{45}Q_{3-\alpha 3}\right] + R_\alpha^A[u_\alpha^*;\gamma], \tag{2.33}$$

$$R_\alpha^A[u_\alpha^*;\gamma] = \frac{3}{2h^3}r_2[\psi_\alpha^A;\gamma] - B_\alpha;$$

$$\begin{aligned} M_\alpha =& -\frac{2h^3}{3}(c_{\alpha\alpha}\partial_\alpha^2 + 2c_{\alpha 6}\partial_{12} + c_{\alpha 3-\alpha}\partial_{3-\alpha}^2)u_3^* \\ &+ \frac{2}{3\delta}h^2(1+2\gamma)\Bigg[\Big[(b_{44}c_{1\alpha} - b_{45}c_{\alpha 6})\partial_1 + (b_{44}c_{\alpha 6} - b_{45}c_{\alpha 2})\partial_2\Big]Q_{13} \\ &+ \Big[(b_{55}c_{\alpha 6} - b_{45}c_{1\alpha})\partial_1 + (b_{55}c_{\alpha 2} - b_{45}c_{\alpha 6})\partial_2\Big]Q_{23} \\ &+ \frac{\delta}{2}b_{\alpha 3}b_{33}^{-1}(g_3^+ - g_3^-)\Bigg] + R_{\alpha+82}^A[M_\alpha;\gamma], \end{aligned} \tag{2.34}$$

$$
\begin{aligned}
&R^A_{\alpha+2}[M_\alpha;\gamma] \\
&\quad = M^{NL}_\alpha \\
&\qquad + r^A_2\left[\frac{2}{\delta}\Big((c_{1\alpha}\partial_1 + c_{\alpha6}\partial_2)\psi^A_1 + (c_{\alpha6}\partial_1 + c_{\alpha2}\partial_2)\psi^A_2\Big) + \frac{b_{23}}{b_{33}}\sigma_{33};\gamma\right]
\end{aligned}
$$

$$
\begin{aligned}
M_{\alpha\beta} = &-\frac{2h^3}{3}(c_{16}\partial_{12} + 2c_{66}\partial_{12} + c_{26}\partial_{22})u^*_3 \\
&+\frac{2h^2}{3\delta}(1+2\gamma)\Big[((b_{44}c_{16} - b_{45}c_{66})\partial_1 + (b_{44}c_{66} - b_{45}c_{26})\partial_2)Q_{13} \\
&\qquad +((b_{55}c_{66} - b_{45}c_{16})\partial_1 + (b_{55}c_{26} - b_{45}c_{66})\partial_2)Q_{23} \\
&\qquad +\frac{\delta}{2}b_{36}b^{-1}_{33}(g^+_3 - g^-_3)\Big] \\
&+R^A_5[M_{\alpha\beta};\gamma],
\end{aligned}
\tag{2.35}
$$

$$
R^A_5 = M^{NL}_{\alpha\beta} + r^A_2\left[\frac{2}{\delta}((c_{16}\partial_1 + c_{66}\partial_2)\psi^A_1 + (c_{66}\partial_1 + c_{26}\partial_2)\psi^A_2) + \frac{b_{36}}{b_{33}}\sigma_{33};\gamma\right].
$$

In their turn, equations corresponding to values $Q_{\alpha3}$ take the following form:

$$
\begin{aligned}
Q_{\alpha3} = &-\frac{2h^3}{3}L_{\alpha3}u^*_3 + \frac{2h^2}{3\delta}(1+2\gamma)S_{\alpha\beta}(\partial_1,\partial_2)Q_{\beta3} - \int_{-h}^{h} tf_\alpha dt \\
&+ 2L_{0\alpha}(\partial_1,\partial_2)\underline{\int_{-h}^{h} tdt\int_0^t \sigma_{\beta3,\beta}dt} + L_{0\alpha}(\partial_1,\partial_2)\int_{-h}^{h} t\sigma_{33}dt + h(g^+_\alpha + g^-_\alpha) \\
&+ \frac{2}{\delta}r_1\left[tS_{\alpha\beta}(\partial_1\partial_2)\int_0^t \sigma_{\beta3}dt;\gamma\right] \\
&- \frac{2}{3\delta}h^2(1+2\gamma)S_{\alpha\beta}(\partial_1,\partial_2)\int_{-h}^{h}\sigma_{k3}u_{\beta,k}dt + M^{NL}_{\beta\alpha,\beta}.
\end{aligned}
$$

Here

$$
\begin{aligned}
S_{\alpha\beta}(\partial_1,\partial_2) &= b_{3+\beta,3+\beta}L_{\alpha\beta} - b_{45}L_{\alpha3-\beta} - \delta\partial_\beta L_{0\alpha}(\partial_1,\partial_2) \\
L_{0\alpha}(\partial_1,\partial_2) &= b^{-1}_{33}(b_{\alpha3}\partial_\alpha + b_{36}\partial_{3-\alpha}).
\end{aligned}
$$

The use of the third equilibrium equation of the system (1.1) in the underlined term gives:

$$
\int_{-h}^{h}(h^2 - t^2)\sigma_{\beta3,\beta}dt = \int_{-h}^{h}(h^2 - t^2)(f_3 - \sigma_{k\beta,\beta}u_{3,k} - \partial_\beta(\sigma_{3\beta}u_{3,3}))dt
$$

$$-\frac{2}{3}h^2(1+2\gamma)[g_3^+ - g_3^- + 2hL[u_3^*, F^*]]$$

$$-2r_1\left[t(\sigma_{33}+\sigma_{k3}u_{k,3}+\int_0^t \sigma_{\beta\delta}u_{\beta,\beta\delta}dt, \gamma\right]$$

$$-\frac{2h^2}{3}\left[(-1)^{\beta+\delta}u^*_{3,\beta\delta}P_{\beta\delta} + \int_{-h}^{h} \sigma_{\beta\delta}\partial_{\beta\delta}(u_3 - u_3^*)dt\right].$$

The account of (2.15) reduces the following nonlocal representations for $Q_{\alpha 3}$:

$$\begin{aligned} Q_{\alpha 3} &- \frac{2h^2}{3\delta}(1+2\gamma)S_{\alpha\beta}(\partial_1,\partial_2)Q_{\beta 3} \\ &= -\frac{2h^3}{3}L_{\alpha 3}u_3^* - \frac{h^2}{3}(1+2\gamma)L_{0\alpha}\left[g_3^+ - g_3^- + 4hL[u_3^*, F^*]\right] + h(g_\alpha^+ - g_\alpha^-) \\ &- \int_{-h}^{h}\left(tf_\alpha - L_{0\alpha}(h^2-t^2)f_3\right)dt + R^A_{5+\alpha}[Q_{\alpha 3};\gamma]. \end{aligned} \tag{2.36}$$

Here

$$\begin{aligned} R^A_{5+\alpha}&[Q_{\alpha 3};\gamma] \\ &= r_1\left[t\left(2\delta^{-1}S_{\alpha\beta}\int_0^t \sigma_{\beta 3}dt - L_{01}\left(\sigma_{33}+\sigma_{k3}u_{3,k} + +2\int_0^t \sigma_{\beta\delta}u_{\beta,\beta\delta}dt\right)\right);\gamma\right] \\ &+ M^{NL}_{\beta\alpha,\beta} - L_{0\alpha}\int_{-h}^{h}(h^2-t^2)(\sigma_{k\beta,\beta}u_{3,k} + \partial_\beta(\sigma_{3\beta}u_{3,3}))dt \\ &- \frac{2h^2}{3}(1+2\gamma)\left[L_{0\alpha}\left((-1)^{\beta+\delta}u_{3,\beta\delta}P_{\beta\delta} + \int_{-h}^{h}\sigma_{\beta\delta}\partial_{\beta\delta}(u_3-u_3^*)dt\right)\right. \\ &\left. + \delta^{-1}S_{\alpha\beta}\int_{-h}^{h}\sigma_{k3}u_{\beta,k}dt\right] - L_{0\alpha}\int_{-h}^{h}t\sigma_{k3}u_{3,k}dt. \end{aligned}$$

By virtue of formulae (2.31), (2.36), equation (1.1) and the above-given reasoning, the first basic relation, corresponding to von Karman's equation for deflection in an anisotropic case, takes the form:

$$\begin{aligned} \frac{2h^3}{3}L_4^0(\partial_1,\partial_2;c)u_3^* &= \frac{2h^2}{3\delta}(1+2\gamma)\partial_\alpha S_{\alpha\beta}Q_{\beta 3} + 2hL[u_3^*, F^*] + h(g^+_{\alpha,\alpha}g^-_{\alpha,\alpha}) \\ &+ g_3^+ - g_3^- - \frac{h^2}{2}(1+2\gamma)L_{33}(g_3^+ - g_3^- + 4hL[u_3^*, F^*]) \end{aligned}$$

$$-\int_{-h}^{h}\Big[tf_{\alpha,\alpha}+(h^2-t^2)(1-L_{33})f_3\Big]dt+R_8^A[u_3^*;\gamma], \quad (2.37)$$

$$R_8^A[u_3^*;\gamma]=R_{5+\alpha,\alpha}^A+(-1)^{\alpha+\beta}u_{3,\alpha\beta}^*P_{\alpha\beta}$$
$$+\int_{-h}^{h}[\partial_\alpha(\sigma_{\alpha 3}u_{3,3}-\sigma_{k3}u_{\alpha,k})+\sigma_{\alpha\beta,\alpha}u_{3,\beta}+\sigma_{\alpha\beta}\partial_{\alpha\beta}(u_3-u_3^*)]\,dt.$$

Thus we have constructed the system of integro-differential equations (2.17) – (2.22) in isotropic and (2.26), (2.33) – (2.37) in anisotropic cases. If we denote

$$U^*=(u_1^*,u_2^*,M_1,M_2,M_{21},M_{12},Q_{13},Q_{23},u_3^*,F^*)^T,$$
$$R=(R_1,R_2,R_3,R_4,R_5'[M_{21};\gamma],R_5''[M_{12};\gamma],R_6,R_7,R_8,R_9)^T,$$

then, for brevity and convenience, we can join in the form of operator equation:

$$2P_h(\partial_1,\partial_2,\gamma)U^*=G(x_1,x_2;\gamma)+R[U^*;\gamma],\quad (x_1,x_2)\in\bar{D},$$

we added the necessary boundary conditions , averaged with respect to the plate thickness, originated from the appropriate data on the lateral surface S (see below subsections 3.5 – 3.7).

We shall rewrite the last system without remainder term R: we shall obtain the two-dimensional parametric system of nonlinear equations:

$$\boldsymbol{P_h(\partial_1,\partial_2;\gamma)U^h=G(x_1,x_2;\gamma),\quad (x_1,x_2)\in\bar{D},} \qquad (\boldsymbol{P_h})$$

where $U_h=(u_1^h,u_2^h,M_1^h,M_2^h,M_{21}^h,M_{12}^h,Q_{13}^h,Q_{23}^h,u_3^h,F^h)^T$ denotes the approximate solution of the initial three-dimensional problem (1.1) – (1.3).

The system of equations (P_h) has the structure typical for refined theories; a part of relations (P_h), corresponding to functions u_3^h, F^h and $Q_{\alpha 3}^h$ forms the closed strictly elliptical nonlinear system of differential equations in D in partial derivatives; the rest of the functions u_α^h, M_α^h and $M_{\alpha\beta}^h$ are expressed by means of obvious formulae.

2.3. Basic results

Let us try to give the formulation of the results, following from exact nonlocal representations ((2.17) – (2.22) in the isotropic case, (2.26), (2.33) – (2.37) in the anisotropic case) and from the scheme (P_h).

Let the initial problem (1.1) – (1.5) be linear.

2.3.1. The vector error R in this case depends on linear combination of remainder terms of quadrature formulae of Simpson, trapezoid, or Gauss (with one node) only. At the same time, as is known, for integrals of type (2.15) weighted t in the case of sufficiently smooth functions σ_{i3} the remainder term of Simpson's rule will have the order of exactness with respect to h, equal to six at most. In the class of functions with generalized derivatives the remainder terms of all quadrature formulae have the same order; moreover, Sard–Nikolski's absolute constants appear to be minimal for Simpson's rule, i.e., when $\gamma = 0$ in the scheme (P_h) (see section 5, where results of the theory of quadrature formulae are used for the evaluation of integrals of type (2.15) and (2.16)).

2.3.2. In the case of (discontinuous) functions with integrable singularity, the optimal parameter value depends on the order of the singularity. So, when the order of singularity is equal to 1/2 the refined Reissner's theory will be the optimal model, that is to say, provided $\gamma = 0.1$ in the scheme (P_h).

2.3.3. The scheme (P_h) with $\gamma = -0.5$ gives the classic theory of Kirchhoff in the isotropic case; (2.21) is transformed into Germen–Lagrange's equation, and coefficients with summands, causing the effect of boundary layer, vanish.

Further, as will be seen below, the known refined theories of plate bending (including Reissner's, Kromm's, Henki's models the calculation of which is given by means of physical hypotheses, e.g., in the monograph of Panč [1975]. p.I, ch. 3–5) and new models, equivalent to them in the sense of point 2.3.1, follows as the variation of parameter γ from the scheme (P_h).

2.3.4. Depending on the physical pattern of stress–strained state $\Omega_h(x_1, x_2, x_3)$ we can construct the corresponding nonhomogeneous models of mixed type with different $\gamma = \gamma_i(x_1, x_2)$, $i = 1, 2, \ldots, 8$, $(x_1, x_2) \in \bar{D}$ by virtue of scheme (P_h), below we shall call them *refined theories of mixed type*.

2.3.5. The exactness of the remainder term R_9 in equation (2.22) or (2.26) is defined by the trapezoid formula. In the case of sufficiently smooth functions σ_{i3}, in order to increase the exactness of order of approximation to h, we should use the one-dimensional interpolated Hermite formula with multiple nodes $x_3 = \pm h$ for the function σ_{33} and the equailibrium equation (1.1) (see also point 2.3.14).

Let us note here that the representation for R_9 given in (2.22) specifically substantiates the correctness of introducing summands $0.5\nu\partial_{\alpha\alpha}(g_3^+ + g_3^-)$ to be put into the righthand parts of von Karman's system, taking into account transverse loads (see Lukasiewicz [1979] ch.2, section 21).

2.3.6. In the linear anisotropic case the following is obtained from the scheme (P_h): with $\gamma = -0.5$ — the classic theory set forth, e.g., in Lekhnitski's paper [80]; with $\gamma = 0.1$ — the theory of Reissner–Ambartsumian (see Ambartsumian [1974]); the theory of Girkmann–Beer in the orthotropic case with $\gamma = 0.1$ (see, e.g., Panč [1975], ch.19).

In the case of a transversal isotropic plate with $\gamma = 0.1$ the model of Reissner–Lekhnitski's type follows from scheme (P_h) and the corresponding system of differential equations is splitted into separate equations with respect to u_3^h and Q_α^h accordingly.

In the anisotropic case (1.4) the fulfilment of the identity below, appears to be the nesessary and sufficient condition for the splitting of scheme (P_h) in D:

$$L_{13}(b_{44}\sigma_{13} - b_{45}\sigma_{23}) + L_{23}(b_{55}\sigma_{23} - b_{45}\sigma_{13}) = (l_1\partial_{11} + 2l_2\partial_{12} + l_3\partial_{22})\sigma_{\alpha 3,\alpha}. \quad (2.38)$$

In its turn, (2.38) is equivalent to the following relations:

$$\begin{aligned}(b_{11}b_{33}-b_{13}^2)b_{44}&=(b_{22}b_{33}-b_{23}^2)b_{55};\\ b_{11}b_{33}-b_{13}^2&=(b_{16}b_{33}-b_{13}b_{36})b_{45}^{-1};\\ b_{16}b_{33}-b_{13}b_{36}&=(b_{26}b_{33}-b_{23}b_{36})b_{44}^{-1}b_{55};\\ b_{12}b_{33}-b_{13}b_{23}+2(b_{33}b_{66}-b_{36}^2)&=(b_{11}b_{33}-b_{13}^2)(b_{44}b_{55}+2b_{45}^2)b_{55}^{-2};\end{aligned} \tag{2.39}$$

$$l_1=\delta(b_{33}b_{55})^{-1}l_4,\ l_2=\delta b_{45}b_{33}^{-1}l_4,\ l_3=\delta b_{44}b_{33}^{-1}b_{55}l_4,\ l_4=b_{11}b_{33}-b_{13}^2.$$

Let us call the fourth equation of (2.39) *the condition of generalized transversality*. Elastic constants, corresponding to transversalisotropic bodies, satisfy these equations; in this case the known relation $2b_{66}=b_{11}-b_{12}$ follows from (2.39) immediately.

Let us consider the nonlinear case.

2.3.7. While estimating the terms, containing differences $u_3-u_3^*$, $L[u_3,u_3]-L[u_3^*,u_3^*]$, it is quite sufficient to use Taylor's formula.

2.3.8. Choosing the γ parameter to be as it was in the linear case, brings us to equivalent models with exactness to R; the model (P_h) with $\gamma=0$ will be the optimal one in the sense of point 2.3.2.

2.3.9. Let $f_\alpha=g_\alpha^+=0$ in problem (1.1) - (1.2) and $\gamma=-0.5$ in scheme (P_h). Then the obtained system (P_h) coincides with von Karman's system of differential equations.

2.3.10. Nonlinear equations, given in ch.2 Lukasiewicz [1979] (see points 2.3.5 and 2.3.12 too) in isotropic case with $\gamma=0.1$ and some obvious simplifications follow from scheme (P_h) (see also Cheng Zhen-giang, Wang Xin-xi, Huang Mao-guang [1993].)

2.3.11. Evidently, nonlinear models, given in Ambartsumian [1967] (ch.II, 8), also follow from (P_h) with $\gamma=0.1$ in an anisotropic case.

2.3.12. The investigation of models with account of summands, containing $\Delta L[u_3^*,F^*]$, $\Delta u_{3,\alpha}^*Q_{\alpha3}$ (see formulae (2.23), (2.24)), evidently, deserves certain attention. At the same time we should notice that $\Delta L[\ ,\ ]$ is the nonlinear differential operator of elliptic type of the fourth order, and the mechanical meaning of $u_3^*Q_{\alpha3}$ product is shown, for example, in Lukasiewicz [1979] (ch.I, section 1.9).

2.3.13. In the monograph of Landau and Lifchitz [1970] they construct two-dimensional models for elastic plates based on *a priori* assumptions of a physical nature. Evidently, all of these models follow from (P_h) scheme easily by correspondingly choosing control parameters γ_i, without using simplifying hypotheses.

2.3.14. The basic equation (2.14) contains arbitrary functions $P_{\alpha\beta}(x_1,x_2)$. When $P_{\alpha\beta}\equiv0$ and plates are homogeneous and isotropic, from (2.14), as we showed above, follow two-dimensional partial differential equations of the generalized plane strain problem (Filon's model as the second equation of von Karman system). In such a particular case the method of constructing equations of (2.14) type (without remainder terms, too) coincides with well known classical methods (compare, e.g., with Muskhelishvili [1966], Timoshenko, Voinovski–Krieger [1959]). But we show that the method considered above (in case when functions $P_{\alpha\beta}$ and remainder term $R_9[F^*]$ are not identical to zero) generalized the last one. We consider formulae (2.12) and (2.13).

If $P_{12} \equiv 0$, from (2.13) follow:

$$P_{\alpha\alpha} = \int^{x_\alpha} \Phi_\alpha[u] dx_\alpha + c_\alpha(x_{3-\alpha}).$$

Then for the expression $(1+\nu)\Phi_{\alpha,\alpha} - \Delta P_{\alpha\alpha}$ we have:

$$(1+\nu)\Phi_{\alpha,\alpha} - \Delta P_{\alpha\alpha}$$

$$= \nu \left[\int_{-h}^{h} (f_{\alpha,\alpha} - \partial_{\alpha\beta}(\sigma_{\beta k} u_{\alpha,k}))dt - (g^+_{\alpha,\alpha} - g^-_{\alpha,\alpha}) \right]$$

$$-\partial^2_{3-\alpha} \left[\int^{x_\alpha} dx_\alpha \left(\int_{-h}^{h} (f_\alpha - \partial_\beta(\sigma_{\beta k} u_{\alpha,k}))dt - (g^+_\alpha - g^-_\alpha) \right) + c_\alpha(x_{3-\alpha}) \right].$$

Substituting the last formula in expression for $\Delta^2 F^*$, we get immediately

$$\Delta^2 F^* = \frac{E}{2} I[A] + \frac{\nu}{2h} \int_{-h}^{h} \Delta\sigma_{33} dt$$

$$+\frac{1}{2h} \left[\nu \left(\int_{-h}^{h} (f_{\alpha,\alpha} - \partial_{\alpha\beta}(\sigma_{\beta k} u_{\alpha,k}))dt - (g^+_{\alpha,\alpha} - g^-_{\alpha,\alpha}) \right) \right.$$

$$-\partial^2_{3-\alpha} \Big(\int^{x_\alpha} dx_\alpha \Big[\int_{-h}^{h} (f_\alpha - \partial_\beta(\sigma_{\beta k} u_{\alpha,k}))dt - (g^+_\alpha - g^-_\alpha) \Big] \Big)$$

$$\left. +\partial^2_{3-\alpha}(c_\alpha(x_{3-\alpha})) \right]$$

or (see p. 18)

$$\Delta^2 F^* = -\frac{E}{2} L[u_3^*, u_3^*] + \frac{\nu}{2} \Delta(g_3^+ - g_3^-)$$

$$+ \frac{1}{2h} \left[\nu \Big(\int_{-h}^{h} f_{\alpha,\alpha} dt - (g^+_{\alpha,\alpha} - g^-_{\alpha,\alpha}) \Big) \right.$$

$$- \partial^2_{3-\alpha} \Big(\int^{x_\alpha} dx_\alpha \Big(\int_{-h}^{h} (f_\alpha dt - \partial_\beta(\sigma_{\beta k} u_{\alpha,k}))dt \Big) - (g^+_\alpha - g^-_\alpha)$$

$$\left. + c_\alpha(x_{3-\alpha}) \Big) \right] + R''_g[F^*]. \tag{2.40}$$

$$R_9''[F^*] = \frac{\nu}{2h}\rho_{tr}(\sigma_{33} + \sigma_{k3}u_{3,k}) - \frac{E}{2h}\int\limits_{-h}^{h}(L[u_3,u_3] - L[u_3^*,u_3^*] + L[u_\alpha,u_\alpha])\,dt$$

$$-\frac{1}{2h}\left[\nu\int\limits_{-h}^{h}\partial_{\alpha\beta}(\sigma_{\beta k}u_{\alpha,k})dt - \partial_{3-\alpha}^2\int\limits^{x_\alpha}dx_\alpha\int\limits_{-h}^{h}\partial_\beta(\sigma_{\beta k}u_{\alpha,k})dt\right]$$

$$-\frac{\nu}{2h}\int\limits_{-h}^{h}\Delta(\sigma_{k3}u_{3,k})dt.$$

Here $c_\alpha(x_{3-\alpha})$ (no sum!) are arbitrary functions .

2.3.15. In the work Tvalchrelidze [1984] there are constructing shell theory using two basic surfaces. Evidently, these models follow from the (P_h) scheme by accordingly choosing unknown functions (defining by $u_i(x_1,x_2,h_2)\pm u_i(x_1,x_2,h_1)$) and parameter γ_i, without using simplifying hypotheses (see also subsection 11.4).

2.4. Variational formulation for refined theories, generalized Hellinger- Reissner variational principle

The variational formulation for constructing of refined theories has very important meaning, (see Chien Wei-zang [1988], Ciarlet [1997], Hellinger [1914], Lukasievicz [1979], Reissner [1950], Washizu [1982]). Here are also historical notes, commentaries, comparisions of different theories with a very rich bibliography. As follows from these works, the variational formulation also has other convinient properties (e.g. Washizu [1982], part A), but I call readers' attention to the following well known destinguishing features such as: the possibility of investigating and approximately solving initial BVPs in the most general Sobolev (functional) spaces defined by its data.

For simplicity and definiteness we consider the linear theory of elasticity, when Ω_h is an isotropic homogeneous elastic medium of constant thickness, but the consideration of more general cases (such as, anisotropies and non-linearities) does not represent any difficulties.

Let the boundary conditions are such:

$$u\big|_{S_1} = \sigma_n\big|_{S_2} = 0, \qquad S_1\cup S_2 = S, \qquad \sigma_3\big|_{S^\pm} = g^\pm \tag{2.41}$$

This conditions we understand in the sense of the trace of functions (e.g., Ciarlet [1988], p.279.)

Follow Mikhlin [1970], let us considering the following functional:

$$\Phi[u] = \int\limits_{\Omega_h}(\frac{1}{2}\sigma e + uf)dv - \int\limits_{S+}u\sigma_n ds - \int\limits_{S-}u\sigma_n ds, \tag{2.42}$$

where n is outnet normal.

If we use the third equality of Hooke's law (1.6) we have

$$\Phi[u]=\frac{1}{2}\int_{\Omega_h}\Big[((\lambda^*+2\mu)u_{\alpha,\alpha}+\lambda^* u_{3-\alpha},3-\alpha)u_{\alpha,\alpha}+\mu(u_{1,2}+u_{2,1})^2$$

$$+\sigma_{\alpha,3}(u_{\alpha,3}+u_{3,\alpha})++\frac{\lambda}{\lambda+2\mu}\sigma_{33}u\alpha,\alpha+\sigma_{33}u_{3,3}\Big]dv$$

$$+\int_{\Omega_h} uf dv-\int_{S\pm} u\sigma_n ds.$$

If we assume that (see, e.g, Berdichevski [1973]):

$$u_i(x_1,x_2,t)=v_i(x_1,x_2)+tw_i(x_1,x_2)+r_i[u], \tag{2.43}$$

for $\Phi[u]$ immediately follows:

$$\Phi[u]=\frac{1}{2}\int_D\Big[(\lambda^*+2\mu)[2hv^2_{\alpha,\alpha}+\frac{2h^3}{3}w_{\alpha,\alpha}]$$

$$+\lambda^*\Big[2hv_{\alpha,\alpha}v_{3-\alpha,3-\alpha}+\frac{2h^3}{3}w_{\alpha,\alpha}w_{3-\alpha,3-\alpha}\Big]$$

$$+\mu\Big(2hv^2_{\alpha,3-\alpha}+\frac{2h^3}{3}w^2_{\alpha,3-\alpha}\Big)+(v_{3,\alpha}+w_\alpha)\int_{-h}^{h}\sigma_{\alpha,3}dt+w_{3,\alpha}\int_{-h}^{h}t\sigma_{\alpha,3}dt$$

$$+\frac{\lambda}{\lambda+2\mu}\Big(v_{\alpha,\alpha}\int_{-h}^{h}\sigma_{3,3}dt+w_{\alpha,\alpha}\int_{-h}^{h}t\sigma_{3,3}dt\Big)+w_3\int_{-h}^{h}\sigma_{3,3}dt\Big]dw$$

$$+\int_D\Big[v_i\int_{-h}^{h}f_i dt+w_i\int_{-h}^{h}tf_i dt\Big]dw-\int_{S\pm}(v_i\pm hw_i)\sigma_{3n}ds+R[u], \tag{2.44}$$

where $r_{ij}[u]$ defined from (2.43) evidently and

$$R[u]=\int_{\Omega_h}(\frac{1}{2}\sigma_{ij}r_{ij}[u]+f_ir_i[u])dv-\int_{S\pm}\sigma_{3n}r_i[u]ds.$$

Now let us consider the difference

$$\Phi[v,w]=\Phi[u]-R[u]. \tag{2.45}$$

This expression we call the **Hellinger–Reissner generalized two-dimensional functional.** Euler–Lagrange conditions of stationarity give us the following two-dimensional system of PDEs:

$$2h[\mu\Delta v_\alpha + (\lambda^* + \mu)\partial_\alpha(v_{\alpha,\alpha} + v_{3-\alpha,3-\alpha})] + \frac{\lambda}{2(\lambda+2\mu)}\int_{-h}^{h}\sigma_{33,\alpha}dt$$

$$= \int_{-h}^{h} f_\alpha dt - (g_\alpha^+ - g_\alpha^-),$$

$$\frac{1}{2}\int_{-h}^{h}\sigma_{\alpha 3,\alpha}dt = \int_{-h}^{h} f_3 dt - (g_3^+ - g_3^-),$$

$$\frac{2h^3}{3}[\mu\Delta w_\alpha + (\lambda^* + \mu)\partial_\alpha(w_{\alpha,\alpha} + w_{3-\alpha,3-\alpha})] \tag{2.46}$$

$$-\frac{1}{2}\int_{-h}^{h}\sigma_{\alpha 3}dt + \frac{\lambda}{2(\lambda+2\mu)}\int_{-h}^{h} t\sigma_{33,\alpha}dt$$

$$= \int_{-h}^{h} tf_\alpha dt - h(g_\alpha^+ + g_\alpha^-),$$

$$\frac{2h^3}{3}\Delta w_3 - \frac{1}{2}\int_{-h}^{h}\sigma_{33}dt = \int_{-h}^{h} tf_3 dt - h(g_3^+ + g_3^-).$$

If now use formulae of type (2.15) and (2.16), assume also $w_3 \equiv 0$, from (2.45) follows parametrical representation of refined theories full identical to (P_h) schemes (see p.25).

Thus the following conclusion are true.

Conclusion 2.1.

The functional $\Phi[u] - R[u]$ corresponds to the refined theories (in wide sense) and scheme (P_h) for this functional represents Euler's equations.

In particular, from equations (2.46_{3-5}) follow:

$$\frac{2h^3}{3}[\mu\Delta w_\alpha + (\lambda^* + \mu)\,\mathrm{grad\,div}\, w_+] - \frac{\mu h}{(1+2\gamma)}(w_\alpha + v_{3,\alpha})$$

$$= \int_{-h}^{h} tf_\alpha dt - h(g_\alpha^+ + g_\alpha^-) - \frac{\lambda}{2(\lambda+2\mu)}\int_{-h}^{h} t\sigma_{33,\alpha}dt = F_\alpha, \tag{2.47}$$

$$\frac{\mu h}{(1+2\gamma}[\Delta v_3 + w_{\alpha,\alpha}] = \int_{-h}^{h} f_3 dt - (g_3^+ + g_3^-) = F_3.$$

This system, of course, gives, e.g., Reissner's refined theory when $\gamma = 0,1$ (as before, see point 2.3.2.), Mindlin's theory [1951] for $\gamma = \frac{12-\pi^2}{2\pi^2}$, for the theory of Washizu–Vekua $\gamma = 0$ (see 2.3.1.) (compare with Washizu [1982], ch.8, §8.8. or Lukasievicz [1982], ch.2, section 2.1; here it is discused physical aspects of such differences).

Remark 2.6. The problems of existence and solvability of general solutions of BVP as well as the construct of fundamental system of solutions for *PDEs* (2.46) investigated immediately by using correspondently results referenced in, e.g., Morozov [1978] (ch.III), Constanda [1990] (ch.3-4.), if $\gamma \geq -0.5$.

The form of system of differential equations of refined theories for anisotropic case, as follows from subsection 2.2, respect to averaged deflection v_3 and components of normals rotation w_α is such:

$$\frac{2h^3}{3}\left[(c_{1\alpha}\partial_1 + c_{\alpha 6}\partial_2)w_{1,\alpha} + (c_{\alpha 6}\partial_1 + c_{26}\partial_2)w_{2,\alpha} + (2c_{16}\partial_1 + c_{66}\partial_2)w_{1,3-\alpha}\right.$$
$$\left. + (c_{66}\partial_1 + 2c_{26}\partial_2)w_{2,3-\alpha}\right]$$
$$-\frac{h}{(1+2\gamma)\delta}\left[a_{3+\alpha,3+\alpha}(w_\alpha + v_{3,\alpha}) - a_{3+\alpha,6-\alpha}(w_{3-\alpha} + v_{3,3-\alpha})\right]$$
$$= F_\alpha,$$

$$\frac{h}{(1+2\gamma)\delta}\left[(a_{44}\partial_{11} - 2a_{45}\partial_{12} + a_{55}\partial_{22})v_3 + (a_{3+\alpha,3+\alpha}\partial_\alpha - a_{3+\alpha,6-\alpha}\partial_{3-\alpha})w_\alpha\right]$$
$$= F_3.$$

where, we recall the modulus of elasticity and rigidity-numbers a, b are defined from Hooke's law (1.5) , for c see subsection 2.2, $\delta = a_{44}a_{55} - a_{45}^2$.

3. On the Construction of Refined Theories for Nonhomogeneous Plates and on the Problems of Boundary Conditions

The method of the construction of refined theories without hypotheses, being discussed in the preceding section, will be extended below to the nonhomogeneous and nonstationary cases.

3.1. Nonhomogeneous case

For clearness, we shall consider the case of an isotropic elastic body Ω_h and write problem (1.1) – (1.6) in the linear formulation as

$$\sigma_{ij,j} = f_i, \quad f = (0,0,-Z)^T, \quad x \in \Omega_h,$$
$$\sigma_{ij} = \lambda(x_1,x_2)e_{kk}\delta_{ij} + 2\mu(x_1,x_2)e_{ij}, \quad e_{ij} = \frac{1}{2}(u_{i,j} + u_{j,i}), \qquad (B)$$
$$\sigma_3 = (0,0,g_3^\pm), \quad x \in S^\pm; \quad l(\partial_1,\partial_2,\partial_3)u = g, \quad x \in S.$$

Thus, we proceed to the inference of the exact nonlocal representations, corresponding to problem (B). Similarly in these formulae, as in a homogeneuos case, we obtain:

$$M_{\alpha\beta,\alpha\beta} = Q_{\alpha,\alpha} = -g_3^+ + g_3^- - \int_{-h}^{h} Z dt.$$

Further, due to formulae (2.7) for moments M_α and M_{12} we have:

$$M_1 = \int_{-h}^{h} t\sigma_{11} dt$$
$$= \lambda^* \int_{-h}^{h} tdt \left[\int_0^t -\partial_{\alpha\alpha} u_3 dt + \partial_1 \left(\mu^{-1} \int_0^t \sigma_{13} dt \right) + \partial_2 \left(\mu^{-1} \int_0^t \sigma_{23} dt \right) \right]$$
$$+ \frac{\lambda}{\lambda + 2\mu} \int_{-h}^{h} t\sigma_{33} dt + 2\mu \int_{-h}^{h} tdt \left[- \int_0^t \partial_{11} u_3 dt + \partial_1 \left(\mu^{-1} \int_0^t \sigma_{13} dt \right) \right],$$

$$M_2 = \int_{-h}^{h} t\sigma_{22} dt$$
$$= \lambda \int_{-h}^{h} tdt \left[\int_0^t -\partial_{\alpha\alpha} u_3 dt + \partial_\alpha \left(\mu^{-1} \int_0^t \sigma_{\alpha 3} dt \right) \right]$$
$$+ \frac{\lambda}{\lambda + 2\mu} \int_{-h}^{h} t\sigma_{33} dt + 2\mu \int_{-h}^{h} tdt \left[- \int_0^t \partial_{22} u_3 dt + \partial_2 \left(\mu^{-1} \int_0^t \sigma_{23} dt \right) \right],$$

$$M_{12} = M_{21} = \int_{-h}^{h} t\sigma_{12} dt$$
$$= \mu \int_{-h}^{h} tdt \left[-2 \int_0^t \partial_{12} u_3 dt + \partial_\alpha \left(\mu^{-1} \int_0^t \sigma_{3-\alpha 3} dt \right) \right].$$

Hence, taking these formulae into account we have the following representations for $M_{\alpha\beta;\alpha\beta}$ in terms of u_3 and σ_{i3}:

$$M_{\alpha\beta,\alpha\beta} = - \int_{-h}^{h} tdt \int_0^t [\partial_{11}(\lambda^* \Delta u_3) + 2\partial_{11}(\mu \partial_{11} u_3) + 4\partial_{12}(\mu \partial_{12} u_3)$$
$$+ \partial_{22}(\lambda^* \Delta u_3) + 2\partial_{22}(\mu \partial_{22} u_3)] \, dt + \partial_{11} \left(\frac{\lambda}{\lambda + 2\mu} \int_{-h}^{h} t\sigma_{33} dt \right)$$

$$+\partial_{22}\left(\frac{\lambda}{\lambda+2\mu}\int_{-h}^{h} t\sigma_{33}dt\right)+\partial_{11}\left((\lambda^*+2\mu)\partial_1\left(\mu^{-1}\int_0^t \sigma_{13}dt\right)\right.$$
$$\left.+\lambda^*\partial_2\left(\mu^{-1}\int_0^t \sigma_{23}dt\right)\right)+2\partial_{12}\left(\mu\partial_\alpha\left(\mu^{-1}\int_0^t \sigma_{3-\alpha 3}dt\right)\right.$$
$$\left.+\partial_{22}\left(\lambda^*\partial_1\left(\mu^{-1}\int_0^t \sigma_{13}dt\right)+(\lambda^*+2\mu)\partial_2\int_0^t \sigma_{23}dt\right)\right).$$

The further simplication of this formula gives us the following:

$$\frac{3}{2h^3}M_{\alpha\beta,\alpha\beta}=-\Delta((\lambda^*+2\mu)\Delta u_3^*)+2L[u_3^*,\mu]+\frac{3}{2h^3}\left[\partial_{\alpha\alpha}\left(\frac{\lambda}{\lambda+2\mu}\int_{-h}^{h} t\sigma_{33}dt\right)\right]$$
$$+\int_{-h}^{h} tdt\int_0^t\left[\Delta\sigma_{\alpha 3,\alpha}+\partial_\alpha((\mu\partial_\alpha\mu^{-1})\sigma_{\alpha 3,\alpha}+\partial_\beta(\mu\partial_\alpha\mu^{-1})\sigma_{\beta 3})\right]dt$$
$$+\Delta\int_{-h}^{h} tdt\int_0^t \lambda^*\partial_\alpha(\mu^{-1}\sigma_{\alpha 3})dt,$$

or

$$\begin{aligned}\Delta[(\lambda^* &+ 2\mu)\Delta u_3^*]\\ &=2(-1)^{\alpha+\beta}\partial_{3-\alpha}\partial_{3-\beta}\mu\partial_{\alpha\beta}u_3^*\\ &\quad+\frac{3}{2h^3}\left[g_3^+-g_3^-+\int_{-h}^{h} Zdt\right.\\ &\quad\left.-\int_{-h}^{h} t\left[\Delta\left(\sigma_{33}+\int_0^t Zdt\right)+\partial_\alpha\left((\mu\partial_\alpha\mu^{-1})\left(\sigma_{33}+\int_0^t Zdt\right)\right)\right]dt\right]\\ &\quad+\partial_{\alpha\alpha}\left(\frac{\lambda}{\lambda+2\mu}\int_{-h}^{h} t\sigma_{33}dt\right)\\ &\quad+\int_{-h}^{h} tdt\int_0^t\left[\Delta(\lambda^*(\partial_\alpha(\mu^{-1}\sigma_{\alpha 3})))+\partial_\alpha(\mu\partial_\alpha\mu^{-1})\sigma_{\beta 3}\right]dt.\end{aligned}$$

To deduce the equation corresponding to the intersecting forces Q_α we shall use the equality:

$$M_{\alpha\beta,\beta}=\int_{-h}^{h} t\sigma_{\alpha\beta,\beta}dt.$$

Further, using the representation of moments $M_{\alpha\beta}$ in terms of u_3^* and σ_{i3} we shall have:

$$Q_\alpha = -\frac{2h^3}{3}[\partial_\alpha(\lambda^*\Delta u_3^*) + 2\partial_{\alpha\beta}(\mu\partial_{\alpha\beta}u_3^*)] + \int_{-h}^{h} t\partial_\alpha\left[\frac{\lambda}{\lambda+2\mu}\sigma_{33}\right]dt$$

$$+\partial_\alpha\left[\lambda^*\partial_\beta\left(\mu^{-1}\int_0^t \sigma_{\beta3}dt\right)\right] + \int_{-h}^{h} t\partial_\alpha\left[\lambda^*\partial_\beta\left(\mu^{-1}\int_0^t \sigma_{\beta3}dt\right)\right]dt$$

$$+2\int_{-h}^{h} t\left[\mu\partial_\alpha\left(\mu^{-1}\int_0^t \sigma_{\alpha3}dt\right)\right]dt + \partial_{3-\alpha}\left[\mu\partial_\beta\left(\mu^{-1}\int_0^t \sigma_{3-\beta,3}dt\right)\right].$$

Finally, using the equilibrium equation (1.1), we represent (2.6) for ψ_α, as well as in homogeneous case; *exact nonlocal representations*, corresponding to the problem (B) and based on formulae of this point, take the form:

$$u_\alpha^* = -u_{3,\alpha}^* + D_1^{-1}\psi_\alpha, \tag{3.1}$$

$$M_\alpha = -D(\partial_\alpha^2 + \nu\partial_{3-\alpha}^2)u_3^* + \lambda^*\mu_{1,\beta}\psi_\beta + 2\mu\partial_\alpha(\mu_1\psi_\alpha)$$
$$-\nu(1-\nu)^{-1}\int_{-h}^{h}\left(t\sigma_{33} - (h^2-t^2)Z\right)dt, \tag{3.2}$$

$$M_{12} = -(1-\nu)Du_{3,12}^* + \psi_{\alpha,3-\alpha} + \mu\mu_{1,3-\alpha}\psi_\alpha, \tag{3.3}$$

$$Q_\alpha = -\partial_\alpha(D\Delta u_3^*) + \partial_\alpha(\lambda^*\mu_{1,\beta}\psi_\beta) - \partial_\beta[\mu(\mu_{1,\beta}\psi_\alpha + \mu_{1,\alpha}\psi_\beta)] + \Delta\psi_\alpha$$
$$-\partial_\alpha\left[(1-\nu)^{-1}\int_{-h}^{h}\left(t\sigma_{33} + (1+\nu)(h^2-t^2)Z\right)dt\right], \tag{3.4}$$

$$\Delta(D\Delta u_3^*) - 2(-1)^{\alpha+\beta}D_{1,\alpha\beta}\partial_{3-\alpha}\partial_{3-\beta}u_3^* + \Delta(D_{2,\alpha}\psi_\alpha)$$
$$-(-1)^{\alpha+\beta}\partial_{\alpha\beta}(D_{3,3-\alpha}\psi_{3-\beta})$$
$$= g_3^+ - g_3^- + \int_{-h}^{h} Z dt - \Delta\left[\frac{1}{1-\nu}\int_{-h}^{h}[(2-\nu)t\sigma_{33} + (h^2-t^2)Z]\,dt\right]. \tag{3.5}$$

In these formulae, as in section 2, we have

$$u_\alpha^* = \frac{3}{2h^3}\int_{-h}^{h} tu_\alpha dt, \quad u_3^* = \frac{3}{4h^3}\int_{-h}^{h}(h^2-t^2)u_3 dt, \quad M_\alpha = \int_{-h}^{h} t\sigma_{\alpha\alpha}dt,$$

$$M_{12} = \int_{-h}^{h} t\sigma_{12}dt, \quad Q_\alpha = \int_{-h}^{h}\sigma_{\alpha3}dt, \quad \psi_\alpha = \frac{1}{2}\int_{-h}^{h}(h^2-t^2)\sigma_{\alpha3}dt,$$

$$\mu_1 = \frac{1}{\mu}, \quad D_1 = \frac{2}{3}h^3\mu, \quad D_{2,\alpha} = \frac{1}{1-\nu}D_{3,\alpha}, \quad D_3 = \ln\mu.$$

Now, if formulae (2.15) and (2.16) are used then from system (3.1) – (3.5) we obtain a one-parameter scheme, corresponding to non-homogeneous case. Let us use the representation (3.1) – (3.5) without remainder terms also called (P_h) scheme. Thus, the following is true:

3.1.1. With $\gamma = -0.5$ the classic theory of plate bending comes from (P_h); in other cases the models with respect to γ obtained either correspond to already known refined theories, or are new ones.

3.1.2. The principles of this point, obviously, are valid for an anisotropic case, having been considered in the previous paragraph.

3.2. Dynamic case

Now, let us consider the dynamic case. It is quite enough, for our purposes (see p.8), to consider Ω_h to be an isotropic homogeneous medium.

Let us use relations (3.1) – (3.5) and let $Z = -f_3(x,t) - \rho\partial_{tt}u_3(x,t)$, where $\partial_t = \partial/\partial t$, $x = (x_1, x_2, z)$, t is time, ρ is material density. Then from (3.5) we shall obtain immediately:

$$\begin{aligned}(D\Delta^2 + 2h\rho\partial_{tt} - 2DE^{-1}(1+\nu)\rho\partial_{tt}\Delta)u_3^h \\ = g_3^+ - g_3^- - \frac{2-\nu}{1-\nu}\frac{h^2(1+2\gamma)}{3}\Delta(g_3^+ - g_3^-) \\ - \int_{-h}^{h}\left(1 - \frac{1}{1-\nu}(h^2 - z^2)\Delta\right) f_3 dz.\end{aligned} \tag{R. – L.}$$

Basing on (3.5), $2\rho(1-\nu)^{-1}$ is the rigorous meaning of the operator coefficient $\partial_{tt}\Delta$ on the class of **Rayleigh-Lamb two-dimensional (R.-L.) models.**

(R.-L.) Models have for a long time been the domain of paramount interest for many scientists, begining with Cree and Pohhammer (see Love [1959]). The detailed analysis of the results, physical aspects and the proper bibliography of the early period are given in the monographs Love [1959], Rayleigh [1977], Timoshenko, Voinovski–Krieger [1959]. The construction of refined theories for the dynamic case (in proper sense) offers no difficulty (e.g., see the deduction of the equation (1.27) (Ambartsumian [1967], ch.IV, §1). Therefore, among the works of the later period, we pay special attention to the papers of Kennard [1953] and Berdichevski [1973, 1983], which by their content are closer to our investigations.

3.3. Problems connected with lateral boundary conditions, paradoxes of classical refined theories

In the case of a nonhomogeneous plate, in order to regulate the strongly elliptical system, corresponding to the equations (3.4) and (3.5), *it is necessary for ∂D to give four linear independent conditions* .

In this connection we shall consider beforehand two examples with their further explanations.

3.3.1. Example 1. Let Ω_h in problem (B) be isotropic and homogeneous, $g = 0$; the part of boundary conditions averaged by ∂D splits with respect to u_3^h and for the definitions takes the form:

$$u_3^h = \frac{\partial}{\partial n} u_3^h = 0, \quad (x_1, x_2) \in \partial D, \tag{3.6}$$

where $\partial/\partial n$ is the normal derivative.

Furthermore, let us assume $g_3^+ = q(x_1, x_2)$, $g_3^- = 0$.

Then the system of differential equations, corresponding to refined theories of plate bending, takes the form

$$D\Delta^2 u_3^h = q - \frac{2-\nu}{1-\nu}\frac{1+2\gamma_1}{3} h^2 \Delta q,$$
$$\frac{h^2(1+2\gamma_2)}{3}\Delta Q_\alpha^h - Q_\alpha^h = D\Delta u_{3,\alpha}^h + \frac{h^2(1+2\gamma_3)}{3(1-\nu)}\partial_\alpha q. \tag{3.7}$$

Let us choose the function $q(x_1, x_2)$ so that it is the nontrivial solution of Helmholtz's equation:

$$\frac{h^2(2-\nu)(1+2\gamma_1)}{3(1-\nu)}\Delta q - q = 0, \quad (x_1, x_2) \in D.$$

Subsequently, if the homogeneous equation (3.7) with conditions (3.6) has a single solution, hence, $u_3^h = 0$ in $\bar{D}$; and this, of course, does not coincide with the problem (B) for the arbitrariness of transversal load $q(x_1, x_2)$.

3.3.2. Example 2. Let Ω_h be isotropic and homogeneous, $g \equiv 0$.

Let us consider a sequence of functions $q_k = q_k(x_1, x_2)$, contained in the right hand parts (3.7) such as

$$\lim_{k\to\infty} q_k(x_1, x_2) = \begin{cases} q^0(x_1, x_2) \not\equiv 0, & (x_1, x_2) \in \partial D, \\ 0, & (x_1, x_2) \in D. \end{cases}$$

Hence, if the problems (3.7) with corresponding boundary conditions averaged by ∂D and with their right hand parts, being formed by means of $q_k(x_1, x_2)$ are solved subsequently, then sequences $(u_3^h)_k$ and $(Q_\alpha^h)_k$ will tend to a zero of a function in $\bar{D}$ in case of unique solution, and will contradict the solution of the boundary three-dimensional problem (B).

Let us, now, return to the problems of boundary conditions, connected with scheme (P_h). Let the lateral surface $S_1 \subset S$ be the part of the plane with normal Ox_1. The classical homogeneous boundary conditions given on S_1 are formulated then as:

a rigidly clamped edge: $u_1 = u_2 = u_3 = 0$,
a free edge: $\sigma_{11} = \sigma_{12} = \sigma_{13} = 0$,
a hinge-supported edge: $\sigma_{11} = u_2 = u_3 = 0$.

In case of rigid clamping for the averaged boundary conditions we have:

$$u_3^* = 0, \tag{3.8}$$

Further, by virtue of $u_1^* = \sigma_{23} = 0$ two more conditions follow:

$$u_{3,1}^* = D_1^{-1}\psi_1, \quad Q_2 = \psi_2 = 0. \tag{3.9}$$

Since conditions $(u_3, 1) = 0$ identify the equality $(u_{3,3}, t) = 0$, we have:

$$u_{1,1}^* = D_3^{-1} \int_{-h}^{h} t\sigma_{33} dt = D_3^{-1}(\sigma_{33}, t).$$

Now using formula (3.1) together with the last relation, we have:

$$u_{3,11}^* = \partial_1(D_1^{-1}\psi_1) - D_3^{-1} \int_{-h}^{h} t\sigma_{33} dt. \tag{3.10}$$

Conditions (3.8) – (3.10) will form four independent boundary conditions, corresponding to equations (3.4) – (3.5), if the remainder terms are neglected.

It should be noted that two linear indepedent conditions $(u_\alpha, 1) = 0$ correspond to the equations derived from (2.22).

Let us pay attention to the integral term from condition (3.10). It can be calculated by virtue of formula (2.15) within the accuracy of refined theories of section 2. Just the same term appears in other basic averaged boundary conditions, too.

Indeed, using equation (3.2) and assuming for convenience that $Z = 0$ we have: in case of a free edge

$$D(u_{3,11}^* + \nu u_{3,22}^*) = \frac{2}{1-\nu}(\psi_{1,1} + \nu\psi_{2,2}) - \nu(1-\nu)^{-1} \int_{-h}^{h} t\sigma_{33} dt, \tag{3.11}$$

in case of a hinge-supported edge

$$Du_{3,11}^* = \lambda^* \mu_{1,1}\psi_1 + 2\psi_{1,1} - \nu(1-\nu)^{-1} \int_{-h}^{h} t\sigma_{33} dt. \tag{3.12}$$

Let us return to examples 1 and 2 and note the following:

3.3.3. The resonance phenomena of Example 1 can be neglected by choosing nonhomogeneous models of mixed type, in terms of subsection 2.7. For that in system (3.7) it is quite enough to assume $\gamma_1 = -0.5$, $\gamma_{4-\alpha} \neq -0.5$. Investigating the problem of models from the scheme (P_h), one must pay attention to the following remarks:

Remark 3.1. According to the papers by Kondratiev [1967], Mazija, Nazarov, Plamenievski [1981] and other investigations of this trend (e.g., see appropriate bibliography in Goldenveizer [1976], Parton, Perlin [1981], Vorovich [1965]), on the class of solution of boundary value problems in domains with angular edges, that all refined theories, the classic one inclusive (it was mentioned in subsection 1.4), have a similar order of exactness with respect to h;

Remark 3.2. In the theory of plates and shells a great deal of information has been accumulated, which *a priori* displays, more or less, peculiar physical phenomena;

Remark 3.3. In problems important for practice and difficult for calculation, the stress–strain state, changing in different parts of the plate, takes, evidently a nonhomogeneous character.

3.3.4. The account of the integral correction term in averaged boundary conditions (3.10) – (3.12), corresponding to system (3.7), reduces to nonhomogeneous boundary value problem and explains the paradox of Example 2.

In this connection, let us note that some of boundary value problems with local tranversal loads of the plate's edge were already investigated in Lukasiewicz [1979] (ch.4). The solution of some important practical problems of this aspect were given there, too.

Remark 3.4. As is known, the influence of interior stress state on the plate's edge and the introduction of some correction in the boundary conditions, corresponding to the classic theory, were the main subject of many investigations (e.g., see Friedrichs, Dressler [1961], Goldenveiser [1976], with vast literature). Within this monograph Goldenveiser [1976] (p.114) *we should notice that edge effects of that kind are not connected with surface loads, as distinct from transversal loads' influences, acting on the plate's edge and introduced by virtue of conditions (3.10) – (3.12).*

Basing on the abovesaid, let us make the following remarks on the problems of satisfying conditions at the boundary ∂D.

As it is seen from system (3.4), it cannot be represented by a single equation of second-order only, as it was in a homogeneous case due to the desired function of uniformal differential form of first-order from Q_α. Although the introduction of such a function makes refined theories of plates more coordinated, its presence contains definite arbitrariness.

Thus, taking into account all the aforementioned and reasoning of this paragraph with respect to boundary conditions (see (3.8) – (3.10)), we should coordinate the quantity and the exactness of approximation of conditions on ∂D with scheme (P_h) (comp. Donnel [1976], §5).

4. On Construction of Refined Theories of Elastic Plates with Variable Thickness

The method of constructing the finite models of group A without any simplification hypotheses, given in section 2, admits generalization in the case of plates with variable thickness, i.e., when

$$\Omega_h(x_1, x_2, x_3) = D(x_1, x_2) \times]h_1(x_1, x_2), h_2(x_1, x_2)[.$$

For clearness, we shall examine the linear isotropic homogeneous case.

4.1. Models corresponding to bending problems

Let us introduce the subsidiary functions, connected with transverse component of displacement vector $u_3(x_1, x_2, x_3)$ and tangent tensions $\sigma_{\alpha_3}(x_1, x_2, x_3)$ of exterior stress vector, acting on the plane with normal Oz:

$$v_\alpha(x_1, x_2, x_3) = u_{3,\alpha} - \mu^{-1}\sigma_{\alpha 3}. \tag{4.1}$$

Then, using formulae, defined from (1.6), we have

$$u_{3,3} = \frac{1}{\lambda + 2\mu}\sigma_{33} - \frac{\lambda}{\lambda + 2\mu}u_{\alpha,\alpha},$$

and

$$u_{\alpha,\alpha}(x_1, x_2, z) = u_{\alpha,\alpha}(x_1, x_2, \hbar) + v_\alpha(x_1, x_2, \hbar)\partial_\alpha \hbar - \int\limits_{\hbar}^{z} v_{\alpha,\alpha}dt,$$

for the values $(t - \hbar, \sigma_{\alpha\beta,\gamma})$ we shall obtain:

$$\int\limits_{h_1}^{h_2}(t-\hbar)\sigma_{\alpha\alpha,\alpha}dt = -\int\limits_{h_1}^{h_2}(t-\hbar)\partial_\alpha\left[\int\limits_{\hbar}^{t}(\lambda^* v_{\beta,\beta} + 2\mu v_{\alpha,\alpha})dt - \frac{\lambda}{\lambda+2\mu}\sigma_{33}\right]dt, \tag{4.2}$$

$$\int\limits_{h_1}^{h_2}(t-\hbar)\sigma_{12,\alpha}dt = -\mu\int\limits_{h_1}^{h_2}(t-\hbar)\partial_\alpha\left[\int\limits_{\hbar}^{t}(v_{1,2}+v_{2,1})dt\right]dt. \tag{4.3}$$

Based on these formulae (4.1) – (4.3) and by means of reasoning from the derivation in section 2 reducing the exact nonlocal representations, we shall obtain:

$$Q_\alpha - \frac{1}{2}\int\limits_{h_1}^{h_2}\left(h^2 - (t-\hbar)^2\right)\Delta\sigma_{\alpha 3}dt$$

$$= -\frac{\lambda^* + 2\mu}{2}\int\limits_{h_1}^{h_2}\left(h^2 - (t-\hbar)^2\right)\Delta u_{3,\alpha}dt$$

$$+h(\sigma_{\alpha 3}^{+} + \sigma_{\alpha 3}^{-}) - \int\limits_{h_1}^{h_2}(t-\hbar)f_\alpha dt - \frac{2(\lambda+\mu)}{\lambda+2\mu}\int\limits_{h_1}^{h_2}(t-\hbar)\sigma_{33,\alpha}dt$$

$$+\frac{3\lambda+2\mu}{2(\lambda+2\mu)}\int\limits_{h_1}^{h_2}\left(h^2-(t-\hbar)^2\right)f_{3,\alpha}dt;$$

$$\frac{1}{2}\partial_\alpha \int\limits_{h_1}^{h_2} \left(h^2 - (t-\hbar)^2\right) \left[(\lambda^2 + 2\mu)\Delta u_{3,\alpha} - \Delta\sigma_{\alpha 3}\right] dt$$

$$= \sigma_{33}^+ - \sigma_{33}^- - \int\limits_{h_1}^{h_2} f_3 dt$$

$$+\partial_\alpha \left[h(\sigma_{\alpha 3}^+ + \sigma_{\alpha 3}^-) - \int\limits_{h_1}^{h_2} (t-\hbar) f_\alpha dt + \frac{2(\lambda+\mu)}{\lambda+2\mu} \int\limits_{h_1}^{h_2} (t-\hbar)\sigma_{33,\alpha} dt - \right.$$

$$\left. - \frac{3\lambda + 2\mu}{2(\lambda+2\mu)} \int\limits_{h_1}^{h_2} \left(h^2 - (t-\hbar)^2\right) f_{3,\alpha} dt \right].$$

Hence there follows:

$$\frac{3}{4h^3} D\partial_\alpha \int\limits_{h_1}^{h_2} \left(h^2 - (t-\hbar)^2\right) \Delta u_{3,\alpha} dt$$

$$= \sigma_{33}^+ - \sigma_{33}^- - \int\limits_{h_1}^{h_2} f_3 dt$$

$$+\partial_\alpha \left[h(\sigma_{\alpha 3}^+ + \sigma_{\alpha 3}^-) - \int\limits_{h_1}^{h_2} (t-\hbar) f_\alpha dt \right] - (h_{2,\alpha}\sigma_{\alpha 3}^+ - h_{1,\alpha}\sigma_{\alpha 3}^-)$$

$$- \frac{h}{1-\nu}(\sigma_{33,\alpha}^+ h_{2,\alpha} + \sigma_{33,\alpha}^- h_{1,\alpha}) + \frac{1}{1-\nu} \int\limits_{h_1}^{h_2} \left(h^2 - (t-\hbar)^2\right) \Delta f_3 dt$$

$$- \frac{1+\nu}{1-\nu} \int\limits_{h_1}^{h_2} (h\partial_\alpha h + (z-\hbar)\partial_\alpha \hbar) f_{3,\alpha} dt - \frac{2-\nu}{1-\nu} \int\limits_{h_1}^{h_2} (t-\hbar)\Delta\sigma_{33} dt$$

$$+ \int\limits_{h_1}^{h_2} (h\partial_\alpha h + (t-\hbar)\partial_\alpha \hbar)\Delta\sigma_{\alpha 3} dt, \tag{4.4}$$

$$Q_\alpha - \frac{1}{2} \int\limits_{h_1}^{h_2} \left(h^2 - (t-\hbar)^2\right) \Delta\sigma_{\alpha 3} dt$$

$$= -\frac{3}{4h^3} D \int\limits_{h_1}^{h_2} \left(h^2 - (t-\hbar)^2\right) \Delta u_{3,\alpha} dt + h(\sigma_{\alpha 3}^+ - \sigma_{\alpha 3}^-) - \int\limits_{h_1}^{h_2} (t-\hbar) f_\alpha dt$$

$$-\frac{1\cdot}{1-\nu}\int_{h_1}^{h_2}(t-\hbar)\sigma_{33,\alpha}dt+\frac{1+\nu}{2(1-\nu)}\int_{h_1}^{h_2}\left(h^2-(t-\hbar)^2\right)f_{3,\alpha}dt. \quad (4.5)$$

For the values, characterizing angular deformations u^*_α and moments $M_{\alpha\beta}$, taking into account Hooke's law and formulae (4.1) – (4.3),we obtain:

$$u^*_\alpha=-\frac{3}{4h^3}\int_{h_1}^{h_2}\left(h^2-(t-\hbar)^2\right)v_\alpha(x_1,x_2,t)dt, \quad (4.6)$$

$$M_\alpha=-\frac{3D}{4h^3}\int_{h_1}^{h_2}\left(h^2-(t-\hbar)^2\right)(v_{\alpha,\alpha}+\nu v_{3-\alpha,3-\alpha})dt+\frac{\nu}{1-\nu}\int_{h_1}^{h_2}(t-\hbar)\sigma_{33}dt, \quad (4.7)$$

$$M_{12}=-\frac{3(1-\nu)D}{4h^3}\int_{h_1}^{h_2}\left(h^2-(t-\hbar)^2\right)(v_{1,2}+v_{2,1})dt \quad (4.8)$$

Now we shall return to formulae (4.4) and (4.5) and construct a scheme of type (P_h) in the case under consideration.

Let us formulate the integrals containing functions u_3 and $\sigma_{\alpha 3}$ by means of Simpson's formula: [1]

$$\begin{aligned}I_1&=\int_{h_1}^{h_2}\left(h^2-(t-\hbar)^2\right)\Delta u_{3,\alpha}dt\\&=\frac{4h^3}{3}\Delta u_{3,\alpha}(x_1,x_2,\hbar)+\rho_{Sm}\left[\left(h^2-(t-\hbar)^2\right)\Delta u_{3,\alpha}\right],\\I_2&=\int_{h_1}^{h_2}\left(h^2-(t-\hbar)^2\right)\Delta\sigma_{\alpha 3}dt\\&=\frac{4h^3}{3}\Delta\sigma_{\alpha 3}(x_1,x_2,\hbar)+\rho_{Sm}\left[\left(h^2-(t-\hbar)^2\right)\Delta\sigma_{\alpha 3}\right],\end{aligned} \quad (4.9)$$

$$Q_\alpha=\int_{h_1}^{h_2}\sigma_{\alpha 3}dt=\frac{4h}{3}\sigma_{\alpha 3}(x_1,x_2,\hbar)+\frac{h}{3}(\sigma^+_{\alpha 3}+\sigma^-_{\alpha 3})+\rho_{Sm}[\sigma_{\alpha 3}],$$

$$I_3=\int_{h_1}^{h_2}(h\partial_\alpha h+(t-\hbar)\partial_\alpha\hbar)\Delta\sigma_{\alpha 3}dt \quad (4.10)$$

---

[1] Using formulae of type (2.15), we have:

$$I_1=\gamma I_1+(1-\gamma)I_1=\frac{4h^3(1+2\gamma)}{3}\Delta u_{3,\alpha}(x_1,x_2,\hbar)+((1-\gamma)\rho_{Sm}+\gamma\rho_G)\left[\left(h^2-(t-\hbar)^2\right)\Delta u_{3,\alpha}\right];$$

analogously, we have same expressions for integrals I_2, Q_α and I_3 too.

$$= \frac{4h^2}{3}\partial_\alpha h\Delta\sigma_{\alpha3}(x_1,x_2,\hbar) + \frac{h^2}{3}\left[(\partial_\alpha h + \partial_\alpha \hbar)\Delta\sigma^+_{\alpha3}\right.$$
$$\left. + \quad (\partial_\alpha h - \partial_\alpha\hbar)\Delta\sigma^-_{\alpha3}\right] + \rho_{Sm}\left[(h\partial_\alpha h + (t-\hbar)\partial_\alpha\hbar)\Delta\sigma_{\alpha3}\right].$$

Taking into account these formulae or representations of type (2.15) when $\gamma = 0$ in equations (4.4) - (4.5), we shall have:

$$\frac{1}{h^3}D\partial_\alpha\left(h^3\Delta u_{3,\alpha}(x_1,x_2,\hbar)\right)$$
$$= \sigma^+_{33} - \sigma^-_{33} - \int_{h_1}^{h_2} f_3 dt$$
$$+\partial_\alpha\left[h(\sigma^+_{\alpha3}+\sigma^-_{\alpha3}) - \int_{h_1}^{h_2}(t-\hbar)f_\alpha dt\right] - (h_{2,\alpha}\sigma^+_{\alpha3} - h_{1,\alpha}\sigma^-_{\alpha3})$$
$$-\frac{h}{1-\nu}(\sigma^+_{33,\alpha}h_{2,\alpha} + \sigma^-_{33,\alpha}h_{1,\alpha}) + \frac{1}{1-\nu}\int_{h_1}^{h_2}\left(h^2-(t-\hbar)^2\right)\Delta f_3 dt$$
$$-\frac{1+\nu}{1-\nu}\int_{h_1}^{h_2}(h\partial_\alpha h + (t-\hbar)\partial_\alpha\hbar)f_{3,\alpha}dt$$
$$-\frac{h^2(2-\nu)}{3(1-\nu)}(1+2\gamma)(\Delta\sigma^+_{33} - \Delta\sigma^-_{33})$$
$$+\frac{h^2}{3}\left[\partial_\alpha(h+\hbar)\Delta\sigma_{\alpha3} + \partial_\alpha(h-\hbar)\Delta\sigma^-_{\alpha3}\right]$$
$$+\frac{4h^2\partial_\alpha h}{3}\Delta\sigma_{\alpha3}(x_1,x_2,\hbar) + R_1[u_3(x_1,x_2,\hbar);\gamma], \qquad (4.11)$$

where

$$R_1[u_3,\gamma] = -\frac{3D}{4h^3}\partial_\alpha\rho_{Sm}\left[\left(h^2-(t-\hbar)^2\right)\Delta u_{3,\alpha}\right]$$
$$+\rho_{Sm}\left[(h\partial_\alpha h + (t-\hbar)\partial_\alpha\hbar)\Delta\sigma_{\alpha3}\right] - \frac{2-\nu}{1-\nu}r_1\left[(t-\hbar)\Delta\sigma_{33};\gamma\right].$$

Further

$$\frac{4h}{3}\sigma_{\alpha3}(x_1,x_2,\hbar) - \frac{2h^3}{3}\Delta\sigma_{\alpha3}(x_1,x_2,\hbar)$$
$$= -D\Delta u_{3,\alpha}(x_1,x_2,\hbar) + \frac{2}{3}h(\sigma^+_{\alpha3}+\sigma^-_{\alpha3})$$
$$-\int_{h_1}^{h_2}(t-\hbar)f_\alpha dt - \frac{(1+2\gamma)h^2}{3(1-\nu)}(\sigma^+_{33,\alpha} - \sigma^-_{33,\alpha})$$
$$+\frac{1+\nu}{2(1-\nu)}\int_{h_1}^{h_2}\left(h^2-(t-\hbar)^2\right)f_{3,\alpha}dt + R_{1+\alpha}[Q_\alpha;\gamma], \qquad (4.12)$$

where

$$R_{1+\alpha}[Q_\alpha;\gamma] = -\rho s_m[\sigma_{\alpha 3}] + \frac{1}{2}\rho s_m\left[\left(h^2-(t-\hbar)^2\right)\Delta\sigma_{\alpha 3}\right] - \\ -\frac{3D}{4h^3}\rho s_m\left[\left(h^2-(t-\hbar)^2\right)\Delta u_{3,\alpha}\right] - \frac{1}{1-\nu}r_1\left[(t-\hbar)\sigma_{33,\alpha};\gamma\right].$$

Similarly, for values u_α^* and $M_{\alpha\beta}$ from formulae (4.6) - (4.8) we obtain:

$$u_\alpha^* = -v_\alpha(x_1,x_2,\hbar) - \frac{3}{4h^3}\rho s_m\left[\left(h^2-(t-\hbar)^2\right)v_\alpha\right], \tag{4.13}$$

$$\begin{aligned} M_\alpha = &-D\left[v_{\alpha,\alpha}(x_1,x_2,\hbar) + \nu v_{3-\alpha,3-\alpha}(x_1,x_2,\hbar)\right] \\ &+\frac{h^2(1+2\gamma)\nu}{3(1-\nu)}(\sigma_{33}^+ - \sigma_{33}^-) \\ &-\frac{3D}{4h^3}\rho s_m\left[\left(h^2-(t-\hbar)^2\right)(v_{\alpha,\alpha}+\nu v_{3-\alpha,3-\alpha})\right] \\ &+\frac{\nu}{1-\nu}r_1[(t-\hbar)\sigma_{33};\gamma], \end{aligned} \tag{4.14}$$

$$\begin{aligned} M_{12} = &-(1-\nu)D\left[v_{1,2}(x_1,x_2,\hbar) + v_{2,1}(x_1,x_2,\hbar)\right] \\ &-\frac{3(1-\nu)}{4h^3}D\rho s_m\left[\left(h^2-(t-\hbar)^2\right)(v_{1,2}+v_{2,1})\right]. \end{aligned} \tag{4.15}$$

Obviously, while uniting these formulae (4.10) – (4.14) we obtain the parametric representation of type of refined theory of plate bending. This system, by analogy with section 2, will be put down in the form of the operator equation, adding boundary conditions given on S. Thus we obtain:

$$P_\hbar(\partial_1,\partial_2;\gamma)U(x_1,x_2,\hbar) = G(x_1,x_2;\gamma) + R(x_1,x_2;\gamma,\hbar).$$

If the remainder vector R is neglected, we have:

$$P_\hbar(\partial_1,\partial_2;\gamma)U^\hbar = G(x_1,x_2;\gamma). \tag{$P_\hbar$}$$

Here the equations with regard to $u_3^\hbar, \sigma_{\alpha 3}^\hbar$ constitute an independent strongly elliptic system of differential equations, and if these values with averaged boundary conditions on ∂D are found, we should obtain the approximate values for components of a rotation of normals $u_\alpha^\hbar$ and for bending and twisting moments $M_{\alpha\beta}^\hbar$.

For plates with variable thickness we shall make the following special remark: as is now seen from exact nonlocal representations (4.10) – (4.11), the summand, depending on $\Delta\sigma_{\alpha 3}(x_1,x_2,\hbar)$ appears in the equation for deflection (4.10). The analogous term appears in the equation (4.11) and the boundary conditions form the boundary layer phenomenon. In this case, as a consequence, the system of differential equations with respect to values $u_3^\hbar, \sigma_{\alpha 3}^\hbar$ and defined from ($P_\hbar$) appears to be linked, unlike the plates of constant thickness. Moreover, similar to the homogeneous case, discussed in section 3, the representation of the system, corresponding to formula (4.11), is possible in the form of a single equation, however, now this procedure is less effective.

4.2. Models corresponding to plane stress-strained problems

Let us construct the system with respect to the averaged horisontal vector components u_α, being in accordance with the problem of generalized plane stress state for the case of plates with constant thickness.

The method of Filon [1903], with respect to the values containing σ_{33}, gives:

$$\sigma_{\alpha\beta,\beta} = \mu\Delta u_\alpha + (\lambda^* + \mu)\partial_\alpha u_{\beta,\beta} + \frac{\lambda}{\lambda + 2\mu}\sigma_{33,\alpha}. \tag{4.16}$$

By virtue of Hooke's law we obtain:

$$\begin{aligned}
&\Delta u_\alpha(x_1, x_2, z) \\
&= \Delta u_\alpha(x_1, x_2, \hbar) - \Delta \int_{\hbar}^{z} v_\alpha dt \\
&= \Delta u_\alpha(x_1, x_2, \hbar) + \Delta \hbar v_\alpha(x_1, x_2, \hbar) + 2\partial_\beta \hbar v_{\alpha,\beta}(x_1, x_2, \hbar) - \int_{\hbar}^{z} \Delta v_\alpha dt.
\end{aligned}$$

$$\begin{aligned}
&\partial_\alpha u_{\beta,\beta}(x_1, x_2, z) \\
&= \partial_\alpha \left[u_{\beta,\beta}(x_1, x_2, \hbar) + \partial_\beta \hbar v_\beta(x_1, x_2, \hbar) - \int_{\hbar}^{z} v_{\beta,\beta} dt \right] \\
&= u_{\beta,\alpha\beta}(x_1, x_2, \hbar) + \partial_{\alpha\beta} \hbar v_\beta(x_1, x_2, \hbar) + \partial_\alpha \hbar v_{\beta,\beta}(x_1, x_2, \hbar) \\
&\quad + \partial_\beta \hbar v_{\beta,\alpha}(x_1, x_2, \hbar) - \int_{\hbar}^{z} v_{\beta,\alpha\beta} dt.
\end{aligned}$$

Then, taking into account the last equalities in (4.15), we have:

$$\begin{aligned}
\frac{1}{2h}\int_{h_1}^{h_2} \sigma_{\alpha\beta,\beta} dt &= \mu\Delta u_\alpha(x_1, x_2, \hbar) + (\lambda^* + \mu)\partial_\alpha u_{\beta,\beta}(x_1, x_2, \hbar) \\
&\quad + \mu\Bigg(\Delta \hbar v_\alpha(x_1, x_2, \hbar) + 2\partial_\beta \hbar v_{\alpha,\beta}(x_1, x_2, \hbar) \\
&\qquad - \frac{1}{2h}\int_{h_1}^{h_2} dt \int_{\hbar}^{t} \Delta v_\alpha dt \Bigg) \\
&\quad + (\lambda^* + \mu)\left[\partial_{\alpha\beta} \hbar v_\beta(x_1, x_2, \hbar) + \right. \\
&\quad \left. + \partial_\alpha \hbar v_{\beta,\beta}(x_1, x_2, \hbar) + \partial_\beta \hbar v_{\beta,\alpha}(x_1, x_2, \hbar) - \frac{1}{2h}\int_{h_1}^{h_2} dt \int_{\hbar}^{t} v_{\beta,\alpha\beta} dt \right]
\end{aligned}$$

$$+\frac{\lambda}{\lambda+2\mu}\frac{1}{2h}\int_{h_1}^{h_2}\sigma_{33,\alpha}dt$$

$$=-\frac{1}{2h}(\sigma^{+}_{\alpha 3}-\sigma^{-}_{\alpha 3})+\frac{1}{2h}\int_{h_1}^{h_2}f_\alpha(x_1,x_2,t)dt.$$

Using Dirichlet's formula (for repeated integrals, containing v) and quadrature formula of trapezoid, after some calculations the last relation will take the form

$$\mu\Delta u_\alpha(x_1,x_2,\hbar)+(\lambda^*+\mu)\partial_\alpha u_{\beta,\beta}(x_1,x_2,\hbar)=f^{\hbar}_\alpha+R_\alpha[u_\alpha]. \tag{4.17}$$

Here

$$\begin{aligned}
&f^{\hbar}_\alpha(x_1,x_2)\\
&\quad=\frac{1}{2h}\left(\int_{h_1}^{h_2}f_\alpha(x_1,x_2,t)dt-\sigma^{+}_{\alpha 3}+\sigma^{-}_{\alpha 3}\right)-\frac{\lambda}{2(\lambda+2\mu)}(\sigma^{+}_{33,\alpha}+\sigma^{-}_{33,\alpha})\\
&\quad-\mu\left[\Delta\hbar v_\alpha(x_1,x_2,\hbar)+2\partial_\beta\hbar v_{\alpha,\beta}(x_1,x_2,\hbar)\right]\\
&\quad-(\lambda^*+\mu)\left[\partial_{\alpha\beta}\hbar v_\beta(x_1,x_2,\hbar)+\partial_\alpha\hbar v_{\beta,\beta}(x_1,x_2,\hbar)+\partial_\beta\hbar v_{\beta,\alpha}(x_1,x_2,\hbar)\right],\\
&R_\alpha[u_\alpha(x_1,x_2,\hbar)]\\
&\quad=\frac{1}{2h}\left[\mu(\rho_{tr}\left[(h_1-t)\Delta v_\alpha\right]+\rho_{tr}\left[(h_2-t)\Delta v_\alpha\right])\right]\\
&\quad+(\lambda^*+\mu)\left[\rho_{tr}\left[(h_1-t)v_{\beta,\alpha\beta}\right]+\rho_{tr}\left[(h_2-t)v_{\beta,\alpha\beta}\right]\right]\\
&\quad-\frac{\lambda}{2h(\lambda+2\mu)}\rho_{tr}[\sigma_{33,\alpha}].
\end{aligned}$$

Let us add equation (4.16) without the remainder vector to scheme (P_h). We should mention that the values $v_\alpha(x_1,x_2,\hbar)$ are approximately defined from system (P_h) by virtue of the equality (4.1).

Thus, in the case of a plate with variable thickness the truth of the following conclusion is proved.

So, the problem of definition of stress–strain state of the plate of variable thickness is split into two problems, which should be solved consecutive: first, we must solve the problem of values from scheme (P_h), characterizing the deflection of average surface $u_3(x_1,x_2,\hbar)$, intersecting forces Q_α, defined by means of $\sigma_{\alpha 3}(x_1,x_2,\hbar)$ horizontal rotation of normals u^*_α and moments $M_{\alpha\beta}$; then in the second part we must solve the problem of finding functions $u_\alpha(x_1,x_2,\hbar)$ of horizontal components of displacement vector $u(x_1,x_2,\hbar)$ by (4.16) (ignoring the remainder term) and taking into account conditions on S. With all this, as it follows from (4.16), the righthand parts of $f^{\hbar}_\alpha$ depend on the values, defined from the first problem.

Remark 4.1. The above given processes for plates with variable thickness are valid when functions σ_{i3} are known on $S^{\pm}$. When the surfaces are not sufficiently shallow let the boundary conditions be:

$$\sigma_n = (\sigma_{1n}, \sigma_{2n}, \sigma_{3n}) = g^{\pm}, \qquad x \in S^{\pm}, \tag{1.2a}$$

where n is an inner normal.

Below we construct an iterative process, by which the above scheme is valid also for this case (if $3\cos^2(n, x_3) > 2$).

By using the expressions (5) §18, ch.I Muskhelishvili [1966], we have:

$$\begin{aligned} \sigma_{\alpha 3}^{[m]} &= -\operatorname{sc}(n, x_3)\left[\cos(n, x_\beta)\sigma_{\alpha\beta}^{[m-1]} - \sigma_{\alpha n}\right], \\ \sigma_{33}^{[m]} &= -\operatorname{sc}(n, x_3)\left[\cos(n, x_\alpha)\sigma_{\alpha 3}^{[m]} - \sigma_{3n}\right]. \end{aligned} \qquad x \in S^{\pm} \tag{4.18}$$

Here m denotes an iteration number.

As $\cos(n, x_\alpha) = h_{\gamma,\alpha}(1 + h_{\gamma,1}^2 + h_{\gamma,2}^2)^{-1/2}$, $\cos(n, x_3) = -(1 + h_{\gamma,1}^2 + h_{\gamma,2}^2)^{-1/2}$, the spherical norm of the matrix corresponding to (4.17) is:

$$2\operatorname{sc}^2(n, x_3)\left(\cos^2(n, x_1) + \cos^2(n, x_2)\right),$$

and for convergence we have

$$2\operatorname{sc}^2(n, x_3)\left(\cos^2(n, x_1) + \cos^2(n, x_2)\right) < 1,$$

or as

$$\cos^2(n, x_1) + \cos^2(n, x_2) + \cos^2(n, x_3) = 1,$$

then

$$3\cos^2(n, x_3) > 2.$$

For the functions $\sigma_{\alpha\beta}^{[m]}(x_1, x_2, h_\gamma)$ by using Simpson's quadrature formula we have:

$$\sigma_{\alpha\beta}^{[m]}(x_1, x_2, h_\gamma) = \frac{3}{2h} T_{\alpha\beta}^{[m]} - 2\sigma_{\alpha\beta}^{[m]}(x_1, x_2, \hbar) + (-1)^\gamma \frac{3}{2h^2} M_{\alpha\beta}^{[m]}. \tag{4.19}$$

Here $T_{\alpha\beta}^{[m]}$, $\sigma_{\alpha\beta}^{[m]}$ and $M_{\alpha\beta}^{[m]}$ are approximate value of surfaces efforts, plane components of stress tensor and bending and twisting moments which are determined by above-mentioned processes for each m.

Remark 4.2. The expressions of type (4.19) when $h_2 = -h_1 = \text{const}$ see, e.g., section 2.1, ch.2 Lukasiewicz [1979]. See also section 9, p. 121.

Remark 4.3. If $\cos^2(n, x_3) \geq 2$, it is evident that the problem of definition of stress–strained state of the plate of variable thickness does not split into two problems. In this case the corresponding systems (4.10)–(4.14) and (4.16) (ignoring the remainder terms) are connected by the functions

$$v_\alpha^{\hbar}(x_1, x_2, \hbar) = u_{3,\alpha}^{\hbar}(x_1, x_2, \hbar) - \mu^{-1}\sigma_{\alpha 3}^{\hbar}(x_1, x_2, \hbar).$$

"Is there — is there balm in Gilead?"
The Raven, Edgar Allan Poe

5. On Unimprovable Estimates on the Class of Functions for Transition Errors for Refined Theories

Within the sections 2–4 the method of construction of two-dimensional models of group A was developed without any hypotheses of geometrical and physical character. With this, for remainder vector R, characterizing the transition of initial problems of three-dimensional theory of elastisity (1.1) – (1.8) to refined theories, we obtained a representation common to all models.

Now, using these representations we shall start with the estimation of error transition.

Deriving from formulae sections 2–4, in the case of a bounded inverse reverse operator, corresponding to scheme (P_h), we can obtain the estimation for error transition without any difficulties. The construction of strict expressions for the remainder vector R gives us the opportunity to prove the truth of the more essential result: the error transition of refined theories on the class of functions is bounded from below.

In this respect we shall consider the problem of definition of the stress–strain state of an isotropic homogeneous plate with constant thickness in the linear formulation, and let the acting forces be on its plane (e.g., see Muskhelishvili [1966], §26). Under these assumptions the three-dimensional problem of the theory of elasticity (1.1) – (1.4), (1.6) which will be for brevity represented in the form

$$L(\partial_1, \partial_2, \partial_3)u = f, \quad x \in \bar{\Omega}_h \tag{5.1}$$

is approximately substituted by the generalized plane strain problem state.

$$L^h(\partial_1, \partial_2)u^h = f^h(x_1, x_2), \quad x \in \bar{D}. \tag{5.2}$$

We should mention that the vector function $u^h = (u_1^h, u_2^h)^T$, unlike the functions introduced in (4.16), is an approximate value for averaged over thickness horizontal components $u_\alpha(x_1, x_2, x_3)$ of displacement vector, $L^h = \mu\Delta + (\lambda^* + \mu)\,\mathrm{grad\,div}$; when $(x_1, x_2) \in \partial D$, averaged boundary conditions, according to the procedure of Muskhelishvili [1966], §26, correspond to the equation (5.2).

Below, based on results of the theory of quadrature formulae (see Sard [1949] or Nikolski [1979]), the exactness of approximation of the problem (5.1) with the problem (5.2) is studied.

Following Nikolski [1979], let us introduce the classes $W^r_{Lp}(\Omega; r-1)$ of function that have absolutely continuous derivatives $D^{r-1}(\cdot)$ – of $(r-1)$ orders in $\bar{\Omega}_h$ and derivatives of order $\partial_3 D^{r-1}(\cdot) \in Lp(-h, h)$ – of order r, satisfying the condition

$$M_p(\cdot) = \left\{ \frac{1}{2h} \int_{-h}^{h} |\partial_3 D^{r-1}(\cdot)|^p dx_3 \right\}^{1/p} \leq M, \quad p \geq 1, \quad x \in D \tag{5.3}$$

and let us consider

$$u_i \in W^r_{Lp}(D \times [-h,h]; M) \bigcap W^{r-1}_{Lp}(S \times [-h,h]; M) = H^r_p \quad (r = 3,4)$$

We define the vector-valued function $\eta^h(x_1, x_2) = (\eta^h_1, \eta^h_2)^T$ in the following way:

$$\eta^h(x_1, x_2) = u^h - (u_1(x_1, x_2, 0), u_2(x_1, x_2, 0))^T = u^h - u^0$$

and call it the mean residual of solution of the problem (5.1) relative to the scheme (5.2).

Evidently, η^h satisfies the system

$$L^h(\partial_1, \partial_2)\eta^h = f^h - L^h(\partial_1, \partial_2)u^0 = \varphi^h(x_1, x_2) \tag{5.4}$$

The right-hand part $\varphi^h = (\varphi^h_1, \varphi^h_2)^T$ will be introduced as the approximation of the problem (5.1) by means of the scheme (5.2) on the solution u (by analogy with, e.g., Marchuk [1977]).

Using the statements of section 2 with respect to the representation of remainder terms for trapezoid formula, as well as the conclusion of equations (4.16) for functions φ^h_α with $(x, y) \in D$, we shall obtain:

$$\begin{aligned}
\varphi^h_\alpha(x,y) &= f^h_\alpha(x,y) - L^h_\alpha(\partial_1, \partial_2)u^0 \\
&= f^h_\alpha - \frac{1}{2h}\int_{-h}^{h} L^h_\alpha[u_1, u_2]dt + \frac{1}{2h}\int_{-h}^{h} F_{r-2,G}(t;h)\partial_3^{r-2} L^h_\alpha[u_1, u_2]dt \\
&= f^h_\alpha - \frac{1}{2h}\int_{-h}^{h} \sigma_{i\alpha,i}dt + \frac{\lambda}{2h(\lambda + 2\mu)}\int_{-h}^{h} \partial_\alpha \sigma_{33}dt \\
&\quad + \frac{1}{2h}\int_{-h}^{h} F_{r-2,tr}(t;h)\partial_3^{r-2} L^h_\alpha[u_1, u_2]dt \\
&= f^h_\alpha - \frac{1}{2h}\int_{-h}^{h} f_\alpha dt \\
&\quad + \frac{1}{2h}\left[\int_{-h}^{h} F_{r-2,G}(t;h)\partial_3^{r-2} L^h_\alpha[u_1, u_2]dt \right. \\
&\qquad \left. - \frac{\lambda}{\lambda + 2\mu}\int_{-h}^{h} F_{r-2,tr}(t;h)\partial_3^{r-2}\sigma_{33,\alpha}dt\right],
\end{aligned} \tag{5.5}$$

since

$$\varphi_\alpha^h = \frac{1}{2h}\left[\int_{-h}^{h} F_{r-2,G}(t;h)\partial_3^{r-2} L_\alpha^h[u_1,u_2]dz - \frac{\lambda}{\lambda+2\mu}\int_{-h}^{h} F_{r-2,tr}(t;h)\partial_3^{r-2}\sigma_{33,\alpha}dt\right], \tag{5.5a}$$

where

$$\begin{aligned}
L_\alpha^h[u_1,u_2] &= L_\alpha^h(\partial_1,\partial_2)(u_1(x_1,x_2,x_3),u_2(x_1,x_2,x_3))^T,\\
F_{r-2,G}(t;h) &= \frac{1}{(r-3)!}\left[\frac{(t-r)^{r-2}}{r-2} - 2hK_{r-2}(-t)\right],\\
F_{r-2,tr}(t;h) &= \frac{1}{(r-3)!}\left[\frac{(h-t)^{r-2}}{r-2} - h(K_{r-2}(h-t)+K_{r-2}(-h-t))\right],\\
K_{r-2}(t) &= \begin{cases} t^{r-3}, & t \ge 0,\\ 0, & t<0.\end{cases}
\end{aligned}$$

Now, evidently, by (5.5) for φ^h the strict estimation with respect to h is true:

$$\begin{aligned}
|\varphi^h| &\le h^{r-2}\left[c_{r-2,q}(1)M_p(\partial_3^{r-2}L^h[u_1,u_2])\right.\\
&\quad \left.+\frac{\lambda}{\lambda+2\mu}\bar{c}_{r-2,q}(1)M_p(\partial_3^{r-2}\operatorname{grad}\sigma_{33})\right]\\
&\le cMh^{r-2},
\end{aligned} \tag{5.6}$$

where

$$c_{r-2,q}(1) = \left\{\frac{1}{2}\int_{-1}^{+1}|F_{r-2,G}(t;1)|^q dt\right\}^{1/q},$$

$$\bar{c}_{r-2,q}(1) = \left\{\frac{1}{2}\int_{-1}^{+1}|F_{r-2,tr}(t;1)|^q dt\right\}^{1/q}, \qquad \frac{1}{p}+\frac{1}{q}=1.$$

Let us show, that $c_{r-2,q}(1)+\lambda(\lambda+2\mu)^{-1}\bar{c}_{r-2,q}(1)$ is an unimprovible constant on the class of $[W_{L_p}(D\times[-h,h];M)]^3$.

In fact, let us consider the systems of differential equations

$$L^h[u_1,u_2] = \operatorname{grad}(x+y)\psi^h(z), \tag{5.7}$$

$$\operatorname{grad}\sigma_{33} = -\operatorname{grad}(x+y)\bar{\psi}^h(z), \tag{5.8}$$

where

$$\psi(z) = \frac{M}{(r-3)!}\int_{-h}^{z}(t-z)^{r-3}[F_{r-2,G}(t;h)]^{q-1}\operatorname{sign}F_{r-2,G}(t;h)dt\left\{\int_{-h}^{h}|F_{r-2,G}(t;h)|^q dt\right\}^{-1/q},$$

$$\bar{\psi}^h(z) = \frac{M}{(r-3)!}\int_{-h}^{h}(t-z)^{r-3}[F_{r-2,tr}(t;h)]^{q-1}\operatorname{sign}F_{r-2,tr}(t;h)dt\left\{\int_{-h}^{h}|F_{r-2,tr}(t;h)|^q dt\right\}^{-1/q}$$

Then, for example, the functions

$$u_\alpha(x,y,z) = \Gamma_1(z)\Gamma_{1\alpha}(x,y) + \operatorname{grad}(x+y)\frac{(x+y)^2}{4(\lambda^*+2\mu)}\psi^h(z),$$

$$\sigma_{33}(x,y,z) = \Gamma_2(z) - (x+y)\bar{\psi}^h(z),$$

where $\Gamma_\alpha(z)$ are continuous in $[-h,h]$ and $\partial_\alpha\Gamma_{1\alpha}(x,y)$ is harmonic in D arbitrary functions , have solutions of systems (5.7) and (5.8) respectively, for which the relation (5.6) becomes an exact equality.

Let now $(x,y)\in\partial D$. For definiteness, let us think, that in problem (5.1) the boundary conditions, which must be averaged, and given on S, take the form (below n is an inner normal):

$$l_1 u = \sigma_{11}\cos(n,x) + \sigma_{12}\cos(n,y) + \sigma_{13}\cos(n,z) = f_1$$

and make the further propositions for this case. Then we shall obtain:

$$l_1^h(\partial_1,\partial_2)u^h = ((\lambda^*+2\mu)u_{1,1}^h + \lambda^* u_{1,2}^h)\cos(n,x_1) + \mu(u_{2,1}^h + u_{1,2}^h)\cos(n,x_2) = f_1^h.$$

or (see (5.5)):

$$\varphi_1^h = l_1^h\eta^h = \frac{1}{2h}\left[\int_{-h}^{h}F_{r-2,G}(t;h)\partial_3^{r-3}l_1^h u dt - \frac{\lambda\cos(n,x_1)}{\lambda+2\mu}\int_{-h}^{h}F_{r-2,tr}(t;h)\partial_3^{r-2}\sigma_{33}[u]dt\right].$$

Then, evidently, the estimates of type (5.6) are true for the case when $(x_1,x_2)\in\partial D$, too:

$$|\varphi_1^h(x,y)| \le c\bar{M}h^2, \tag{5.6a}$$

where

$$\bar{M} = \begin{cases} M, & r=3 \\ \sup\limits_{\partial D}\left\{\dfrac{1}{2h}\displaystyle\int_{-h}^{h}|D^{(2)}(\cdot)|^p dt\right\}^{1/p}, & r=4 \end{cases}$$

For other classical conditions the truth of values of type (5.6) is checked analogously.

After this procedure we shall return to the boundary value problem (5.4) again, and prove, that *the order of approximation relative to h of the mean residual in the class of functions H_p^r cannot be better than $r-2$.*

In fact, let us consider the boundary value problem :

$$L^h(\partial_1,\partial_2)\eta^h = \alpha\,\mathrm{grad}(x+y), \quad D: x^2+y^2<1,$$
$$\eta^h|_{\partial D} = \beta\,\mathrm{grad}(x+y), \quad \partial D: x^2+y^2=1.$$

Then, evidently, its solution has the form:

$$\eta^h = \left(\frac{x_2+y_2-1}{2(\lambda^*+3\mu)}\alpha+\beta\right)\mathrm{grad}(x+y).$$

If we now choose α and β according to the abovegiven procedure, developed in studying the inequality (5.6), then the last statement will be proved; with this, the last equality, satisfying the error η, will be strict, i.e., attainable on class $[H_p^r]^3$ not only with h-order, but with constants, appearing in the inequality (5.6) as well.

The above given substantiation of attainable estimates (on the class of solutions of initial problems) for error transition, evidently, has the general character for all refinedf theories, derived from (P_h). In this respect we should give essential remarks.

Remark 5.1. The realization of the above-given procedure for anisotropic case (1.8) represents a definite interest, although there is not any principal need.

In this case the operator of generalized plane stress state with $(x_1,x_2)\in D$ by virtue of (1.3) takes the form:

$$L^h(\partial_1,\partial_2) = \left\| \begin{matrix} c_{11}\partial_{11}+2c_{10}\partial_{12}+c_{66}\partial_{22}, & c_{16}\partial_{11}+(c_{12}+c_{66})\partial_{12}+c_{26}\partial_{12} \\ c_{16}\partial_{11}+(c_{12}+c_{66})\partial_{12}\partial_{22}, & c_{66}\partial_{11}+2c_{26}\partial_1\partial_2+c_{22}\partial_{22} \end{matrix} \right\|,$$

where

$$c_{ij} = b_{ij} - b_{i3}b_{j3}/b_{33} \quad (i,j=1,2,6)$$

For approxiomation of φ^h the following representation is true:

$$\varphi^h = \frac{1}{2h}\left[\int_{-h}^{h} F_{r-2,G}(t;h)L^h[u_1,u_2]dt - \int_{-h}^{h} F_{r-2,tr}(t;h)\partial_3^{r-2}T^h(\partial_1,\partial_2)\sigma_{33}dt\right],$$

$$T^h(\partial_1,\partial_2) = b_{33}^{-1}(b_{13}\partial_1+b_{36}\partial_2;b_{36}\partial_1+b_{23}\partial_2)^T.$$

We represent the solution of systems, similar to (5.7) and (5.8) in the form:

$$u_\alpha = \Gamma_1(z)\Gamma_{1\alpha}(x_1,x_2) - \Gamma_\alpha(x_1,x_2)\psi^h(z),$$
$$\sigma_{33} = \Gamma_2(z) - q(x_1,x_2)\bar{\psi}(z).$$

Here $\Gamma=(\Gamma_{11},\Gamma_{12})^T$ – is the general solution of the equation

$$L^h(\partial_1,\partial_2)\Gamma = 0,$$

coefficients of homogeneous polynomials $(p_1, p_2)^T = p$ of second-order and q of first-order, follow from the conditions:

$$L^h(\partial_1, \partial_2)p = 1, \quad p|_{\partial D} = 0.$$

$$T^h(\partial_1, \partial_2)q = 1.$$

All the abovesaid may be summarized in the following statement:

Theorem 5.1. *Let $u_i \in H_p^r (r = 3, 4)$ and*

$$|\varphi^h| = ch^{r-2}M, \quad (x_1, x_2) \in \bar{D}.$$

Then: 1) $\exists$ such functions as u_i and σ_{33}, on which the last inequality transforms into the exact equality;

2) the boundary value problem with respect to η^h is such that

$$\eta^h = ch^{r-2}(p+1)M \operatorname{grad}(x_1 + x_2).$$

Remark 5.2. Evidently, it is not necessary to display the procedure of obtaining the desirable values for other models of the scheme (P_h) (e.g., in the case of plate bending). Nevertheless, we note the following. In the example considered the admissible class of solution was

$$H_p^r = W_{Lp}^r(D \times [-h, h]; M) \bigcup W_{Lp}^{r-1}(S \times [-h, h]; M) \ni u_i \quad (r = 3, 4)$$

As follows from the above procedure, in case of smoother functions, evidently, we cannot obtain an improved result, for the remainder term of model (5.2) cannot have the order more than 2 (by virtue of usage of quadrature formula of trapezoid and Gauss's formula with one node). At the same time in the class of functions less smooth than H_p^3 for the values, characterizing the approximation of initial problems by two-dimensional models of (P_h), as the order of smoothness of solutions u of the class of admissible functions defines the order with respect to h and the values of Sard-Nikolski's constants. The possibility for estimations follows from the following reasoning.

Let $u_i \in W_1^r(\Omega_h)$ be the function with summable to $r \geq 1$ order generalized derivatives inclusive. If we use theorem 2.7.1. (Mikhlin [1977], p.41), then the functions $\partial_3^k u_i$ $(k = \overline{0, r-1})$ will be absolutely continuous with respect to z in $D \times [-h, h]$. Now let us introduce the classes of functions $W_1^r(D \times [-h, h]; M)$ with the elements of absolutely continuous derivatives to order $r-1$ with z inclusive in $D \times [-h.h], \partial_3^r u_i \in L_p(-h, h)$ and satisfy the inequality (5.3) with $p = 1$. Then instead of $H_p^r (r = 3, 4)$ it is possible to estimate the error transition within the class of functions, defined in the following way:

$$H_1^r = W_1^r(D \times [-h, h]; M) \bigcup W_1^{r-1}(S \times [-h, h]; M) \quad (r \geq 1).$$

In the case of investigation of the problem of error transition for other models of (P_h) different from (5.2), it is necessary to introduce error functions of the type

$$\eta^*(x_1, x_2) = u^h(x_1, x_2) - u^*(x_1, x_2),$$

being connected with the representation of function by virtue of averaging, according to Reissner, for example, formulae (2.5) or (2.6), when u^h, as is accepted, denotes the exact solution of scheme (P_h).

Chapter II. Theories with Regular Processes

6. The Construction and Investigation of Vekua's Two-Dimensional Models

Within this and the next section some methods of group B will be investigated (see subsection 1.2) when the initial boundary value problem (1.1)–(1.5) is linear.

As is known, the expansion of desired fields in incomplete series with respect to the variable z or in asymptotic series appears to be the starting point for this group's methods. Further, application of the projective method, for example, Bubnov–Galiorkin's or minimization of the quadratic functional, corresponding to the initial problem with respect to expansion, results in infinite systems of two-dimensional differential problems. Choosing a finite number of problems from this succession results the initial problem to two-dimensional models, which are usually called reduced system or approximations of high order.

Vekua's method, as is assumed in literature, appears to be typical and handy, at the same time, within the class of these group methods; the essence of this projective method lies in the application of Legendre's polynomials as the coordinate (basis) system.

6.1. Review of related works and basic equations

For more than 30 years (1942–1977) Vekua has systematically studied the problem of the construction of the theory of plates and shells. So, results of this investigation were summarized in the monograph Vekua [1985], with an appropriate bibliography of all his papers.

Vekua's method is to construct the consistent theory of plates and shells from the point of view of boundary value problems theory for differential equations in partial derivatives; it is rather a universal way within the class of direct methods of reduction of multi-dimensional problems to the problems of lower dimension, which was introduced by Kantorovich [1933] and investigated in the works of the following authors such as Kantorovich, Krilov [1962], Keldish [1942], Krasnoselski, Vainikko, Zabreiko, Rutitski, Stetsenko [1969], Krilov, Bogoliubov [1929], Mikhlin [1970], Polski [1955], Sobolev [1950], Vogelius, Babuška [1981] and others.

Vekua's method was succesfully adopted by many scientists in their investigations and calculations of problem of the theory of elastic plates and shells. Besides the works quoted in review articles Galinich [1967], Vorovich [1965,1975], who used the basic system of polynomials, we should note the following, such as the works by Fellers, Soler [1970], Gordeziani [1974,1975], Gotsuliak, Guliaev, Tchibiriakov [1973], Guliaev, Bagenov, Lizunov [1978], Hutchins, Soler [1973], Kaldani [1983], Khvoles [1971], Khoma [1985], Magnaradze [1984], Meunargia [1991], Mindlin,

Medik [1959], Pelekh, Lazko [1982], Poniatovski [1962], Savin, Khoma [1971], Shleniov [1975], Soler [1968], Tskhadaia [1977]. Zgenti [1988], Zgenti, Khvoles [1982]. The further bibliography connected with Vekua's method and his applications are given in these works (see also section 11).

One more advantage of Vekua's reduction method lies in the fact that introduced basic characteristics, according to this theory, "make it possible to have a rather concrete interpretation in terms of classic mechanics" (Vekua [1985], p.9). In this respect the development of reduction method for the problems of theory of elasticity in case of bodies with small thickness has a principal meaning, as well as the method of integral relations on problems with flow about solid bodies, proposed by Belotserkovski, Dorodnitsin [1960], method of invariant immersion (see, e.g., Casti, Kalaba [1973]), Vlasov's method in problems of constructive mechanics (Vlasov [1959]) and method of spherical harmonics in problems of transport theory (Vladimirov [1961]).

While investigating projective methods of Vekua type, depending on boundary conditions , given at $S^{\pm}$ we distinguish two cases:

1) with linear combination of exterior tension vector and displacement vector on the surfaces;

2) with displacement vector on the surfaces.

In the first case with an exterior tension vector, i.e., $d = 0$ only, given on $S^{\pm}$, Vekua obtained equlibrium equation and corresponding boundary conditions at lateral surface S. The case with $d \geq 0$ is considered within subsection 6.4.

In the second case with formal $d = +\infty$ (as is accepted in the theory of differential equations), (1.2) goes over into Dirichlet's conditions. This case is considered in subsection 6.5.

Thus, the following boundary condition is to be considered:

$$[-\mu \operatorname{rot}\operatorname{rot} + (\lambda + 2\mu)\operatorname{grad}\operatorname{div}]u = f, \quad x \in \Omega_h, \tag{6.1}$$

$$\sigma_{i3} - du = g_i^{\pm}, \qquad x \in S^{\pm} \tag{6.2}$$

$$l(\partial_1, \partial_2, \partial_3)u = g, \qquad x \in S \tag{6.3}$$

Let us assume that boundary conditions and domains Ω_h are such, that boundary value problem (6.1)–(6.3) has a unique solution.

6.2. Convergence of incomplete Fourier–Legendre series for domains with angles and edges. Pollard's result to limited density

Below, we shall often use Fourier–Legendre incomplete series:

$$u(x,y,z) = \sum_{k=0}^{\infty} \overset{k}{u}(x,y) P_k\left(\frac{z}{n}\right) = \sum_{k=0}^{N} \overset{k}{u}(x,y) P_k\left(\frac{z}{n}\right) + R_N(u;z). \tag{6.4}$$

Here and below N is the positive integer; it defines the order of approximation in Vekua's method, $R_N(u_1; z)$ is the remainder term, characterizing approximation of function u by aggregate of (6.4) form.

Let us assume that

$$u_i(x,y,z) \in W_p^{2+\alpha}(\Omega_h), \quad \alpha \geq 0, \ p \geq 1, \tag{6.5}$$

i.e., the solution of initial problem (6.1) – (6.3) has generalized derivatives of $2+\alpha$-order, all variables inclusive and derivatives $D^{2+\alpha}(u_i) \in L_p(\Omega_h)$, i.e., summable in degree p .

The set $W_p^k(\Omega_h)$ $(k = \alpha + 2)$ becomes Sobolev linear normed space by means of different equivalent norms (see Sobolev [1975]). Below we follow concepts and construction designations, assumed in work Mikhlin [1977]. In case of noninteger α it is possible to introduce weight spaces according to Kondratiev [1967], or spaces of Hölder type, introduced by Morrey [1959], Besov, Iliin, Nikolski [1975].

Before starting the investigation of the initial problem, based on Vekua method, let us take condition (6.5) and discuss the question of correctness of an assumption of this inclusion.

As follows from the general theory of differrential equations by Agmon, Douglis, Nirenberg [1959] in case of the single solution of elliptic differential problems with infinite differentiability of their right hand parts, coefficients and domain boundaries, the solution will also be infinitely differentiable.

Differentiability of a solution of the same basic problem of the theory of elasticity, concerning the differentiability of dyad $[0, g]$ and domain boundary, were studied in the works by Gegelia [1962] (see also Kupradze, Gegelia, Basheleishvili, Burchuladze [1976], ch. VI, §3) to a certain (fixed) order. At the same time the apparatus of multi-dimensional singular integral equations was developed, differential properties for corresponding elastic potentials were settled, embedding theorems of functional spaces were used.

We should pay attention to the paper Thompson [1969], too; there the general theory of differential equations in partial derivatives Agmon, Douglis, Nirenberg [1959] is used for the case with smooth boundary for investigation properties of the solution of some boundary value problems of the theory of elasticity.

A somewhat different matter is in the case with nonsmooth boundary domains, which can be proved by simple examples. The availability of *angular points and ribs* at domain boundary stipulates discontinuity of derivatives of the solution of the differential problem with analyticity of the remaining data, if additional relations are not fulfilled, i.e., the so called "conditions on dyad agreement" $[f, g]$ (e.g., see Veindiner [1973], Volkov [1965]), are not fulfilled.

As is known from the literature, as far back as Magnaradze [1937] paid attention to the importance of investigation of boundary value problems with angular points. Further, Williams [1952] made successive steps in investigation of problems of theory of elasticity.

With investigation of value functions, given on the domain boundary with angular (conic) points, the works by Kondratiev [1967] and Mazija, Nazarov, Plamenievski [1981] appeared to be basic ones. They became the source of many successive investigations, devoted to differential values of elliptic differential problems with angular lines and points. The appropriate bibliography together with a number of principal results for problems of the theory of elasticity are given in the following work Parton and Perlin [1981] (ch.III, §17).

Let us mention some results, interpreted in Parton and Perlin [1981] and based on the works by Aksentian [1967], Kalandia [1973], Koldorkina [1973]. In the case, which we considered, Ω_h domain contain angular lines with solid angle $\alpha = \pi/2$.

Roots of corresponding transcendental equations, defining asymptotic value of the solution in the vicinity of angular (conic) points and ribs, have the order no less than 1/2 for all boundary conditions defined by virtue of problems (6.1) – (6.3).

If we now use theorems of Mikhlin [1977] (ch. 3, §3), then the next lemma will be correct, concerning the correctness of inclusion (6.5).

Lemma 6.1.

1) in the case that tensions vanish on S near the rib,

$$u_i \in W_p^{2+\alpha}(\Omega_h), \quad \alpha \geq 0,$$

if $1 \leq p < 3$;

2) in the case near the rib when displacements or tangents tensions and normal displacement are equal to zero,

$$u_i \in W_p^{2+\alpha}(\Omega_h), \quad \alpha \geq 0,$$

if $1 \leq p \leq 6/(3+2\alpha) \leq 2$.

Below, we shall take into account inclusion (6.5) being fulfilled in the case $0 \leq d \leq +\infty$.

The condition (6.5) makes it possible to study the question of convergence of (6.4) series and to get an estimation for theremainder term R_N. The following lemma will be valid in this respect:

Lemma 6.2. (On convergence). *Let inclusion (6.5) be fulfilled. Then the next inequality is correct*

$$|R_N| \leq c_{\alpha,p} M_{\alpha+2,p} h^{\alpha+2} N^{-(1+2\alpha)/2} \rho_N, \quad c_{\alpha,p} > 0, \tag{6.6}$$

where $c_{\alpha,p}$ is a constant, independent of u, N and h, ρ_N is a sequence of non-negative n umbers, tending to zero,

$$\operatorname*{ess\,sup}_D ||\partial_3^{\alpha+2} u_i||_{L_p(-h,h)} \leq M_{\alpha+2,p}.$$

Proof. According to theorem 2.7.1 Mikhlin [1970] u_i will have almost everywhere in $D \times [-h, h]$ by z to $\alpha+1$-order *absolute continuous derivatives* inclusive by virtue of (6.5).

Further, if reasoning from the book by Jackson [1941] is used, we assume that

$$R_N(u_i; z) = o\left(N^{-(1+2\alpha)/2}\right). \tag{6.7}$$

Now let us prove that R_N with respect to h has $\alpha + 2$ infinitesimal order. For this purpose in equality (37) (Jackson [1941]) (ch. II), representing expression for R_N we shall use Taylor's formula with remainder term in integral form. Assuming $\alpha+2 < N$ and temporarily excluding index i and dependence on x_α we shall have

$$R_N(u; z) = \frac{N+1}{2(\alpha+1)!} \int_{-h}^{h} dt Q_N\left(\frac{z}{h}, \frac{t}{h}\right) \frac{1}{z-t} \int_t^z (z-s)^{\alpha+1} \partial_3^{\alpha+2} u(s) ds$$

$$= \frac{(N+1)h^{\alpha+2}}{2(\alpha+1)!}\int_{-1}^{1} dt \frac{Q_N(\zeta,\tau)}{\zeta-\tau}\int_{\tau}^{\zeta}(\zeta-\sigma)^{\alpha+1}[\partial_3^{\alpha+2}u]_{s=\sigma h}d\sigma, \tag{6.8}$$

$$(-1 \leq \zeta \leq 1).$$

Here $Q_N(\zeta,\tau) = P_{N+1}(\zeta)P_N(\tau) - P_{N+1}(\tau)P_N(\zeta)$ is Christoffel kernel, P_N is Legendre's polynomial of degree N.

Let us consider the integral:

$$I(\zeta,\tau) = \frac{1}{\zeta-\tau}\int_{\tau}^{\zeta}(\zeta-\sigma)^{\alpha+1}[\partial_3^{\alpha+2}u(z)]_{z=\sigma h}d\sigma.$$

By virtue of Hölder's inequality we obtain

$$|I(\zeta,\tau)| \leq \frac{1}{\zeta-\tau}\left(\int_{\tau}^{\zeta}|\zeta-\sigma|^q d\sigma\right)^{1/q}\left(\int_{\tau}^{\zeta}|\partial_3^{\alpha+2}u(z)|^p_{z=\sigma h}d\sigma\right)^{1/p}$$
$$\leq (\alpha q+q+1)^{1/q}|\zeta-\tau|^{\alpha+1/q}M_{\alpha+2,p}.$$

Taking into account the last value and relation (6.7) in equality (6.8), we have

$$|R_N(u;z)| \leq \frac{h^{\alpha+2}M_{\alpha+2,p}(N+1)}{2(\alpha+1)!\sqrt[q]{\alpha q+q+1}}\int_{-1}^{+1}|Q_N(\zeta,\tau)||\zeta-\tau|^{\alpha+1/q}d\tau$$
$$\leq c_{\alpha,p}h^{\alpha+2}M_{\alpha+2,p}N^{-(1+2\alpha)/2}\rho_N.$$

In case of non-integer α, as was observed above, the estimates (6.6) are valid, if norms of Hölder type according to Besov, Ilin, Nikolski [1975] are introduced for functions $u_{i,33}$.

Inequality (6.6), which gives the estimation by h and N simultaneously, generalizes corresponding results Gordeziani [1974], Vekua [1965, 1985]. In particular, from (6.6) the convergence of series in usual sense follows.

In substantiation of projective methods, as it is known, (Courant, Hilbert [1961], Mikhlin [1966]), the quality of limit density for linear shell according to system $\{P_i(z/h), i=\overline{0,N}\}$ is quite essential in energetic space H_{-L} for admissible values of p, appearing in expressions (6.5) – (6.6). In this respect, the lemma, being the corollary of results by Pollard [1947], referring to the theory of Jacobi orthogonal polynomials (see Szegö [1959]) is correct.

Lemma 6.3. (On limited density) . *The linear shell according to system $P_i(z/h)$ is of maximum density in H_{-L} by virtue of system basis in $L_p(-h,h)$ in case of* $4/3 < p < 4$.

If now regions of admissible values for p are compared in lemma 6.1 and 6.3, then we shall obtain $p \in (4/3, 3)$ for the first case, and $p \in (4/3, 2)$ for the second one.

6.3. Construction and justification Vekua type systems in the case when plates' surfaces are free. Korn type inequality

Let us return to the initial problem (6.1) – (6.3) and consider the case with $0 \le d < +\infty$ in (6.2).

In the above-mentioned works, based on Vekua's system, there was considered a case where components of exterior vector tension σ_3 were given at $S^{\pm}$, i.e., $d = 0$. The problem of satisfying these conditions or any approximation was the difference among proposed systems, constructed according to Vekua: for some models they are natural, for others they appear to be the main ones, in the sense of variational methods (see, e.g., Courant, Hilbert [1961], Rektorys [1980]).

First, we shall assume $d = 0$ and construct a class of operator equations, in fact, coinciding with system (8.33) (Vekua [1965]), for brevity, *we shall denote it as* (V).

Let us use expansion into Fourier–Legendre for incomplete series components of stress tensor. By virtue of boundary conditions by $S^{\pm}$ we have:

$$\sigma_{\alpha\beta} = \sum_{k=0}^{\infty} \overset{s}{\sigma}_{\alpha\beta} P_s\left(\frac{z}{h}\right), \tag{6.4a}$$

$$\sigma_3 = \frac{(h+z)g^+ - (h-z)g^-}{2h} + \sum_{s=1}^{\infty} \overset{s}{\sigma}_{3j} \left[P_{s+1}\left(\frac{z}{h}\right) - P_{s-1}\left(\frac{z}{h}\right)\right]. \tag{6.4b}$$

Then equilibrium equations in terms of components of stress tensor will be equivalent to the following infinite system as

$$\begin{aligned} &c_m h \overset{m}{\sigma}_{\alpha\beta,\beta} + (2m+1) c_m \overset{m}{\sigma}_{\alpha\beta} = f_\alpha^m - h \frac{g_\alpha^+ - g_\alpha^-}{2} c_0 \delta_{m0}, \\ &c_m h (\overset{m-1}{\sigma}_{\alpha\beta,\alpha} - \overset{m+1}{\sigma}_{\alpha\beta,\alpha}) + (2m+1) c_m \overset{m}{\sigma}_{33} \\ &\qquad = \overset{m}{f}_3 - h c_0 \delta_{m0} \frac{g_{\alpha,\alpha}^+ + g_{\alpha,\alpha}^-}{2} \\ &\qquad\quad - h c_1 \delta_{m1} \frac{g_{\alpha,\alpha}^+ - g_{\alpha,\alpha}^-}{2} - h c_0 \delta_{m0} \frac{g_3^+ - g_3^-}{2}, \end{aligned} \tag{6.9}$$

where

$$\overset{m}{f} = \int_{-h}^{h} f(x_1, x_2, t) P_m\left(\frac{t}{h}\right) dt, \quad c_m = \frac{2}{2m+1}, \quad m = 0, 1, 2, \dots$$

Hooke's law takes the following form:

$$c_m h\overset{m}{\sigma}_{11} = (\lambda+2\mu)c_m h\overset{m}{u}_{1,1} + \lambda c_m h\overset{m}{u}_{2,2} + \lambda(2m+1)c_m \sum_{k\geq m(2)} \overset{k+1}{u}_3,$$

$$c_m h\overset{m}{\sigma}_{12} = \mu h c_m(\overset{m}{u}_{1,2} + \overset{m}{u}_{2,1}),$$

$$c_m h\overset{m}{\sigma}_{22} = \lambda c_m h\overset{m}{u}_{1,1} + (\lambda+2\mu)c_m h\overset{m}{u}_{2,2} + \lambda(2m+1)c_m \sum_{k\geq m(2)} \overset{k+1}{u}_3,$$

$$\begin{aligned} c_m h(\overset{m-1}{\sigma}_{3\alpha} - \overset{m+1}{\sigma}_{3\alpha}) &= \mu h c_m \overset{m}{u}_{3,\alpha} + \mu(2m+1)c_m \sum_{k\geq m(2)} \overset{k+1}{u}_\alpha \\ &\quad - hc_0\delta_{m0}\frac{g_\alpha^+ + g_\alpha^-}{2} - hc_1\delta_{m1}\frac{g_\alpha^+ - g_\alpha^-}{2}, \end{aligned} \tag{6.10}$$

$$\begin{aligned} c_m h(\overset{m-1}{\sigma}_{33} - \overset{m+1}{\sigma}_{33}) &= \lambda h c_m \overset{m}{u}_{\alpha,\alpha} + (\lambda+2\mu)(2m+1)c_m \sum_{k\geq m(2)} \overset{k+1}{u}_3 \\ &\quad - hc_0\delta_{m0}\frac{g_3^+ + g_3^-}{2} - hc_1\delta_{m1}\frac{g_3^+ - g_3^-}{2}. \end{aligned}$$

Here and (often) below the following note is used:

$$\sum_{k\geq i(s)} \overset{k}{u} = \overset{i}{u} + \overset{i+s}{u} + \overset{i+2s}{u} + \ldots, \qquad \sum_{k\leq i(s)} \overset{k}{u} = \overset{i}{u} + \overset{i-s}{u} + \overset{i-2s}{u} + \ldots.$$

Formulae (6.9) and (6.10) make it possible to obtain explicit form of Vekua's system in displacement components. For this purpose we use Hooke's law for values σ_{3i} and condition (6.2) (with $d=0$). We shall have:

$$g_\alpha^+ = \mu\sum_{k=0}^{\infty}\left(\overset{k}{u}_{3,\alpha} + \frac{k(k+1)}{2h}\overset{k}{u}_\alpha\right),$$

$$g_\alpha^- = \mu\sum_{k=0}^{\infty}(-1)^k\left(\overset{k}{u}_{3,\alpha} - \frac{k(k+1)}{2h}\overset{k}{u}_\alpha\right),$$

and

$$g_3^+ = \sum_{k=0}^{\infty}\left(\lambda\overset{k}{u}_{\alpha,\alpha} + (\lambda+2\mu)\frac{k(k+1)}{2h}\overset{k}{u}_3\right),$$

$$g_3^- = \sum_{k=0}^{\infty}(-1)^k\left(\lambda\overset{k}{u}_{\alpha,\alpha} - (\lambda+2\mu)\frac{k(k+1)}{2h}\overset{k}{u}_3\right).$$

We define values $g^+ \pm g^-$, entering (6.9). We shall have:

$$
\begin{aligned}
g_\alpha^+ + g_\alpha^- &\doteq 2\mu \sum_{k=0}^{\infty} \left(\overset{2k}{u}_{3,\alpha} + \frac{(k+1)(2k+1)}{h} \overset{2k+1}{u}_{\alpha} \right), \\
g_\alpha^+ - g_\alpha^- &= 2\mu \sum_{k=0}^{\infty} \left(\overset{2k+1}{u}_{3,\alpha} + \frac{k(2k+1)}{h} \overset{2k}{u}_{\alpha} \right), \\
g_3^+ + g_3^- &= 2 \sum_{k=0}^{\infty} \left(\lambda \overset{2k}{u}_{\alpha,\alpha} + (\lambda+2\mu) \frac{(k+1)(2k+1)}{h} \overset{2k+1}{u}_{3} \right), \\
g_3^+ - g_3^- &= 2 \sum_{k=0}^{\infty} \left(\lambda \overset{2k+1}{u}_{\alpha,\alpha} + (\lambda+2\mu) \frac{k(2k+1)}{h} \overset{2k}{u}_{3} \right).
\end{aligned}
\tag{6.11}
$$

From equations (6.10), summing up the three last formulae, for values $\overset{m}{\sigma}_{3\alpha}$ we obtain:

$$
\begin{aligned}
\sum_{s \le m(2)} (\overset{s-2}{\sigma}_{3\alpha} - \overset{s}{\sigma}_{3\alpha}) = -\overset{s}{\sigma}_{3\alpha} \\
= \mu \sum_{s \le m(2)} \overset{s-1}{u}_{3,\alpha} + \frac{\mu}{h} \sum_{s \le m(2)} (2s-1) \sum_{k \ge s(2)} \overset{k}{u}_{\alpha} \\
-\frac{1}{2}(g_\alpha^+ + g_\alpha^-) \sum_{s \le m(2)} \delta_{s-1,0} - \frac{1}{2}(g_\alpha^+ - g_\alpha^-) \sum_{s \le m(2)} \delta_{s-1,1}.
\end{aligned}
$$

Similarly

$$
\begin{aligned}
-\overset{m}{\sigma}_{33} = \lambda \sum_{s \le m(2)} \overset{s-1}{u}_{\alpha,\alpha} + \frac{\lambda+2\mu}{h} \sum_{s \le m(2)} (2s-1) \sum_{k \ge s(2)} \overset{k}{u}_{3} \\
-\frac{1}{2}(g_3^+ + g_3^-) \sum_{s \le m(2)} \delta_{s-1,0} - \frac{1}{2}(g_3^+ - g_3^-) \sum_{s \le m(2)} \delta_{s-1,1}.
\end{aligned}
$$

In these expressions $\overset{-1}{\sigma}_{3\alpha} = \overset{0}{\sigma}_{3\alpha} = 0$ is assumed.

Now, using formulae (6.11) from latter representations, after some computations, we get:

$$
\begin{aligned}
\overset{m}{\sigma}_{3\alpha} &= \mu \sum_{s \ge (m+1)(2)} \left[\overset{s}{u}_{3,\alpha} + \frac{1}{2h}((s+1)(s+2) - m(m+1)) \overset{s+1}{u}_{\alpha} \right], \\
\overset{m}{\sigma}_{33} &= \sum_{s \ge (m+1)(2)} \left[\lambda \overset{s}{u}_{\alpha,\alpha} + \frac{1}{2h}(\lambda+2\mu)((s+1)(s+2) - m(m+1)) \overset{s+1}{u}_{3} \right].
\end{aligned}
$$

Taking into account the last formulae, as well as (6.10), after obvious simplifications with respect to components of displacement vector we obtain the following infinite system of Vekua's differential equations:

$$
\begin{aligned}
& l_2\overset{m}{u}_+ + (\lambda+\mu)h^{-1}(2m+1)\sum_{k\geq m(2)} \operatorname{grad}\overset{k+1}{u}_3 + \mu h^{-2}\frac{2m+1}{2} \\
& \quad \times \sum_{k\geq m(2)} [k(k+1)-m(m+1)]\overset{k}{u}_+ = \frac{1}{c_m h}\left[\overset{m}{f}_+ - \frac{g_+^+ - g_+^-}{2}\delta_{m0}\right], \\
& \mu\Delta\overset{m}{u}_3 + (\lambda+\mu)h^{-1}(2m+1)\sum_{k\geq m(2)} \operatorname{div}\overset{k}{u}_+ + (\lambda+2\mu)h^{-2}\frac{2m+1}{2} \\
& \quad \times \sum_{k\geq m(2)} [k(k+1)-m(m+1)]\overset{k}{u}_3 = \frac{1}{c_m h}\left[\overset{m}{f}_3 - \frac{g_3^+ - g_3^-}{2}\delta_{m0}\right].
\end{aligned}
\tag{6.12}
$$

Here

$u_+ = (u_1, u_2)^T,\ f_+ = (f_1, f_2)^T,\ (l_2 u_+, u_+) = \mu(\Delta u_\alpha, u_\alpha) + (\lambda+\mu)(\operatorname{grad}\operatorname{div} u_+, u_+)$.

For clearness, we write out some particular cases, assuming $N = 0, 1, 2$ in system (6.12). We shall have:

$N = 0$

$$l_2\overset{0}{u}_+ = \frac{1}{2h}\left[\overset{0}{f}_+ - g_+^+ + g_+^-\right],$$

$$\mu\Delta\overset{0}{u}_3 = \frac{1}{2h}\left[\overset{0}{f}_3 - g_3^+ + g_3^-\right],$$

$N = 1, \quad m = 0, 1:$

$$l_2\overset{0}{u}_+ + (\lambda+\mu)h^{-1}\operatorname{grad}\overset{1}{u}_3 = \frac{1}{2h}\left[\overset{0}{f}_+ - (g_+^+ - g_+^-)\right] = \overset{0}{F}_+,$$

$$\mu\Delta\overset{0}{u}_3 + (\lambda+\mu)h^{-1}\overset{1}{u}_{\alpha,\alpha} = \frac{1}{2h}\left[\overset{0}{f}_3 - (g_3^+ - g_3^-)\right] = \overset{0}{F}_3,$$

$$l_2\overset{1}{u}_+ = (hc_1)^{-1}\overset{1}{f}_+ = \overset{1}{F}_+,$$

$$\mu\Delta\overset{1}{u}_3 = (hc_1)^{-1}\overset{1}{f}_3 = \overset{1}{F}_3,$$

$N = 2, \quad m = 0, 1, 2:$

$$l_2\overset{0}{u}_+ + (\lambda+\mu)h^{-1}\operatorname{grad}\overset{1}{u}_3 + 3\mu h^{-2}\overset{2}{u}_+ = \overset{0}{F}_+,$$

$$\mu\Delta\overset{0}{u}_3 + (\lambda+\mu)h^{-1}\operatorname{div}\overset{1}{u}_+ + 3h^{-2}(\lambda+2\mu)\overset{2}{u}_3 = \overset{0}{F}_3,$$

$$l_2\overset{1}{u}_+ + 3(\lambda+\mu)h^{-1}\operatorname{grad}\overset{2}{u}_3 = \overset{1}{F}_+,$$

$$\mu\Delta\overset{1}{u}_3 + 3(\lambda+\mu)h^{-1}\operatorname{div}\overset{2}{u}_+ = \overset{1}{F}_3,$$

$$l_2\overset{2}{u}_+ = \overset{2}{F}_+,$$

$$\mu\Delta\overset{2}{u}_3 = \overset{2}{F}_3.$$

Below, we shall return to system (6.12) again, but now let us give another form to Vekua's system.

From system (6.10), evidently, for values $\overset{m}{\sigma}_{\alpha 3}$ we have:

$$\overset{m-1}{\sigma}_{\alpha 3} = \overset{m+1}{\sigma}_{\alpha 3} + \mu \overset{m}{u}_{3,\alpha} + \mu \frac{2m+1}{h} \sum_{k \geq m(2)} \overset{k+1}{u}_{\alpha} - \frac{1}{2}\delta_{m0}(g_\alpha^+ + g_\alpha^-) - \frac{1}{2}\delta_{m1}(g_\alpha^+ - g_\alpha^-)$$

$$= \overset{m+3}{\sigma}_{\alpha 3} + \mu \overset{m+2}{u}_{3,\alpha} + \mu h^{-1}(2m+5) \sum_{k \geq m(2)} \overset{k+3}{u}_{\alpha} + \mu \overset{m}{u}_{3,\alpha} + \mu h^{-1}(2m+1) \sum_{k \geq m(2)} \overset{k+1}{u}_{\alpha} - \sum_{k \geq m(2)} \left(\frac{g_\alpha^+ + g_\alpha^-}{2}\delta_{k0} + \frac{g_\alpha^+ - g_\alpha^-}{2}\delta_{k1} \right),$$

$$\overset{m}{\sigma}_{\alpha 3} = \mu \sum_{k \geq m(2)} \overset{k+1}{u}_{3,\alpha} + \mu h^{-1} \sum_{s \geq (m+1)(2)} (2s+1) \sum_{k \geq m(2)} \overset{k+2}{u}_{\alpha} - \frac{1}{2} \sum_{k \geq (m+1)(2)} (g_\alpha^+ + g_\alpha^-)\delta_{k0} + (g_\alpha^+ - g_\alpha^-)\delta_{k1}, \quad m = 1, 2, \ldots$$

$$\overset{m}{\sigma}_{\alpha 3} = \mu \sum_{k \geq (m+1)(2)} \left(\overset{k}{u}_{3,\alpha} + \frac{1}{2h}((k+1)(k+2) - m(m+1))\overset{k+1}{u}_{\alpha} \right).$$

Analogously,

$$\overset{m}{\sigma}_{33} = \sum_{k \geq (m+1)(2)} \left(\lambda \overset{k}{u}_{\alpha,\alpha} + \frac{1}{2h}(\lambda + 2\mu)((k+1)(k+2) - m(m+1))\overset{k}{u}_{3} \right).$$

Taking into account these formulae we obtain

$$\frac{1}{2}(g_\alpha^+ + g_\alpha^-) \sum_{m=1}^{\infty} \delta_{m-1,0} + \frac{1}{2}(g_\alpha^+ - g_\alpha^-) \sum_{m=1}^{\infty} \delta_{m-1,1} - \overset{m}{\sigma}_{3\alpha} = \mu \sum_{k < (m+1)(2)} \left(\overset{k}{u}_{3,\alpha} + \frac{1}{2h}(k+1)(k+2)\overset{k+1}{u}_{\alpha} \right).$$

Hense for values $\overset{m}{\sigma}_{\alpha 3}$ we have:

$$\overset{m}{\sigma}_{\alpha 3} = \frac{1}{2}(g_\alpha^+ + g_\alpha^-) \sum_{m \geq 1(1)} \delta_{m-1,0} + \frac{1}{2}(g_\alpha^+ - g_\alpha^-) \sum_{m \geq 1(1)} \delta_{m-1,1} - \mu \sum_{k \leq (m-1)(2)} \left[\overset{k}{u}_{3,\alpha} + \frac{1}{2h}(k+1)(k+2)\overset{k+1}{u}_{\alpha} \right].$$

Similarly for $\overset{m}{\sigma}_{33}$ we shall have:

$$\overset{m}{\sigma}_{33} = \frac{1}{2}(g_3^+ + g_3^-) \sum_{m \geq 1(1)} \delta_{m-1,0} + \frac{1}{2}(g_3^+ - g_3^-) \sum_{m \geq 1(1)} \delta_{m-1,1} - \sum_{k \leq (m-1)(2)} \left[\lambda \overset{k}{u}_{\alpha,\alpha} + \frac{\lambda + 2\mu}{2h}(k+1)(k+2)\overset{k+1}{u}_{3} \right].$$

Taking into account these expression in (6.10) we obtain the infinite system of differential equations according to Vekua's system (V) in the following equivalent form:

$$
\begin{aligned}
& l_2 \overset{n}{u}_+ + h^{-1}(2n+1)\,\mathrm{grad}\left(\lambda \sum_{i\geq n(2)} \overset{i+1}{u} - \mu \sum_{i\leq n(2)} \overset{i-1}{u}_3\right) \\
& \qquad -\mu h^{-2}\frac{2n+1}{2} \sum_{i\leq n(2)} i(i+1)\overset{i}{u}_+ \\
& = \frac{1}{hc_n}\left[\overset{n}{f}_+ - \left(\frac{g_+^+ - g_+^-}{2}\delta_{n0} + \right.\right. \\
& \qquad \left.\left. +(g_+^+ + g_+^-)\sum_{i\geq 1(1)} \delta_{i-1,0} + (g_+^+ - g_+^-)\sum_{i\geq 1(1)} \delta_{i-1,1}\right)\right], \\
& \mu\Delta \overset{n}{u}_3 + h^{-1}(2n+1)\,\mathrm{div}(\mu \sum_{i\geq n(2)} \overset{i+1}{u}_+ - \lambda \sum_{i\leq n(2)} \overset{i-1}{u}_+) \\
& \qquad -(\lambda+2\mu)h^{-2}\frac{2n+1}{2}\sum_{i\leq n(2)} i(i+1)\overset{i}{u}_3 \\
& = \frac{1}{hc_n}\left[\overset{n}{f}_3 - \frac{1}{2}(g_3^+ - g_3^-)\delta_{n0} + \right. \\
& \qquad \left. +(g_3^+ + g_3^-)\sum_{i\geq 1(1)} \delta_{i-1,0} + (g_3^+ - g_3^-)\sum_{i\geq 1(1)} \delta_{i-1,1}\right],
\end{aligned}
\tag{6.13}
$$

$n = 0, 1, 2, \ldots, 2N+1.$

The comparison of these equations (6.13) with those of Vekua (V) proves their identity. In Vekua [1965] there are given cases with $N = 0, 1, 2$.

Evidently, in order to obtain effective values *a priori* in the form of energy inequalites for Vekua's operator with fixed N together with highest derivatives, we should pay attention to the explicit form of summands with derivatives of zero and first order from unknown moments $\overset{n}{u}_i(x_1, x_2)$ $(n = 0, 1, 2, \ldots)$ appearing in systems (6.12) or (6.13).

Here we should note that the explicit form of system (V) for the first approximation $N = 1$ was used in the paper Gordeziani [1975] to obtain the effective energy value *a priori* for this case.

Thus we constructed systems (6.12) and (6.13), corresponding to equation (6.1). Reduced boundary conditions, originated by data at S, the construction of which is of no difficulty, should be added to these systems, too. For this purpose we should multiply equalities (6.3) by Legendre polynomials $P_i(z/h)$ and integrate them between $-h$ and h. If Hooke's law and representations of (6.4) form are used, then we come up to finite reduced boundary conditions , defined by ∂D (see below subsection 6.4).

We should note that systems (6.12) and (6.13) were obtained as distinct from method of construction of any systems (V) (see, e.g., Vekua [1955,1965], Vekua, Muskhelishvili [1962], Zgenti [1988], Zgenti, Khvoles [1982]) with the account of satisfying boundary conditions (6.2) in case with $d = 0$ for any N (we shall discuss this question again).

Let us pay attention to the structure of system (6.12). First, the system of such a form was obtained in Vekua [1955], it was systematically investigated by Vekua, by his pupils and followers. If we bound ourselves to the first $N+1$ vector equations, then the reduced system will transform to recurrent relation, the solution of which lies in the solution of nonhomogeneous boundary value problems corresponding to differential operator of plane elasticity theory l_2 and the Laplacian Δ $(N+1)$-times with right hand parts, formed recursively, by means of the obtained functions. All the abovesaid makes this scheme attractive.

In spite of explicit substantiation of reduced systems, following from (6.12) or from more general systems of analogous structure (see Magnaradse [1984]), their solutions are not able to give full characteristics to qualitative vision of the solution of initial problems. For instance, we *cannot catch the effect of boundary layer or clear out any peculiar values of the solution* (on qualitative behaviour of solutions of boundary value problems of the theory of elasticity for bodies with small thickness there are numerous investigations e.g., see Friedrichs [1950], Friedrichs, Dressler [1961], Goldenveizer [1976], Lourie [1942], Shleniov [1975], Vorovich [1965]).

Probably, this reasoning caused the fact, that Vekua never returned to system of (6.12) type.

Now we shall consider condition (6.5) again. As we can see from the aforesaid reasoning, inclusions of (6.5) type appear to be essential characteristics for the solution (6.1)–(6.3). The existense of derivatives (in some general sense) of the second order and higher in Ω_h domain is connected with the strict fulfilment of strict limitations, imposed on dyad $[f, g]$. In this respect, for instance, the work Volkov [1965], devoted to investigation of differential properties of solution of Dirichlet's problem for Laplacian and Poisson equations in domains with piecewise-smooth boundaries, is a clear example of the abovesaid.

The abovementioned lemmas 6.2. and 6.3. (on convergence and limited density) make it possible to substantiate the rightfulness of Vekua's method with respect to reduced systems, originating from scheme (6.13), with corresponding boundary conditions on ∂D. A somewhat different matter is with the scheme (6.12). The problem of application of reduced system, corresponding to scheme (6.12) is connected with the problem of appropriate smoothness of the solution of initial problems.

The following theorem on application of reduced system, originated from scheme (6.12), is correct.

Theorem 6.1. *The reduced system (6.12) is applicable if and only if the series*

$$s = \sum_{k=1}^{\infty} k(k+1)\overset{k}{u}(x_1, x_2),$$

is absolutely convergent for $\forall (x_1, x_2) \in D$.

Proof. Each vector equation of (6.12), corresponding to index m, has summands of the form

$$s_{m+1} = (2m+1) \sum_{k\geq m(2)} (k(k+1) - m(m+1))\overset{k}{u}(x_1, x_2),$$

$$s_1 = \sum_{k\geq 2(2)} k(k+1)\overset{k}{u}, \quad s_2 = 3 \sum_{k\geq 3(2)} (k(k+1) - 2)\overset{k}{u}.$$

Evidently, from the absolute convergence of the last series follows absolute convergence of s and vice versa. Hence there follows the absolute convergence of series s_m. The validity of theorem is proved.

Nevertheless, we should note that for absolute convergence of s type series it is necessary that $\forall (x_1, x_2) \in D$

$$\left| k(k+1)\overset{k}{u} \right| = o(k^{-1-\alpha}), \quad (\alpha > 0).$$

The last condition is guaranteed (as is known) by the theory of orthogonal series Szëgo [1959] in the case of the solution of the initial problem u_i is continuously differentiable to the fourth order inclusive in domain $D \times [-h, h]$ and $\partial_3^5 u_i \in L_p(\Omega_h)$. The rigidity of conditions of function u_i is, evidently, stipulated by the necessity of absolute convergence of the series

$$\sum_{k=0}^{\infty} k(k+1)\overset{k}{u}(x_1, x_2) P_k\left(\frac{z}{h}\right),$$

at the end points $z = \pm h$.

Now let us return to the case with $d \geq 0$ in conditions (6.2), and use the variational principle of Vekua [1970], when system (V) or (6.13) with conditions (6.3) taken into account, is the identity of Euler–Lagrange equation for the initial problem with vector components σ_3, given on surfaces $S^{\pm}$.

For the case with $d \geq 0$ the following lemma is true.

Lemma 6.4. *Vekua's system (V) or (6.13) together with reduced boundary conditions, generated by (6.3) appears to be the Euler–Lagrange equation for energy functional, corresponding to the initial problem.*

Proof. Let us consider a tensor τ, connected with the stress tensor in the following way:

$$\tau_{3i} = \sigma_{3i} - d_i u, \quad \tau_{\alpha\beta} = \sigma_{\alpha\beta}.$$

Evidently, equation(1.1) in linear case in components τ has the form:

$$\tau_{\alpha i,i} + \partial_3(d_\alpha u) = f_\alpha,$$
$$\tau_{3i,i} + \partial_i(d_i u) = f_3.$$

Taking into account conditions (1.5) (with $A_{ij} = 0$) from Hooke's law (1.4) we obtain:

$$\tau_{\alpha\beta} = \sigma_{\alpha\beta} = \overline{B}_{\alpha\beta}(\partial_1, \partial_2, \partial_3)u,$$
$$\tau_3 = \sigma_3 - du = B_3 e - du = (\overline{B}_3(\partial_1, \partial_2, \partial_3) - d)u.$$

Here $\overline{B}_{\alpha\beta}$ and $\overline{B}_3$ are homogenous linear operators of the first order, stipulated by Hooke's law and Cauchy's formulae (see (1.4), (1.5).

The estimation of the last formulae in equilibrium equations gives the invariant form of equation (6.1) in components of displacement vector.

Lemma is proved.

By virtue of all this and the remark on p.271 of Kantorovich, Krilov [1962], concerning natural and main boundary conditions, it is more suitable to represent σ_3 in the form of (see (6.4)):

$$\tau_3 = \frac{(h+z)g^+ - (h-z)g^-}{2h} + \sum_{k=1}^{\infty} \overset{k}{\tau}_3(x_1,x_2)\left[P_{k+1}\left(\frac{z}{h}\right) - P_{k-1}\left(\frac{z}{h}\right)\right].$$

Values $\overset{k}{\tau}_3$ and coefficients $\overset{k}{\tau'}_3$ of the Fourier–Legendre series of vector τ_3 are connected with the following relation by virtue of Abel's relation

$$\overset{0}{\tau'}_3 = \frac{1}{2}(g^+ + g^-) - \overset{1}{\tau}_3, \qquad \overset{2k+1}{\tau}_3 = \frac{1}{2}(g^+ + g^-) - \sum_{i=0}^{k} \overset{2i}{\tau'}_3,$$

$$\overset{1}{\tau'}_3 = \frac{1}{2}(g^+ - g^-) - \overset{2}{\tau}_3, \qquad \overset{2k}{\tau}_3 = \frac{1}{2}(g^+ - g^-) - \sum_{i=0}^{k} \overset{2i+1}{\tau'}_3,$$

$$\overset{m}{\tau'}_3 = \overset{m-1}{\tau}_3 - \overset{m+1}{\tau}_3;$$

Now substituting $\overset{n}{\sigma}_3$ by $\overset{n}{\tau}_3$ in formulae (6.9) and (6.10), we obtain Vekua's system, corresponding to the problem (6.1) – (6.3).

In the problem of substantiation for group B methods, as is known, the essential singularity is investigation of the problem of unique solvability for the corresponding reduced system and the possibility of transition to the limit, with respect to the parameter, stipulating the order of approximation. There exists, as is known, another scheme for investigation (see Kantorovich, Akilov [1959], ch.XIV) it proves the solvability of reduced systems and convergence of the process by means of data on the initial problem.

The problem (6.1) – (6.3) was the subject of investigation for many authors, as concerned the substantiation of its *a priori* estimates. Let us note that after Korn's works [1907,1914], energy inequalities and the questions of their unique solvability for more general domains, not necessarily cylindrical ones (as in our case), were investigated in works by the following authors: Ciarlet [1988], Courant and Hilbert [1961], Dafermos [1968] Duvaut and Lions [1972], Eidus [1951], Fichera [1973], Friedrichs [1947], Koselev [1958], Mikhlin [1952], Mosolov, Miasnikov [1971], Ovtchinnikov, Xantis [1997], Rodrigues [1987]. The appropriate bibliography is given there, too. While comparing these approaches, we should stress that projective methods of group B and Vekua's technique, in particular, appear to be contrary methods in the following sense: on the one hand, the initial problem possesses qualities which can provide unique solvability of reduced systems and convergence of the process at a certain approximation, on the other hand, initial approximations in Vekua's theory ($N = 0, 1, 2$) are important examples of the theory of plates and shells; they have significant and rich geometrical and physical interpretation; in this respect it is essential to investigate the questions of unique solvability and effective solution for reduced systems of any approximation.

Further, in §§12–14 of the paper Vekua [1965] devoted to the investigation of the problem of unique solvability of Vekua's theory of plates and shells, for the first time, important energy relations (12.12) and (12.14) Vekua [1965] (p.38) were obtained; the non-negative bilinear form for tensor of strain, on the basis of which the uniqueness of regular solution for reduced system was proved. As to the problem of existence of such a solution, it can be proved with application of classic results Vekua [1965]. By mapping domain D on circle method Muskhelishvili [1968] is used, after that by means of explicit formulae, constructed by Kiknadze [1960], the system of integral equations should be constructed with the applicable Fredholm alternative.

Further investigations of the reduced Vekua system were realized in the paper Gordeziani [1974]. To obtain *a priori* estimates in the form of Korn's inequalities, the form defined according to (12.12) (Vekua [1970]) was split into N independent problems; for each of them, the procedure described in the book Duvaut, Lions [1972] was used as a result, *the existence of non-negative constant in energetic inequalities of Korn's type with each fixed N* was proved. Some results for the system (V) in the case of anisotropic elastic shellow shells *(also for fixed N)* was proved by Khoma [1985, 1986]. But, in this connection, **the proof of restriction for consequent constants from below with $N \to \infty$ has not been solved yet. Evidently, the solution of this problem is directly connected with the proof of convergent process, defined by virtue of Vekua's method.**

Let us consider reduced systems, generated by system (6.13) if we bound ourselves to $2N+1$ vector equations, and prove the following theorem.

Theorem 6.2. *Let the boundary conditions on the lateral boundary S corresponding to the problem (6.1) - (6.3), be homogeneous and such that the equalities hold:*

$$(\overset{n}{u}_{\alpha,\beta}, \overset{m}{u}_i) = \int\limits_D \overset{n}{u}_{\alpha,\beta}, \overset{m}{u}_i dx_1 dx_2 = -(\overset{n}{u}_\alpha, \overset{m}{u}_{i,\beta}) \tag{6.14}$$

where $\overset{n}{u}_i$ are desired coefficients of expansion function u_i.

Then the operator of the theory of plates corresponding to the reduced systems (6.13), satisfies an inequality of Korn's type with a constant, independent of N.

Proof. Let us multiply the reduced system (6.13) by $\overset{n}{u}$ according to (6.14) and sum from 0 to $2N+1$. After some calculations, taking into account equalities (6.14), we shall obtain:

$$(L_N^1 U_N^1, U_N^1) = \sum_{n=0}^{N} \left[\frac{1}{4n+1}(l_2 \overset{2n}{u}_+, \overset{2n}{u}_+) + \mu \frac{1}{4n+3}(\Delta \overset{2n+1}{u}_3, \overset{2n+1}{u}_3) \right.$$
$$-\lambda h^{-1} \left(\left(\sum_{m=0}^{n} \operatorname{div} \overset{2m}{u}_+, \overset{2n+1}{u}_3 \right) - \left(\sum_{m=n}^{N} \operatorname{grad} \overset{2m+1}{u}_3, \overset{2n}{u}_+ \right) \right)$$

$$-\mu h^{-1}\left(\left(\sum_{m=0}^{n-1}\operatorname{grad}\overset{2m+1}{u}_{3},\overset{2n}{u}_{+}\right)-\left(\sum_{m=n+1}^{N}\operatorname{div}\overset{2m}{u}_{+},\overset{2n+1}{u}_{3}\right)\right)$$

$$-h^{-2}\left(\mu(4n+3)\left(\sum_{m=n+1}^{N}\overset{2m}{u}_{+},\sum_{m=n+1}^{N}\overset{2m}{u}_{+}\right)\right.$$

$$\left.+(\lambda+2\mu)(4n+1)\left(\sum_{m=n}^{N}\overset{2m+1}{u}_{3},\sum_{m=n}^{N}\overset{2m+1}{u}_{3}\right)\right)\Bigg]; \qquad (6.15a)$$

$$(L_N^2U_N^2,U_N^2)=\sum_{n=0}^{N}\Bigg[\frac{1}{4n+3}(l_2\overset{2n+1}{u}_{+},\overset{2n+1}{u}_{+})+\mu\frac{1}{4n+1}\left(\Delta\overset{2n}{u}_{3},\overset{2n}{u}_{3}\right)$$

$$-\lambda h^{-1}\left(\left(\sum_{m=0}^{n-1}\operatorname{div}\overset{2m+1}{u}_{+},\overset{2n}{u}_{3}\right)-\left(\sum_{m=n+1}^{N}\operatorname{grad}\overset{2m}{u}_{3},\overset{2n+1}{u}_{+}\right)\right)$$

$$-\mu h^{-1}\left(\left(\sum_{m=0}^{n}\operatorname{grad}\overset{2m}{u}_{3},\overset{2n+1}{u}_{+}\right)-\left(\sum_{m=0}^{N}\operatorname{div}\overset{2m+1}{u}_{+},\overset{2n}{u}_{3}\right)\right)$$

$$-h^{-2}\left(\mu(4n+1)\left(\sum_{m=n}^{N}\overset{2m+1}{u}_{+},\sum_{m=n}^{N}\overset{2m+1}{u}_{+}\right)\right.$$

$$\left.+(\lambda+2\mu)(4n+3)\left(\sum_{m=n+1}^{N}\overset{2m}{u}_{3},\sum_{m=n+1}^{N}\overset{2m}{u}_{3}\right)\right)\Bigg]. \qquad (6.15b)$$

We should remind, that here $L_N=L_N^1+L_N^2$ is a linear operator, corresponding to reduced systems (6.13), $U_N=U_N^1+U_N^2$ an approximate solution , $\overset{n}{u}=0$ $(n\geq 2N+2)$,

$$(\operatorname{div}u_{+},u_3)=(u_{\alpha,\alpha},u_3),\quad (u_{+},\operatorname{grad}u_3)=(u_\alpha,u_{3,\alpha})^T,$$

$$(l_2u_{+},u_{+})=\mu(\Delta u_\alpha,u_\alpha)+(\lambda+\mu)(\operatorname{grad}\operatorname{div}u_{+},u_{+}).$$

Based on formulae (6.14) and using partial summation, with respect to summands with interior sum from 0 to $n+1$ or n , from the last relations (6.15) we have:

$$-(L_N^1U_N^1,U_N^1)=\sum_{n=0}^{N}\Bigg[\frac{1}{4n+1}\left(\mu(-\Delta\overset{2n}{u}_{+},\overset{2n}{u}_{+})+(\lambda+\mu)(\operatorname{div}\overset{2n}{u}_{+},\operatorname{div}\overset{2n}{u}_{+})\right)$$

$$+\frac{\mu}{4n+3}(-\Delta\overset{2n+1}{u}_{3},\overset{2n+1}{u}_{3})+2\lambda h^{-1}\left(\operatorname{div}\overset{2n}{u}_{+},\sum_{m=n}^{N}\overset{2m+1}{u}_{3}\right)$$

$$+2\mu h^{-1}\left(\sum_{m=n+1}^{N}\overset{2m}{u}_{+},\operatorname{grad}\overset{2n+1}{u}_{3}\right)$$

$$+h^{-2}\left(\mu(4n+3)\left(\sum_{m=n+1}^{N}\overset{2m}{u}_{+}\right),\sum_{m=n+1}^{N}\overset{2m}{u}_{+}\right)$$

$$+(\lambda+2\mu)(4n+1)\left(\sum_{m=n}^{N}\overset{2m+1}{u}_{3},\sum_{m=n}^{N}\overset{2m+1}{u}_{3})\right)\Bigg]$$

$$=\sum_{n=0}^{N}\Bigg[\int\limits_{D}\Bigg(\lambda\left(\frac{1}{\sqrt{4n+1}}\operatorname{div}\overset{2n}{u}_{+}+h^{-1}\sqrt{4n+1}\sum_{m=n}^{N}\overset{2m+1}{u}_{3}\right)^{2}$$
$$+\mu\left(\frac{\operatorname{grad}\overset{2n+1}{u}_{3}}{\sqrt{4n+3}}+h^{-1}\sqrt{4n+3}\sum_{m=n+1}^{N}\overset{2m}{u}_{+}\right)^{2}\Bigg)dw$$
$$+\frac{\mu}{4n+1}\Big(|\operatorname{grad}\overset{2n}{u}_{+}|^{2}+(\operatorname{div}\overset{2n}{u}_{+},\operatorname{div}\overset{2n}{u}_{+})$$
$$+2h^{-2}(4n+1)^{2}\left(\sum_{m=n}^{N}\overset{2m+1}{u}_{3},\sum_{m=n}^{N}\overset{2m+1}{u}_{3}\right)\Big)\Bigg].$$

Here

$$|\operatorname{grad}u_{+}|^{2}=\int\limits_{D}\left[|\partial_{1}u_{1}|^{2}+|\partial_{2}u_{1}|^{2}+|\partial_{1}u_{2}|^{2}+|\partial_{2}u_{2}|^{2}\right]dw,\quad dw=dxdy$$

Analogously

$$-(L_{N}^{2}U_{N}^{2},U_{N}^{2})=\sum_{n=0}^{N}\Bigg[\int\limits_{D}\Bigg(\lambda\left(\frac{\operatorname{div}\overset{2n+1}{u}_{+}}{\sqrt{4n+3}}+h^{-1}\sqrt{4n+3}\sum_{m=n+1}^{N}\overset{2m}{u}_{3}\right)^{2}$$
$$+\mu\left(\frac{\operatorname{grad}\overset{2n}{u}}{\sqrt{4n+1}}+h^{-1}\sqrt{4n+1}\sum_{m=n}^{N}\overset{2m+1}{u}_{+}\right)^{2}\Bigg)dw$$
$$+\frac{\mu}{4n+3}\Big(|\operatorname{grad}\overset{2n+1}{u}_{+}|^{2}+(\operatorname{div}\overset{2n+1}{u}_{+},\operatorname{div}\overset{2n+1}{u}_{+})$$
$$+2h^{-2}(4n+3)^{2}\left(\sum_{m=n+1}^{N}\overset{2m}{u}_{3},\sum_{m=n+1}^{N}\overset{2m}{u}_{3}\right)\Big)\Bigg].$$

Combining these formulae, we obtain:

$$-(L_{N}U_{N},U_{N})=\sum_{n=0}^{N}\Bigg[\frac{\lambda}{2n+1}\left\|\operatorname{div}\overset{n}{u}_{+}+\frac{2n+1}{h}\sum_{i\geq n(2)}\overset{i+1}{u}_{3}\right\|^{2}$$
$$+\frac{\mu}{2n+1}\left\|\operatorname{grad}\overset{n}{u}_{3}+h^{-1}(2n+1)\sum_{i\geq n(2)}\overset{i+1}{u}_{+}\right\|^{2}$$
$$+\frac{\mu}{2n+1}\Big(\|\operatorname{grad}\overset{n}{u}_{+}\|^{2}+\|\operatorname{div}\overset{n}{u}_{+}\|^{2}$$

$$+2h^{-2}(2n+1)^2 \left\| \sum_{i \geq n(2)} \overset{i+1}{u}{}_3 \right\|^2 \Bigg) \Bigg] . \tag{6.16}$$

Now, let us introduce scalar product with weight:

$$(\overset{m}{U}, \overset{n}{V})_1 = \frac{1}{\sqrt{(2m+1)(2n+1)}} (\overset{m}{U}, \overset{n}{V}),$$

$$(\overset{m}{U}, \overset{n}{V})_2 = h^{-2}\sqrt{(2m+1)(2n+1)} \left(\sum_{i \geq m(2)} \overset{i+1}{u}, \sum_{i \geq n(2)} \overset{i+1}{v} \right).$$

Then, evidently, from relation (6.16) the energetic inequality of low form follows:

$$-(L_N U_N, U_N) \geq \mu \left(\| \operatorname{grad} \overset{+}{U}_N \|_1^2 + \| \operatorname{div} \overset{+}{U}_N \|^2 + 2\| \overset{3}{U}_N \|_2^2 \right), \tag{6.17}$$

where

$$U_N = (\overset{+}{U}_N, \overset{3}{U}_N)^T = (\overset{1}{U}_N, \overset{2}{U}_N, \overset{3}{U}_N)^T.$$

The theorem is proved.

Now, based on the last three formulae (6.15), (6.16) and (6.17), let us investigate the questions, connected with problem II, formulated at the end of section 1.

As is known, the boundary conditions (6.3), given in their classic sense, will be reduced in eight basic modifications (Vekua [1965]), §11). It is mentioned, too, that in considering more general boundary conditions when, at one part $S' \subset S$ we have one condition and when $S'' = S/S'(S' \bigcap S'' = 0)$ we have other conditions, etc. In spite of this, as one can easily see, of principal difference are cases, when either the displacement vector, that is to say, Dirichlet's problem is considered for reduced system $(6.13)_N$, or the external stress vector is given.

Thus, let us formulate the basic problems, which will be the subject of our further investigation.

Problem I. *It is required to find the solution for the reduced system (6.13) in the domain $D(x_1, x_2)$, satisfying homogeneous Dirichlet boundary conditions at ∂D and being a trace of functions, defined by inclusion (6.5).*

Problem II. *It is required to find the solution for the reduced system (6.13) in $D(x_1, x_2)$ when the boundary of ∂D domain is free and when (6.5) is fulfilled.*

Let us use theorem 6.2 and investigate problem I in the sense of its solvability. By virtue of homogeneous Dirichlet conditions, (6.14), evidently, is fulfilled. Therefore, inequality (6.17) is valid. Further, for $\| \operatorname{grad} \overset{n}{U}_+ \|_1^2$ let us use Friedrich's inequality and let the following formula

$$(\overset{3}{U}_N, \overset{3}{U}_N)_2 = \frac{1}{2h} \int_{\Omega_h} \left(\sum_{n=0}^{2N+1} \overset{n}{u}_3(x,y) P_{n,3}\left(\frac{z}{h}\right) \right)^2 dz dx dy, \tag{6.18}$$

be the result of the representation (6.13).

Then from inequality (6.17) the *positive definiteness of Vekua's operator for problem (6.1) – (6.3) will follow in case of Dirichlet's boundary conditions* on S

$$(-L_N U_N, U_N) \geq \mu \left(\kappa^2 \|\overset{+}{U}_N\|_1^2 + 2\|\overset{3}{U}_N\|_2^2 \right), \tag{6.19}$$

where κ^2 is a constant in Friedrich's inequality.

Evidently, (6.19) is valid in the limit case when $N \to \infty$. In fact, by virtue of formula (6.18) and definition of scalar product with weight $(\ ,\ )_1$ we have:

$$-(Lu, u)_{L(\Omega_h)} \geq \frac{\mu}{2h} \int\limits_{\Omega_h} (|\partial_\alpha u_\alpha|^2 + u_{3,3}^2) d\Omega, \quad d\Omega = dxdydz.$$

One can easily see that by virtue of Parseval's equality

$$\frac{1}{2h} \int\limits_{\Omega_h} (\partial_\alpha u_\alpha)^2 d\Omega = \sum_{n=0}^{\infty} \frac{1}{2n+1} \|\operatorname{grad} \overset{n}{u}_+\|^2 = \|u_+\|_1^2,$$

and

$$\frac{1}{2h} \int\limits_{\Omega_h} u_{3,3}^2 d\Omega = \sum_{n=0}^{\infty} (2n+1) \left(\sum_{i \geq n(2)} \overset{i+1}{u}_3, \sum_{i \geq n(2)} \overset{i+1}{u}_3 \right) = \|u_3\|_2^2.$$

The comparison of the latter formulae assures us of the possibility of transition to limit in (6.19).

Thus, problem I in the case of $[f, g] = [0, 0]$ has the single zero solution, for $\overset{n}{u}_i = 0, \ \forall n \leq 2N+1$, follows from the anisotropic norm, induced with the right hand part of inequality (6.19). The abovegiven procedure on the construction of energetic functional and application of results, set forth in the works of Courant, Hilbert [1961] (or Friedrichs [1947]), make it possible to investigate problem II.

Based on formulae (6.10) and by virtue of boundary conditions on S in the case of problem II we have:

$$\begin{aligned}
&\frac{1}{2h} \int\limits_S u\sigma_{(n)} ds \\
&= \frac{1}{2h} \int\limits_S (u_\alpha \sigma_{\alpha\beta} n_\beta + u_3 \sigma_{\beta 3} n_\beta) ds \\
&= \sum_{i=0}^{\infty} \frac{1}{2i+1} \int\limits_{\partial D} d\tau n_\alpha \Big[(\lambda + 2\mu) \overset{i}{u}_\alpha \overset{i}{u}_{\alpha,\alpha} + \lambda \overset{i}{u}_\alpha \overset{i}{u}_{3-\alpha,3-\alpha} \\
&\qquad + h^{-1} \lambda (2i+1) \overset{i}{u}_\alpha \sum_{k \geq i(2)} \overset{k+i}{u}_3 \\
&\qquad + \mu \overset{i}{u}_{3-\alpha} (\overset{i}{u}_{1,2} + \overset{i}{u}_{2,1}) + \mu \overset{i}{u}_3 \overset{i}{u}_{3,\alpha}
\end{aligned}$$

$$+\mu h^{-1}(2i+1)\overset{i}{u}_3 \sum_{k\geq i(2)} \overset{i+1}{u}_{\alpha}\Big] = 0. \qquad (6.20)$$

Here ds is the surface element of S, $d\tau$ is the linear element of ∂D, $n_\alpha = \cos(n, x_\alpha)$, n is the exterior normal to the curve ∂D, $\sigma_{(n)} = (\sigma_{n1}, \sigma_{n2}, \sigma_{n3})$, $\sigma_{n\alpha} = \sigma_{\alpha\beta}n_\beta$, $\sigma_{n3} = \sigma_{\beta 3}n_\beta$.

Let L_{v1} denote Vekua's operator of the first kind, corresponding to the system (6.1) with homogeneous conditions (6.2), and $U = (\overset{0}{u}, \overset{1}{u}, \dots, \overset{N}{u}, \dots)^T$ coefficients of expansion into series of (6.4) functions $u_i(x,y,z)$. Let us use the equality:

$$\int_{\Omega_h} u[(\lambda+2\mu)\,\mathrm{grad}\,\mathrm{div} - \mu\,\mathrm{rot}\,\mathrm{rot}]u d\Omega_h = \int_D U L_{v1} U dw = (U, L_{v1}U).$$

Evidently, (6.15) transforms into the scalar product $\frac{1}{2h}(U, L_{v1}U)$ when $N \to \infty$, and for summands, according to the known procedure, we shall have:

$$\begin{aligned}
(l_2 u_+, u_+) &= \mu(\Delta u_\alpha, u_\alpha) + (\lambda+\mu)(\partial_\alpha \,\mathrm{div}\, u_+, u_\alpha) \\
&\quad -\mu(u_{1,11}u_1 + u_{1,22}u_1 + u_{2,11}u_2 + u_{2,22}u_2) \\
&\quad +(\lambda+2\mu)(u_{1,11}u_1 + u_{2,12}u_1 + u_{1,12}u_2 + u_{2,22}u_2) \\
&= (\lambda+2\mu)[\partial_1(u_1, u_{1,1}) - u_{1,1}^2 + \partial_2(u_2, u_{2,2}) - u_{2,2}^2] \\
&\quad +\lambda[\partial_1(u_1, u_{2,2}) - u_{1,1}u_{2,2} + \partial_2(u_2, u_{1,1}) - u_{2,2}u_{1,1}] \\
&\quad +\mu[\partial_1(u_2(u_{1,2}+u_{2,1})) + \partial_2(u_1(u_{1,2}) + u_{2,1})) - (u_{1,2}+u_{2,1})^2] \\
&= \int_{\partial D} d\tau n_\alpha((\lambda+2\mu)u_\alpha u_{\alpha,\alpha} + \lambda u_\alpha u_{3-\alpha,3-\alpha} + \mu u_{3-\alpha}(u_{1,2}+u_{2,1})) \\
&\quad - \int_D [(\lambda+2\mu)(u_{1,1}^2 + u_{2,2}^2) + 2\lambda u_{1,1}u_{2,2} + \mu(u_{1,2}+u_{2,1})^2]dw,
\end{aligned}$$

$$\mu(\Delta u_3, u_3) = \mu \int_{\partial D} u_3 n_\alpha u_{3,\alpha} d\tau - \mu \int_D (u_{3,1}^2 + u_{3,2}^2) dw,$$

$$\begin{aligned}
&\lambda h^{-1} \int_D \overset{i}{u}_+ \sum_{k\geq i(2)} \mathrm{grad}\,\overset{k+1}{u}_3 dw \\
&\qquad = \lambda h^{-1} \int_{\partial D} n_\alpha \overset{i}{u}_\alpha \sum_{k\geq i(2)} \overset{k+1}{u}_3 d\tau - \lambda h^{-1} \int_D u_{\alpha,\alpha} \sum_{k\geq i(2)} \overset{k+1}{u}_3 dw,
\end{aligned}$$

$$\int_D u_3 u_{\alpha,\alpha} dw = \int_{\partial D} u_3 n_\alpha u_\alpha d\tau - \int_D u_\alpha u_{3,\alpha} dw.$$

The consideration of these formulae (in the limit case with $N \to \infty$) and (6.20) gives:

$$-\frac{1}{2h}(L_{v_1}U, U)$$

$$= \sum_{n=0}^{\infty} \frac{1}{2n+1} \int_D \Big[(\lambda + 2\mu)(\overset{n}{u}{}_{1,1}^2 + \overset{n}{u}{}_{2,2}^2) + 2\lambda \overset{n}{u}_{1,1} \overset{n}{u}_{2,2} + \mu(\overset{n}{u}_{1,2} + \overset{n}{u}_{2,1})^2$$

$$+\mu(\overset{n}{u}{}_{3,1}^2 + \overset{n}{u}{}_{3,2}^2) + 2(2n+1)\lambda h^{-1} \overset{n}{u}_{\alpha,\alpha} \sum_{k \geq n(2)} \overset{k+1}{u}_3$$

$$+2(2n+1)\mu h^{-1} \overset{n}{u}_{3,\alpha} \sum_{k \geq n(2)} \overset{k+1}{u}_\alpha$$

$$+h^{-2}(2n+1)^2 \left(\mu \left(\sum_{k \geq n(2)} \overset{k+1}{u}_\alpha, \sum_{k \geq n(2)} \overset{k+1}{u}_\alpha \right) \right.$$

$$\left. \left. +(\lambda + 2\mu) \left(\sum_{k \geq n(2)} \overset{k+1}{u}_3, \sum_{k \geq n(2)} \overset{k+1}{u}_3 \right) \right) \right] dw$$

$$-\frac{1}{2h} \int_S u\sigma_{(n)} ds$$

$$= \sum_{n=0}^{\infty} \frac{1}{2n+1} \int_D \left[\lambda \left(\overset{n}{u}_{\alpha,\alpha} + h^{-1}(2n+1) \sum_{k \geq n(2)} \overset{k+1}{u}_3 \right)^2 \right.$$

$$+\mu \left(\overset{n}{u}_{3,\alpha} + h^{-1}(2n+1) \sum_{k \geq n(2)} \overset{k+1}{u}_\alpha \right)^2$$

$$+2\mu(\overset{n}{u}{}_{1,1}^2 + \overset{n}{u}{}_{2,2}^2) + \frac{1}{2}(2 \overset{n}{u}_{1,2} + \overset{n}{u}_{2,1})^2 +$$

$$\left. +h^{-2}(2n+1)^2 \left(\sum_{k \geq n(2)} \overset{k+1}{u}_3, \sum_{k \geq n(2)} \overset{k+1}{u}_3 \right) \right] dw. \quad (6.21)$$

Representation (6.21) is the desired one, generating with certain definite conditions, covered on domain D and $\overset{n}{u}$, the second Korn's type inequality in case of cylindrical domain Ω_h.

In fact, relations $(u, 1) = (\operatorname{rot} u, 1)$ are equivalent to the following equalities:

$$\int_D \overset{n}{u} dw = \int_D \operatorname{rot} \overset{n}{u}_+ dw = \int_D \left(\overset{n}{u}_{3,\alpha} - h^{-1}(2n+1) \sum_{k \geq n(2)} \overset{k+1}{u}_\alpha \right) dw = 0. \quad (6.22)$$

Further, we shall assume that the domain D is such that it satisfies known restrictions formulated in the works Courant, Hilbert [1961], Friedrichs [1947]. Below, for brevity, we shall consider D as a star domain (or finite sum of star domains) with regard to a certain circle belonging to D. The starness with respect to the set denotes the starness with respect to each point of this set. The last remark means that any ray from the point, has one and only one point common with ∂D (e.g., see Mikhlin [1977]).

Now, word for word, let us repeat Friedrichs' [1947] reasoning (see also Courant, Hilbert [1961], as well as Mikhlin [1952], §1) proofs of Korn's inequalities for two-dimensional problem on the theory of elasticity with free boundaries for summands $\overset{n}{u}{}^2_{1,1}+\overset{n}{u}{}^2_{2,2}+\frac{1}{2}(\overset{n}{u}_{1,2}+\overset{n}{u}_{2,1})^2$ appearing in left hand parts of equality (6.21) and satisfying conditions (6.22). Moreover, let us use theorem 2 (see §41 Mikhlin [1952]), its analogous substantiation (see Fichera [1973], comment on Korn's inequality, p. 80-81) with regard to summands $\overset{n}{u}{}^2_{1,1}+\overset{n}{u}{}^2_{2,2}+\frac{1}{2}(\overset{n}{u}_{1,2}+\overset{n}{u}_{2,1})^2$,

$$\left(\overset{n}{u}_{3,\alpha}+h^{-1}(2n+1)\sum_{k\geq n(2)}\overset{k+1}{u}_{\alpha}\right)^2$$
$$=\left(\overset{n}{u}_{3,1}+h^{-1}(2n+1)\sum_{k\geq n(2)}\overset{k+1}{u}_{1}\right)^2+\left(\overset{n}{u}_{3,2}+h^{-1}(2n+1)\sum_{k\geq n(2)}\overset{k+1}{u}_{2}\right)^2,$$

satisfying conditions (6.22). Then we shall have:

$$-\frac{1}{2h}(L_{v_1}U,U)=2\mu(\kappa_1^2\|\operatorname{grad}U\|_1^2+\kappa_2^2\|U\|_2^2). \tag{6.23}$$

Here $\kappa_1^2=\min\{c_1,c_2\}$, $\kappa_2^2=\min\{1,c_2\}$, where $c_\alpha>0$ are constants, depending only on D.

By virtue of condition (6.22_1) we have:

$$\left(\int\limits_D U dw\right)^2=\sum_{n=0}^{\infty}\frac{1}{2n+1}\left(\int\limits_D \overset{n}{u}dw\right)^2=0.$$

Therefore, with respect to the Poincaré inequality (see, e.g., Courant, Hilbert [1961]):

$$\int\limits_D(\overset{n}{u}{}^2_{i,1}+\overset{n}{u}{}^2_{i,2})dw+c_3\left(\int\limits_D\overset{n}{u}_i dw\right)^2\geq c_4\int\limits_D\overset{n}{u}{}^2_i dw\quad(c_3>0,c_4>0),$$

from *a priori* estimations (6.23) follow:

$$-\frac{1}{2h}(L_{v_1}U,U)\geq 2\mu(\kappa_3^2\|U\|_1^2+\kappa_2^2\|U\|_2^2),\quad \kappa_3^2=\kappa_1^2c_3, \tag{6.24}$$

as well as the positive definition of Vekua's operator for problem II with any N, and then, by virtue of this, the single valued solvability of differential problems, corresponding to reduced systems (6.13).

From the positive definiteness of Vekua's operator follow its boundedness from below and exitense of the unique bounded operator (see Lax–Milgramm's lemma:

Lax, Milgramm [1954] or Koltunov, Kravchuk, Maiboroda [1983]). Hence, in it's turn, based on lemmas 6.2 and 6.3 *the weak convergence* of Vekua's process follows for the error of displacement vector.

Further, with respect to inequalities (6.19) and (6.23) the weak convergence of first derivatives follows from the difference $V - V_N$. With regard to this and representation of a function by means of its derivative, according to the procedure for variational methods, *the convergence of Vekua's process in uniform metric follows for the error of displacement vector.*

6.4. Construction and justification of Vekua type systems in the case when the boundary conditions on surfaces are Newtonian type. Korn type inequalities

Let us return to system $(6.13)_N$ and prove, that with definite restrictions, imposed on boundary conditions (6.3), Vekua's operator $L^N_{v_1}$, corresponding to the system, can be found by means of the inversion of uniform operators according to factorized scheme.

As it follows from system (6.13) ($N < +\infty$), it can be represented in the following form

$$L^N_{v_1} U_N = Q_m \overset{m}{v^-} + P_m \overset{m}{v} + R_m \overset{m}{v^+} - T\overset{m}{v'} = \overset{m}{G} \quad (m = 0, 1, 2, \dots, 2N+1). \quad (6.25)$$

The following.notation is introduced here:

$$Q = h^{-1} \begin{pmatrix} 0 & -\mu \operatorname{grad} \\ -\lambda \operatorname{div} & 0 \end{pmatrix}, \qquad R = h^{-1} \begin{pmatrix} 0 & \lambda \operatorname{grad} \\ \mu \operatorname{div} & 0 \end{pmatrix},$$

$$T = \frac{1}{2h^2} \begin{pmatrix} \mu I_2 & 0 \\ 0 & (\lambda + 2\mu) I_1 \end{pmatrix}, \qquad I_1 = I, \qquad I_2 = (I, I), \quad (6.26)$$

$$P_m = \begin{Vmatrix} (2m+1)^{-1} l_2 - \frac{1}{2} h^{-2} \mu m(m+1) I_2, & h^{-1} \lambda \operatorname{grad} \\ -h^{-1} \lambda \operatorname{div}, & (2m+1)^{-1} \mu \Delta - \frac{1}{2} h^2 (\lambda + 2\mu)(m+1)(m+2) I_1 \end{Vmatrix}, \quad (6.27a)$$

$$\overset{m}{v^-} = \sum_{i \le m(2)} (\overset{i-2}{u}_+, \overset{i-1}{u}_3)^T, \quad \overset{m}{v^+} = \sum_{i \ge m(2)} (\overset{i+2}{u}_+, \overset{i+3}{u}_3)^T,$$

$$\overset{m}{v} = (\overset{m}{u}_+, \overset{m+1}{u}_3)^T, \quad \overset{m}{v'} = \frac{1}{2} \sum_{i \le m(2)} \left((i-2)(i-1) \overset{i-2}{u}_+, i(i-1) \overset{i-1}{u}_3 \right)^T,$$

$$\overset{0}{v^-} = \overset{2k}{v^+} = 0, \quad \overset{m}{G} = \left((2m+1)^{-1} \overset{m}{F}_+, (2m+3) \overset{m+1}{F}_3 \right)^T, \quad (6.28a)$$

$$T_m = T \times \{m(m+1), m(m+1), (m+1)(m+2)\},$$

for even values of m;

$$P_m = \left\| \begin{array}{cc} (2m+1)^{-1} l_2 - \frac{1}{2} h^{-2} \mu m(m+1) I_2, & -h^{-1} \mu \,\mathrm{grad} \\ h^{-1} \mu .\mathrm{div}, & (2m+1)^{-1} \mu \Delta - \frac{1}{2} h^{-2} (\lambda + 2\mu)(m-1) m I_1 \end{array} \right\|, \quad (6.27b)$$

$$\overset{m}{v}{}^{-} = \sum_{i \le m(2)} (\overset{i-2}{u}{}_{+}, \overset{i-3}{u}{}_{3})^T, \quad \overset{m}{v}{}^{+} = \sum_{i \ge m(2)} (\overset{i+2}{u}{}_{+}, \overset{i+1}{u}{}_{3})^T, \quad \overset{m}{v} = (\overset{m}{u}{}_{+}, \overset{m-1}{u}{}_{3},)^T,$$

$$\overset{m}{v}{}' = \frac{1}{2} \sum_{i \le m(2)} \left((i-2)(i-1) \overset{i-2}{u}{}_{+}, (i-3)(i-2) \overset{i-3}{u}{}_{3} \right)^T, \quad (6.28b)$$

$$\overset{m}{G} = \left((2m+1)^{-1} \overset{m}{F}{}_{+}, (2m-1) \overset{m-1}{F}{}_{3} \right)^T,$$

$$T_m = T \times \{m(m+1), m(m+1), (m-1)m\},$$

for odd values of m.

In operators T_m, as assumed, $\{i, j, k\}$ denotes the diagonal matrix.

From operator equation (6.25) the summands, containing $v^{\pm}, v'$ can be represented in the form

$$Q\overset{m}{v}{}^{-} - T\overset{m}{v}{}' = \sum_{i \le m(2)} (Q - T_{i-2}) \overset{i-2}{v}, \qquad R\overset{m}{v}{}^{+} = \sum_{i \ge m(2)} R \overset{i+2}{v}.$$

The following matrix operators, based on these formulae, correspond to the system (6.25):

$$M_0 = \left\| \begin{array}{cccccc} P_0 & R & R & \cdots & R & R \\ Q - T_0 & P_2 & R & \cdots & R & R \\ Q - T_0 & Q - T_2 & P_4 & \cdots & R & R \\ \cdots & \cdots & \cdots & \cdots & \cdots & \cdots \\ Q - T_0 & Q - T_2 & \cdots & \cdots & Q - T_{2N-2} & P_{2N} \end{array} \right\|, \text{ for even values of } m,$$

$$M_1 = \left\| \begin{array}{ccccc} P_1 & R & \cdots & R & R \\ Q - T_1 & P_3 & \cdots & R & R \\ \cdots & \cdots & \cdots & \cdots & \cdots \\ Q - T_1 & Q - T_3 & \cdots & Q - T_{2N-1} & P_{2N+1} \end{array} \right\|, \text{ for odd values of } m.$$

Evidently, matrix operators M_0 and M_1 can be represented in the form:

$$
\begin{aligned}
M_0 &= \left\| \begin{matrix} A_0^+ & -B_0^+ & 0 & \cdots & 0 \\ 0 & A_2^+ & -B_2^+ & \cdots & 0 \\ \cdots & \cdots & \cdots & \cdots & \cdots \\ 0 & 0 & \cdots & A_{2N-2}^+ & -B_{2N-2}^+ \\ Q-T_0 & Q-T_2 & \cdots & Q-T_{2N-2} & P_{2N} \end{matrix} \right\| \\
&= \left\| \begin{matrix} P_0 & R & R & \cdots & R & R & R \\ -B_2^- & A_2^- & 0 & \cdots & 0 & 0 & 0 \\ \cdots & \cdots & \cdots & \cdots & \cdots & \cdots & \cdots \\ 0 & 0 & 0 & \cdots & -B_{2N-2}^- & A_{2N-2}^- & 0 \\ 0 & 0 & 0 & \cdots & 0 & -B_{2N} & A_{2N} \end{matrix} \right\| \\
M_1 &= \left\| \begin{matrix} A_1^+ & -B_1^+ & 0 & \cdots & 0 \\ \cdots & \cdots & \cdots & \cdots & \cdots \\ 0 & 0 & \cdots & A_{2N-1}^+ & -B_{2N-1}^+ \\ Q-T_1 & Q-T_3 & \cdots & Q-T_{2N-1} & P_{2N+1} \end{matrix} \right\| \\
&= \left\| \begin{matrix} P_1 & R & \cdots & R & R \\ -B_3^- & A_3^- & \cdots & 0 & 0 \\ \cdots & \cdots & \cdots & \cdots & \cdots \\ 0 & 0 & \cdots & -B_{2N+1}^- & A_{2N+1}^- \end{matrix} \right\|
\end{aligned}
\tag{6.29}
$$

The following notation is introduced here:

$$A_m^+ = P_m - Q + T_m, \qquad B_m^+ = P_{m+2} - R \qquad (m = 0, 1, 2, \ldots, 2N-1)$$

$$A_m^- = P_m - R, \qquad B_m^- = P_{m-2} - Q + T_{m-2} \qquad (m = 2N, 2N-1, \ldots, 2)$$

Expressions (6.26) – (6.28) give the explicit form for operators A and B. We have:

$$A_m^+ = \left\| \begin{matrix} (2m+1)^{-1} l_2 & h^{-1}(\lambda+\mu)\,\mathrm{grad} \\ 0 & (2m+3)^{-1}\mu\Delta \end{matrix} \right\|,$$

$$B_m^+ = \left\| \begin{matrix} (2m+5)^{-1} l_2 - \frac{1}{2} h^{-2}\mu(m+2)(m+3) I_2, & 0 \\ -h^{-1}(\lambda+\mu)\,\mathrm{div}, & (2m+7)^{-1}\mu\Delta - \frac{1}{2} h^{-2}(\lambda+2\mu)(m+3)(m+4) I_1 \end{matrix} \right\|$$

for even m;

$$A_m^+ = \left\| \begin{matrix} (2m+1)^{-1} l_2 & 0 \\ h^{-1}(\lambda+\mu)\,\mathrm{div} & (2m-1)^{-1}\mu\Delta \end{matrix} \right\|$$

$$B_m^+ = \left\| \begin{matrix} (2m+5)^{-1} l_2 - \frac{1}{2} h^{-2}\mu(m+2)(m+3) I_2, & h^{-1}(\lambda+\mu)\,\mathrm{grad} \\ 0, & (2m+3)^{-1}\mu\Delta - \frac{1}{2} h^{-2}(\lambda+2\mu)(m+1)(m+2) I_1 \end{matrix} \right\| \tag{6.30}$$

for odd m;

$$A_m^- = \left\| \begin{matrix} (2m+1)^{-1}l_2 - \frac{1}{2}h^{-2}m(m+1)I_2, & 0 \\ -h^{-1}(\lambda+\mu)\,\mathrm{div}, & (2m+3)^{-1}\mu\Delta - \frac{1}{2}h^{-2}(\lambda+2\mu)(m+1)(m+2)I_1 \end{matrix} \right\|,$$

$$B_m^- = \left\| \begin{matrix} (2m-3)^{-1}l_2 & 0 \\ h^{-1}(\lambda+\mu)\,\mathrm{div}, & (2m-5)^{-1}\mu\Delta \end{matrix} \right\|,$$

for even m;

$$A_m^- = \left\| \begin{matrix} (2m+1)^{-1}l_2 - \frac{1}{2}h^{-2}\mu m(m+1)I_2, & -h^{-1}(\lambda+\mu)\,\mathrm{grad} \\ 0, & (2m-1)^{-1}\mu\Delta - \frac{1}{2}h^{-2}(\lambda+2\mu)m(m-1)I_1 \end{matrix} \right\|,$$

$$B_m^- = \left\| \begin{matrix} (2m-1)^{-1}l_2 & 0 \\ h^{-1}(\lambda+\mu)\,\mathrm{div}, & (2m-5)^{-1}\mu\Delta \end{matrix} \right\|,$$

for odd m.

Let us assume now (and prove below) that reduced conditions, corresponding to conditions (6.3), satisfy the relation of (6.25) form

$$L_{v_1}^N U_N = q_m \overset{m}{v}{}^- + p_m \overset{m}{v} + r_m \overset{m}{v}{}^+ = \overset{m}{g}, \qquad (x_1, x_2) \in \partial D \tag{6.31}$$

Then differential operators A_m, B_m and P_m, defined on the set of functions $\{\overset{m}{v}\}$, the trace of which satisfies (6.31) will be operators which we identify with them and designate by the same letters A_m, B_m and P_m.

By virtue of explicit operations on $M_{\alpha-1}$, right hand parts G will change accordingly; we will have:

$$\overset{m}{G}{}^+ = \overset{m+2}{G} - \overset{m}{G}, \quad 0 \le m < 2N, \qquad \overset{2N}{G}{}^+ = \overset{2N}{G}, \quad \overset{2N+1}{G}{}^+ = \overset{2N+1}{G},$$

$$\overset{\alpha-1}{G}{}^- = \overset{\alpha-1}{G} \quad (\alpha = 1,2), \qquad \overset{m}{G}{}^- = \overset{m}{G} - \overset{m-2}{G} \quad (m \ge 2)$$

Two-point operator equation

$$A_m^+ \overset{m}{v} = B_m^+ \overset{m+2}{v} + \overset{m}{G}{}^+, \qquad A_m^- \overset{m}{v} = B_m^- \overset{m-2}{v} + \overset{m}{G}{}^-,$$

defined by the chain equalities from (6.25) - (6.29), gives

$$\overset{m}{v} = C_m^+ \overset{m+2}{v} + \overset{m}{G}{}^+, \qquad C_m^+ = (A_m^+)^{-1} B_m^+, \qquad \overset{m}{G}{}^+ = (A_m^+)^{-1} \overset{m}{G}{}^+,$$

$$\overset{m}{v} = C_m^- \overset{m-2}{v} + \overset{m}{G}{}^-, \qquad C_m^- = (A_m^-)^{-1} B_m^-, \qquad \overset{m}{G}{}^- = (A_m^-)^{-1} \overset{m}{G}{}^-.$$

The last equalities are recurrent relations, from which

$$\begin{aligned} \overset{m}{v} &= C_m^+ (C_{m+2}^+ \overset{m+4}{v} + \overset{m+2}{G}{}^+) \\ &= C_m^+ C_{m+2}^+ \overset{m+4}{v} + C_m^+ \overset{m+2}{G}{}^+ + \overset{m}{G}{}^+ \end{aligned}$$

$$
\begin{aligned}
&= \dots \\
&= C^+_m C^+_{m+2}\cdots C^+_{2N-3+\alpha}\overset{N_s}{v} + \overset{m}{G^+} + C^+_m \overset{m+2}{G^+} + \dots + C^+_m C^+_{m+2}\cdots C^+_{2N-4+\alpha}\overset{2N-3+\alpha}{G^+} \\
&= \prod_{j\geq m(2)} C^+_j \overset{N_s}{v} + \sum_{i\geq m(2)} \prod_{j\geq m(2)} C^+_j \overset{i}{G^+},
\end{aligned}
\tag{6.32a}
$$

$$
\begin{aligned}
\overset{m}{v} &= \prod_{j\geq m(2)} C^+_j \overset{2N-\alpha+1}{v} + \sum_{i\geq m(2)} \prod_{j\geq m(2)}^{i-2} C^+_j \overset{i}{G^+} \\
&= \prod_{j\geq m(2)} C^+_j \overset{2N-\alpha+1}{v} + \overset{m}{f^+},
\end{aligned}
$$

where

$$
\prod_{j\geq m(2)}^{m-2} C_j = 1.
$$

Similarly

$$
\begin{aligned}
\overset{m}{v} &= C^-_m \left(C^-_{m-2} \overset{m-4}{v} + \overset{m-2}{G^-} \right) + \overset{m}{G^-} \\
&= \dots \\
&= C^-_m C^-_{m-2} \cdots C^-_{\alpha+1} \overset{\alpha+1}{v} + \sum_{i\leq m(2)} \prod_{j\leq m(2)}^{i+2} C^-_j \overset{i}{G^-} \\
&= \prod_{j\leq m(2)} C^-_j \overset{\alpha-1}{v} + \overset{m}{f^-}.
\end{aligned}
\tag{6.32b}
$$

$\alpha = 1$ corresponds to even values of m, $\alpha = 2$ to odd ones.

Let us substitute these expressions into the last ones (corresponding to positive sign) and into first (corresponding to minus) equations of system (6.25). We shall obtain :

$$
\sum_{m=(\alpha-1)(2)}^{cN-3+\alpha} (Q - T_m) \left(\prod_{j\geq m(2)} C^+_j \overset{2N+1+\alpha}{v} + \overset{m}{f^+} \right) + P_{2N+(\alpha-1)} \overset{2N+1+\alpha}{v} = \overset{2N-1+\alpha}{G},
\tag{6.33}
$$

and

$$
R \sum_{m\geq 2N+\alpha-1} \left(\prod_{j\leq m(2)} C^-_j \overset{\alpha-1}{v} + \overset{m}{f^-} \right) + P_{\alpha-1} \overset{\alpha-1}{v} = \overset{\alpha-1}{G}.
$$

From equations (6.33) we define $\overset{2N+\alpha-1}{v}$ and $\overset{\alpha-1}{v}$ accordingly, by means of which the rest of $\overset{m}{v}$ are defined from equations (6.32).

Now, let us return to the relation (6.31) for all admissible classic conditions, investigated, for example, in Kupradze, Gegelia, Basheleishvili, Burchuladze [1976], operator l is represented in the form of $l = a_1(\partial_1, \partial_2) + a_0(\partial_3)$, where $a_{\alpha-1}$ are differential operators of dimension 3×3, with elements independent of z and carry derivatives of order $\alpha - 1$ at most.

Therefore, Galiorkin's projective method lengthwise generators Ω_h according to system $\{P_k(z/h)\}$ gives in case of classic boundary conditions

$$q = 0, \quad p_m = \frac{1}{2m+1} a_1(\partial_1, \partial_2), \quad r = a_0(\partial_3)$$

At the same time, in each scalar equation (6.31) u_+ and u_3 enter it with indices differing by an even number (see representations (6.10) and (6.20)).

6.5. Three point operator equations and the method of differential factorization of such systems, plates of variable thickness

Let us consider now the case with a displacement vector, given a surfaces $S^\pm$, i.e., in condition (6.2) formally assuming the right hand part to be equal $-dg^\pm$ and then admitting $d = +\infty$.

$$u(x_1, x_2, \pm h(x_1, x_2)) = g^\pm(x_1, x_2), \qquad x \in S^\pm. \tag{6.2'}$$

In the paper Vashakmadze [1973] the numerical process of the solution of system (6.13) was proposed, when (6.13) carries six scalar equations, if we use this process for three-dimensional problems of the theory of elasticity (6.1), (6.2), (6.3), then in case of elastic plates with variable thickness

$$\Omega_h(x) = D(x_1, x_2) \times] - h(x_1, x_2), h(x_1, x_2)[,$$

system (6.1) takes the form of the following three-point operator equation:

$$Q_m \overset{m-2}{v}(x_1, x_2) + P_m \overset{m}{v}(x_1, x_2) + R_m \overset{m+2}{v}(x_1, x_2) = \overset{m}{F}, \quad m = 1, 2, 3, \ldots \tag{6.34}$$

System (6.34) was obtained on the basis of Galiorkin's method, in the case $g^\pm = 0$ and for the displacement vector u the expansion below was used instead of (6.4)

$$u(x_1, x_2, z) = w(x_1, x_2, z) = \sum_{m=1}^{\infty} \overset{m}{w} [P_{m+1}(z/h) - P_{m-1}(z/h)]. \tag{6.35}$$

Evidently the case of homogeneous boundary conditions (6.2′) changes only right hand part of system (6.34).

System (6.34) is, as has already been mentioned, a three-point operator equation with elements defined by the following equalities:

$$\overset{m}{v} = \left(\overset{m}{w}_1, \overset{m}{w}_2, \overset{m-1}{w}_3\right)^T, \qquad \overset{m}{F} = \left(\overset{m+1}{f}_+ - \overset{m-1}{f}_+, \overset{m}{f}_3 - \overset{m-2}{f}_3\right)^T; \tag{6.36}$$

matrices

$$P_m(\partial_1, \partial_2) = \{P_{ij}^m\}, \quad Q_m(\partial_1, \partial_2) = \{Q_{ij}^m\}, \quad R_m(\partial_1, \partial_2) = \{R_{ij}^m\}, \tag{6.37}$$

are differential operators of 3×3, the elements of which are calculated by the following algebraic expressions:

$$\begin{aligned}
P_{\alpha\alpha}^m &= c_m h \left[(\lambda+2\mu)(\partial_{\alpha\alpha} + g_{m,3}\partial_\alpha + l_{m,3}) + \mu(\partial_{3-\alpha,3-\alpha} \right. \\
&\quad \left. + g_{m,3-\alpha} + l_{m,3-\alpha})\right] - 2(2m+1)h^{-1}\mu, \\
P_{\alpha\beta}^m &= (\lambda+\mu)c_m h \left[\partial_{\alpha\beta} + \frac{1}{2}(g_{m,\alpha}\partial_\beta + g_{m,\beta}\partial_\alpha) + d_m\right], \\
P_{\alpha 3}^m &= -(\lambda+\mu)\left[\partial_\alpha - (m-1+h)\partial_\alpha \ln h\right] = -P_{3\alpha}^m, \\
P_{33}^m &= \mu c_{m-1} h \left[\Delta + g_{m-1,\alpha}\partial_\alpha + l_{m-1,\alpha}e_\alpha\right] - 2(2m+1)h^{-1}(\lambda+2\mu), \\
Q_{\alpha\alpha}^m &= -a_{m-1} h \left[(\lambda+2\mu)(\partial_{\alpha\alpha} + g_{m,\alpha}^-\partial_\alpha + l_{m,\alpha}^-) + \mu(\partial_{3-\alpha}^2 \right. \\
&\quad \left. + g_{m,3-\alpha}^-\partial_{3-\alpha} + l_{m,3-\alpha}^-)\right], \\
Q_{\alpha\beta}^m &= -(\lambda+\mu)a_{m-1} h \left[\partial_{\alpha\beta} + (g_{m,\alpha}^-\partial_\alpha + g_{m,\beta}^-\partial_\beta) + d_m^-\right], \quad Q_{\alpha 3}^m = 0, \\
Q_{3\alpha}^m &= -2(\lambda+\mu)\left[\partial_\alpha + (m-2+h)\partial_\alpha \ln h\right], \\
Q_{33}^m &= -\mu a_{m-2} h \left[\Delta + g_{m-1,\alpha}^-\partial_\alpha + l_{m-1,\alpha}^- e_\alpha\right], \\
R_{\alpha\alpha}^m &= -a_{m+1} h \left[(\lambda+2\mu)(\partial_{\alpha\alpha} + g_{m,\alpha}^+\partial_\alpha + l_{m,\alpha}^+) + \mu(\partial_{3-\alpha}^2 \right. \\
&\quad \left. + g_{m,3-\alpha}^+\partial_{3-\alpha} + l_{m,3-\alpha}^+)\right], \\
R_{\alpha\beta}^m &= -(\lambda+\mu)a_{m+1} h \left[\partial_{\alpha\beta} + (g_{m,\alpha}^+\partial_\alpha + g_{m,\beta}^+\partial_\beta) + d_m^+\right], \\
R_{\alpha 3}^m &= 2(\lambda+\mu)\left[\partial_\alpha - (m+2-h)\partial_\alpha \ln h\right], \quad R_{3\alpha}^m = 0, \\
R_{33}^m &= -\mu a_m h \left[\Delta + g_{m-1,\alpha}^+\partial_\alpha + l_{m-1,\alpha}^+ e_\alpha\right].
\end{aligned}$$

Here

$$a_m = \frac{2}{2m+1}, \qquad c_m = \frac{4(2m+1)}{(2m-1)(2m+3)} = a_{m-1} + a_{m+1}, \qquad a_0 = a_{-1} = 0,$$

$$\begin{aligned}
d_m &= 2hc_m^{-1}\left[-\frac{1}{4}c_m\partial_{12}\ln h + \frac{1}{2}c_m h\partial_1\frac{1}{h}\partial_2\frac{1}{h} + \left(\frac{(m+1)^2}{2m+3} + \frac{m^2}{2m-1}\right)\right. \\
&\quad \left. \times h\partial_1\frac{1}{h}\partial_2\frac{1}{h}\right], \\
d_m^+ &= \frac{2h}{a_{m+1}}\left[-\frac{m+2}{2m+3}\partial_{12}\ln h + \frac{(m+2)(m+3)}{2m+3}h\partial_1\frac{1}{h}\partial_2\frac{1}{h}\right], \\
d_m^- &= \frac{2h}{a_{m-1}}\left[-\frac{m-1}{2m-1}\partial_{12}\ln h - \frac{(m-1)(m+2)}{2m-1}h\partial_1\frac{1}{h}\partial_2\frac{1}{h}\right], \\
l_{m,\alpha} &= 2hc_m^{-1}\left[-\frac{1}{4}c_m\partial_{\alpha\alpha}\ln h + \frac{1}{2}c_m h(\partial_\alpha \ln h)^2 - \left(\frac{(m+2)^2}{2m+3} + \frac{m^2}{2m-1}\right)\right. \\
&\quad \left. \times h(\partial_\alpha \ln h)^2\right],
\end{aligned}$$

$$l^+_{m,\alpha} = \frac{2h}{a_{m+1}}\left[-\frac{m+2}{2m+3}\partial_{\alpha\alpha}\ln h - \frac{(m+2)(m+3)}{2m+3}h(\partial_\alpha \ln h)^2\right],$$
$$l^-_{m,\alpha} = \frac{2h}{a_{m+1}}\left[-\frac{m-1}{2m-1}\partial_{\alpha\alpha}\ln h - \frac{(m+2)(m-1)}{2m-1}h(\partial_\alpha \ln h)^2\right],$$
$$g_{m,\alpha} = -\frac{m-1}{2m+1}h\partial_\alpha \ln h,$$
$$g^+_{m,\alpha} = 2(m+2)h\partial_\alpha \ln h,$$
$$g^-_{m,\alpha} = -2(m-1)h\partial_\alpha \ln h,$$
$$e_\alpha = (1,1)^T.$$

In case of $h =$ const operators P_m, Q_m and R_m take the following form:

$$P_m = \begin{Vmatrix} hc_m l_2 - 2\mu h^{-2}(2m+1)I_2, & -2(\lambda+\mu)\,\mathrm{grad} \\ 2(\lambda+\mu)\,\mathrm{div}, & hc_{m-1}\mu\Delta - 2(2m-1)h^{-2}(\lambda+2\mu)I_1 \end{Vmatrix}, \tag{6.38}$$

$$Q_m = -\begin{Vmatrix} ha_{m-1}l_2, & 0 \\ 2(\lambda+\mu)\,\mathrm{div}, & ha_{m-2}\mu\Delta \end{Vmatrix}, \quad R_m = -\begin{Vmatrix} ha_{m+1}l_2, & -2(\lambda+\mu)\,\mathrm{grad} \\ 0, & ha_m\mu\Delta \end{Vmatrix}.$$

Here, as before, l_2 is the operator of the plane theory of elasticity, Δ is a two-dimensional Laplacian, I_α are α-dimensional unit matrices; in addition, grad $= (\partial_1, \partial_2)^T$, div $= (\partial_1, \partial_2)$, the dimension of zero operators is in agreement with Q and R. As follows from representations (6.35), condition (6.3), in the case where the boundary operator l is independent by z, goes over into the usual three point relation:

$$-a_{m-1}l\overset{m-2}{w}_i + c_m l\overset{m}{w}_i - a_{m+1}l\overset{m+2}{w}_i = \overset{m+1}{g}_i - \overset{m-1}{g}_i, \qquad (x_1,x_2)\in\partial D. \tag{6.39}$$

Evidently, by virtue of the representation (6.34) the consideration of more general boundary condition does not offer any difficulty.

From forms of (6.34) and (6.39) we can see that they are a linear combination of the system (6.13) or (V), if representation (6.35) is taken into account and the following differences of Legendre polynomials is choosen for the basis:

$$\chi P_m(z/h) = P_{m+1}(z/h) - P_{m-1}(z/h), \quad P_{-2} = P_{-1} = 0, \quad (m = -1, 0, 1, 2, \ldots).$$

Evidently, with $m = 0, 1$ the system (6.34) should coincide with equations (6.13). With respect to explicit equalities, following from (6.4) and (6.35), we have:

$$\overset{m}{u} = \overset{m-1}{w} - \overset{m+1}{w}, \qquad \overset{m}{w} = \overset{m+1}{u} + \overset{m+2}{u} + \ldots,$$

or

$$u = Aw, \qquad w = Bu = A^{-1}u, \tag{6.40}$$

from (6.13) we obtain:

$$\begin{aligned}
&2hl_2\overset{1}{w}_+ = \overset{0}{f}_+,\\
&2h\mu\Delta\overset{1}{w}_3 = \overset{0}{f}_3,\\
&\frac{2}{3}hl_2\overset{2}{w}_+ - 2(\lambda+\mu)\operatorname{grad}\overset{1}{w}_3 - 2\mu h^{-1}\overset{2}{w}_+ = \overset{1}{f}_+,\\
&\frac{2}{3}h\mu\Delta\overset{2}{w}_3 - 2(\lambda+\mu)\operatorname{div}\overset{1}{w}_+ - 2(\lambda+\mu)h^{-1}\overset{2}{w}_3 = \overset{1}{f}_3.
\end{aligned} \tag{6.41}$$

Taking into account the boundary conditions at $S^{\pm}$ for $\overset{1}{w}$ and $\overset{2}{w}$, we have:

$$2hl\,\overset{1}{w} = \overset{0}{g}, \qquad \frac{2}{3}hl\,\overset{1}{w} = \overset{1}{g}. \tag{6.39$'$}$$

Obviously, $\overset{1}{w}$ and $\overset{2}{w}$ are defined from (6.41) and (6.39′).

By extension of the definition of operators P, Q, R and functions v and f for parameter values $m = 1,2,3$ we can give the common form to system (6.34) (for clearness we limit ourselves to the case with $h = \text{const}$). We have:

$$Q_m \overset{m-2}{v} + P_m \overset{m}{v} + R_m \overset{m+2}{v} = \overset{m}{F}, \qquad (x_1,x_2)\in D, \qquad m = 1,2,\ldots, \tag{6.42}$$

where

$$\overset{-1}{v} = \overset{0}{v} = 0, \quad \overset{1}{v} = (\overset{1}{w}_+, 0)^T, \quad \overset{m}{v} = (\overset{m}{w}_+, \overset{m-1}{w}_3)^T, \quad Q_\alpha = R_\alpha = 0,$$

$$P_1 = \begin{pmatrix} 2hl_2, & 0\\ 0, & 0\end{pmatrix}, \qquad P_2 = \begin{pmatrix} \frac{2}{3}hl_2 - 2\mu h^{-1}I_2, & -2(\lambda+\mu)\operatorname{grad}\\ 0, & 2h\mu\Delta\end{pmatrix},$$

$$P_3 = \begin{pmatrix} hc_3l_2 - 14\mu h^{-1}I_2, & -2(\lambda+\mu)\operatorname{grad}\\ 0, & \frac{2}{3}h\mu\Delta - 2(\lambda+2\mu)h^{-1}I_1\end{pmatrix},$$

$$Q_3 = -\begin{pmatrix} ha_2l_2, & 0\\ 2(\lambda+\mu)\operatorname{div}, & 0\end{pmatrix}, \qquad R_3 = -\begin{pmatrix} ha_4l_2, & -2(\lambda+\mu)\operatorname{grad}\\ 0, & 0\end{pmatrix},$$

$$\overset{0}{F} = (\overset{0}{f}_+, 0)^T, \qquad \overset{1}{F} = (\overset{1}{f}_+, \overset{0}{f}_3)^T, \qquad \overset{2}{F}_3 = \overset{1}{f}_3.$$

For the rest of the admissible values of indices appearing in (6.42) expressions are calculated by means of formulae (6.36) and (6.38).

Now we consider the reduced system $(6.42)_N$ with conditions $(6.39)_N$. For convenience and uniformity of recording let us define operators P_{N+1}, Q_{N+1}, R_{N+1}, as well as functions v and F in the following way:

$$P_{N+1} = \begin{pmatrix} 0, & 0 \\ 0, & hc_N\mu\Delta - 2(2N+1)h^{-1}(\lambda+2\mu)I_1 \end{pmatrix},$$

$$Q_{N+1} = -\begin{pmatrix} 0, & 0 \\ 2(\lambda+\mu)\,\mathrm{div}, & ha_{N-1}\mu\Delta \end{pmatrix},$$

$$R_{N+1} = 0, \qquad \overset{N+1}{v} = (0,0,\overset{N}{w}_3)^T, \qquad \overset{N+1}{F} = (0,0,\overset{N}{f}_3 - \overset{N-2}{f}_3)^T.$$

By virtue of (6.40) and (6.39) boundary conditions take the following form:

$$l\overset{m+1}{w} = -\sum_{k\leq m(2)} \frac{2k+1}{2h}\int_{-h}^{h} gP_k(z/h)dz, \qquad (m = 0,1,\ldots). \tag{6.43}$$

Obviously $l\overset{m}{v} = l(\overset{m}{w}_+, \overset{m-1}{w}_3)^T$ are defined from this equation.

The formulae of one-dimensional factorization for Jacobi three-point matrices are immediately acceptable for reduced system $(6.42)_N$ with appropriate boundary conditions being taken into account, which were induced to $(6.43)_N$. The solution of such a system is found by the inversion of operators (comp.Keldish [1985]):

$$H_m = P_m - Q_m H_{m-2}^{-1} R_{m-2}, \qquad H_{-1} = H_0 = 0, \qquad m = 1,2,\ldots,N+1, \tag{6.44}$$

on the class of vector-functions, satisfying conditions $(6.43)_N$.

In fact, for $m = 1,2$ we have:

$$H_\alpha = P_\alpha$$

Let us assume that (6.44) is correct for $m-2 > 2$. Then we should have:

$$H_{m-2}\overset{m-2}{v} = -R_{m-2}\overset{m}{v} + \overset{m-2}{F^*}$$

or, in case of existense H_{m-2}^{-1} (the inverse operator of H_{m-2}) it follows

$$-QH_{m-2}^{-1}R_{m-2}\overset{m}{v} + H_{m-2}^{-1}\overset{m-2}{F^*} + P_m\overset{m}{v} + R_m\overset{m+2}{v} = \overset{m}{F}.$$

Hence

$$H_m = P_m - Q_m H_{m-2}^{-1} R_{m-2}.$$

Thus finding approximate solution $\bar{v}_N = \bar{w}_N$ of problem (6.1), (6.2), (6.3) is reduced to the solution of uniform problems of type

$$H_m\overset{m}{v} = -R_m\overset{m+2}{v} + \overset{m}{F^*}, \qquad (x_1,x_2)\in\bar{D}, \tag{6.45*}$$

where

$$\overset{m}{F}{}^* = \overset{m}{F} - H_{m-2}^{-1} \overset{m-2}{F}{}^*,$$

in classes satisfying conditions $(6.43)_N$.

We should mention that in the case of symmetry of operator P_m according to Lagrange there follows the symmetry of operator H_m as $R_{m-2} = Q_m^T$ in the same sense.

Similarly, the formula of differential factorization with decreasing value of index m is defined. If we limit ourselves to $N+1$ equation we should obtain:

$$\bar{H}_m \overset{m}{v} = -Q_m \overset{m-2}{v} + \overset{m}{F}_*, \qquad (x_1, x_2) \in \bar{D}, \tag{6.45$_*$}$$

where

$$\bar{H}_m = P_m - R_m \bar{H}_{m+2}^{-1} Q_{m+2}, \qquad \overset{m}{F}_* = \overset{m}{F} - \bar{H}_{m+2}^{-1} \overset{m+2}{F}_*, \qquad m = N+1, N, \cdots .$$

The summation of formulae (6.45^*) and (6.45_*) gives

$$(H_m + \bar{H}_m) \overset{m}{v} = -Q_m \overset{m-2}{v} - R_m \overset{m+2}{v} + 2 \overset{m}{F} - H_{m-2}^{-1} \overset{m-2}{F}{}^* - \bar{H}_{m+2}^{-1} \overset{m+2}{F}_*.$$

The comparison of these formulae with initial ones — $(6.42)_N$ — gives:

$$H_m + \bar{H}_m = P_m, \qquad \overset{m}{F} = H_{m-2}^{-1} \overset{m-2}{F}{}^* + \bar{H}_{m+2}^{-1} \overset{m+2}{F}_*,$$

$$P_m = Q_m H_{m-2}^{-1} R_{m-2} + R_m \bar{H}_{m+2}^{-1} Q_{m+2}.$$

Let us now consider the boundary value problem $(6.42)_N$ with reduced by ∂D homogeneous boundary conditions generated by (6.3)

The following theorem on the value *a priori* will be true:

Theorem 6.3. *Let D be a star region with respect to a certain circle, boundary ∂D is a piecewise-smooth curve, and the domain of definition $\mathfrak{D}(L_{v_1}) \subset L_2(D)$.*

Then $\exists N_0$ natural, and a number $c_{3+\alpha} > 0$ are constants, independent of N_0, then in $\forall N > N_0$ the inequality given below will be valid

$$(-L_{v_1}^N U_N, U_N) \geq c_4 \| \operatorname{grad} U_N \|^2 \geq c_5 \| U_N \|^2_{\overset{0}{W}_2}, \tag{6.46}$$

where $\overset{0}{W}_2 \subset W_2$ are Sobolev spaces, induced by conditions at ∂D.

Proof. Let L_{v1} be Vekua's operator of the first kind, corresponding to the equation of elastic plates in domain D and reduced boundary conditions by ∂D with $h = 1$ (see subsection 6.4).

Thus we have:

$$L_{v_1} u(x_1, x_2) = f(x_1, x_2), \qquad (x_1, x_2) \in \bar{D} \tag{6.47}$$

or

$$L_{v_1}^k u(x_1, x_2) = f^k(x_1, x_2), \qquad k = 0, 1, 2, \ldots \tag{6.47a}$$

Three-point operator equation, corresponding to the initial problem according to (6.42) will be rewritten for convenience as

$$L_{v_2} w(x_1, x_2) = g(x_1, x_2), \qquad (x_1, x_2) \in \bar{D}; \tag{6.48}$$

or

$$L_{v_2}^k w(x_1, x_2) = g^k(x_1, x_2), \qquad k = 1, 2, \ldots . \tag{6.48a}$$

Here L_{v_2} denotes of Vekua's operator of the second kind.

Let us recall that R_N, according to (6.4) is a remainder vector, u and w are connected by relation (6.40). Moreover, U_N and W_{N+1} denote partial sums of series according to polynomials $P_k(z/h)$ and $\chi P_k(z/h)$ respectively.

For convenience we shall put the differential operator corresponding to equation (6.1) in the form

$$L = A_0 \partial_{33} + A_1 \partial_3 + A_2, \qquad x \in \Omega,$$

and let the number of equations in systems (6.47a) and (6.48a) be equal to $3N+3$.

In case of Dirichlet homogeneous boundary conditions on $S^{\pm}$, equations $(6.47)_N$ and $(6.48)_N$ will be connected with relations:

$$\begin{aligned} L_{v_1}^{k-1} U_N - L_{v_1}^{k+1} U_N & \\ = L_{v_2}^k W_{N+1} &- \frac{2}{2N+3} a_2 \overset{N}{w} \delta_{k,N} - \frac{2}{2N+5} A_2 \overset{N+1}{w} \delta_{k,N+1} + 2A_1 \overset{N}{w} \delta_{k,N+1} \\ &+ (2k+1) A_0 \left[\left(i + (-1)^{N+k}\right) \overset{N}{w} + \left(1 - (-1)^{N+k}\right) \overset{N+1}{w} \right], \\ k = \overline{1, N+1}. & \end{aligned} \tag{6.49}$$

Here $\overset{k}{w}$ is the k component of vector w. If we multiply scalarly (6.49) by $\overset{k}{w}$ and take the sum termwise we shall have:

$$\begin{aligned} -(L_{v_1} U_N, U_N) = {} & (-L_{v_2} W_{N+1}, W_{N+1}) \\ & - \frac{2}{2N+3} (A_2 \overset{N}{w}, \overset{N}{w}) - \frac{2}{2N+5} (A_2 \overset{N+1}{w}, \overset{N+1}{w}) + 2(A_1 \overset{N}{w}, \overset{N+1}{w}) \\ & + \sum_{i=1}^{N+1} (2i+1) \left(A_0 \left[\left(1 + (-1)^{i+N}\right) \overset{N}{w} + \left(1 - (-1)^{i+N}\right) \overset{N+1}{w} \right], \overset{i}{w} \right) \\ = {} & -((L_{v_2} - R) W_{N+1}, W_{N+1}). \end{aligned}$$

Now let us use the explicit equality

$$(-L W_{N+1}, W_{N+1})_{L_2^3(\Omega)} = (-L_{v_2} W_{N+1}, W_{N+1})_{L_2^3(D)},$$

and Korn inequality for basic problems of linear theory of elasticity for an isotropic body, according to Mikhlin [1952](§§42, 43). Hence, using values for remainder vector R_N (lemma 6.2), we shall have:

$$\begin{aligned}(-L_{v_1}U_N, U_N) &= (-L_{v_2}W_{N+1}, W_{N+1})_{L_2^3(D)} + (RW_{N+1}, W_{N+1})_{L_2^3(D)}\\ &\geq c_4\|\operatorname{grad} W_{N+1}\|^2_{L_2^3(\Omega)} - \eta_N\|\operatorname{grad} W_{N+1}\|^2_{L_2^3(\Omega)}.\end{aligned} \tag{6.50}$$

Here $c_4 > 0$ is a constant from Korn's inequality for the Dirichlet problem or in basic mixed problems and depending only on Ω,

$\eta_N = o(N^{-(1+2\alpha)/2})$.

Let us return now to relation (6.40) ($N < +\infty$). It is easy to see that eigenvalues of BB^T matrix, which we denote as $\rho(BB')$, are equal to $4\sin^2 \dfrac{k\pi}{2(2N+1)}$ ($k = 1, 3, ..., 2N-1$) as BB^T is Jacobi (three-diagonal) matrix. Therefore

$\|A\|_3 = \max\sqrt{\rho(A'A)} \geq \dfrac{1}{2}.$

By virtue of the last inequality, from (6.50) follows:

$$\begin{aligned}-(L_{v_1}U_N, U_N)_{L_2^3(D)} &\geq (c_6 - \eta_N)\|\operatorname{grad} W_{N+1}\|_{L_2^3(\Omega)}\\ &\geq \frac{1}{2}(c_6 - \eta_N)\|\operatorname{grad} U_N\|^2_{L_2^3(\Omega)}.\end{aligned}$$

Further, repeating the reasoning set forth at the end of the subsection 6.3, with respect to application of Friedrichs' or Poincare's inequalities, depending on boundary conditions (6.3), from the last inequality we shall obtain the proof of theorem 6.3.

It is easy to see that inequality (6.46) is valid also in the case of the Ω_h domain; by virtue of lemmas 6.2 and 6.3 and the convergence of Vekua's process follows in case of Dirichlet's conditions, given at the surfaces $S^\pm$. *At the same time, for the error of displacement vector, as well as in case of $d < +\infty$, the convergence is uniform.*

6.6. The case of mixed boundary conditions

Let us consider the case of other boundary conditions given in their classic form. As was mentioned above, the investigation of the corresponding reduced systems has come to the cases already studied. In this connection it is not necessary to study boundary value problems for equations (6.1) with other (mixed) boundary conditions . Nevertheless, in order to make our reasoning more complete, we shall introduce below Vekua's systems which correspond to other boundary conditions.

Let normal displacement u_3 and tangents of stress $\sigma_{\alpha 3}$ be given on $S^\pm$. Then in terms of operator $L = A_0\partial_{33} + A_1\partial_3 + A_2$ and vector W_{N+1} we shall have ($g^\pm = 0$) :

$$\begin{aligned}hA_2(a_{m-1}\overset{m-2}{w} - c_m\overset{m}{w} + a_{m+1}\overset{m+2}{w}) - 2A_1(\overset{m-1}{w} - \overset{m+1}{w})&\\ +2(2m+1)h^{-1}A_0(\overset{0}{w} - \overset{N_m}{w}) = \overset{m-1}{f} - \overset{m+1}{f},&\end{aligned} \tag{6.51}$$

where the designation of the following form is introduced:

$$\overset{N_m}{w}_\alpha(x_1,x_2) = \begin{cases} \overset{N}{w}_\alpha, & m+N \text{ even}, \\ \overset{N+1}{w}_\alpha, & m+N \text{ odd}, \end{cases} \qquad \overset{N_m}{w}_3 = 0. \tag{6.42a}$$

Operator equation (6.51) in case of an isotropic body disintegrates into two independent subsystems:

$$Q_m \overset{m-2}{v} + P_m \overset{m}{v} + R_m \overset{m+2}{v} = \overset{m}{F} \qquad (m = 1,2,\dots,N+1),$$

where

$$P_m = \begin{Vmatrix} -hc_m l_2 + 2(2m+1)h^{-1}A_{0,2} & -2A_{12} \\ 2A_{12}^T\alpha_m & -hc_{m-1}\alpha_m l_1 + 2(2m+1)h^{-1}\alpha_m A_{0,1} \end{Vmatrix},$$

$$Q_m = \begin{Vmatrix} ha_{m-1}l_2 & 0 \\ -2\alpha_m A_{12}^T & ha_{m-2}\alpha_m l_1 \end{Vmatrix}, \qquad R_m = \begin{Vmatrix} ha_{m+1}l_2 & 2A_{12} \\ 0 & ha_m\alpha_m l_1 \end{Vmatrix}.$$

$$A_2 = \begin{pmatrix} l_2 & 0 \\ 0 & l_1 \end{pmatrix}, \quad A_1 = \begin{pmatrix} 0 & A_{12} \\ A_{12}^T & 0 \end{pmatrix}, \quad A_0 = \begin{pmatrix} A_{02} & 0 \\ 0 & A_{01} \end{pmatrix},$$

$$\alpha_m = \begin{cases} 0, & m = 1, \\ 1, & m > 1, \end{cases} \qquad \overset{m}{v} = (\overset{m}{w}_+, \overset{m-1}{w}_3)^T,$$

$$\overset{m}{F} = \left(\overset{m-1}{f}_+ - \overset{m+1}{f}_+, \overset{m-2}{f}_3 - \overset{m}{f}_3\right)^T + 2(2m+1)h^{-1}A_0 \overset{N_m}{w}.$$

In operators A_i the matrix operators on the main diagonal have dimensions of 2×2 and 1×1 accordingly, the remaining elements appearing in A_i are defined by them.

The case (when normal stress σ_{33} and tangent components u_α of displacement vector are given by $S^\pm$) differs from the previous one by the application of the relation given below for definition $\overset{N_m}{w}$:

$$\overset{N_m}{w}_\alpha = 0, \quad \overset{N_m}{w}_3 = \begin{cases} \overset{N}{w}_3, & m+N \text{ even}, \\ \overset{N+1}{w}_3, & m+N \text{ odd}, \end{cases} \tag{6.42b}$$

Now we should take into account (6.42b) in (6.51) instead of (6.42a).

6.7. Vekua system for an anisotropic nonhomogeneous elastic plate

As we denote in introduction of this section, dissemination of Vekua's method for more general cases does not represent a difficalty.

Let us consider when Hooke's law (1.5) has the form:

$$\sigma = B(x_3)e = B\sum_{i=0}^{\infty}\eta_i P_i\left(\frac{x_3}{h}\right)e. \tag{6.52}$$

i.e., B is a symmetric nonhomogeneous matrix.

If we use this procedure developing in subsection 6.3, for constructing of Vekua's system, it is necessary to calculate explicity the following integrals of such forms:

$$I_1[m,n,s] = \int_{-1}^{+1} P_m(t)P_n(t)P_s(t)dt, \tag{6.53}$$

$$I_2[m,n,s] = \int_{-1}^{+1} P_m(t)P_n(t)P_s'(t)dt, \tag{6.54}$$

For this purpose using the well known formula of Adams:

$$P_m(t)P_n(t) = \sum_{r=0}^{m} b_{m+n-2r}P_{m+n-2r}(t) \tag{6.55}$$

Here

$$b_{m+n-2r} = \frac{A_{m-r}A_rA_{n-r}}{A_{m+n-r}} \cdot \frac{2(m+n-2r)+1}{2(m+n-r)+1}$$

$$A_m = \frac{(2m-1))!!}{m!}$$

(see, e.g., Whittaker, Watson [1935])

Then, using (6.55), for I_1 we will have:

$$\begin{aligned} I_1[m,n,s] &= \sum_{r=o}^{m} \int_{-1}^{+1} b_{m+n-2r}P_{m+n-2r}P_s(t)dt \\ &= \sum_{r=o}^{m} \frac{2}{2s+1} b_{m+n-2r}\delta_{m+n-2r,s}, \end{aligned} \tag{6.56}$$

The calculation of integrals type (6.54), if we take into accont the well known formula (Szegö [1959]):

$$P_s'(t) = (2s-i)P_{s-1}(t) + (2s-5)P_{s-3}(t) + \ldots,$$

gives:

$$I_2[m,n,s] = 2\sum_{r=0}^{m} b_{m+n-2r}(\delta_{m+n-2r,s-1} + \delta_{m+n-2r,s-3} + \ldots)$$

If now we multiply (6.52) on $h^{-1}\left(s+\frac{1}{2}\right)P_s(t)$ and then integrate from $-h$ till h we have finally:

$$\overset{s}{\sigma} = 2B\sum_{h=0}^{\infty}\left\{\overset{n}{e'}\left[\sum_{m=0}^{n}\sum_{r=0}^{m}\eta_m b_{m+n-2r}\delta^s_{m+n-2r}\right.\right.$$

$$+\sum_{m=n+1}^{\infty}\sum_{r=0}^{n}\eta_n b_{m+n-2r}\delta^s_{m+n-2r}\Bigg]$$

$$+\frac{2s+1}{2h}\overset{n}{e''}\Bigg[\sum_{m=0}^{n}\sum_{r=0}^{m}\eta_m b_{m+s-2r}(\delta^{n-1}_{m+s-2r}+\delta^{n-3}_{m+s-2r}+\ldots)$$

$$+\sum_{m=s+1}^{\infty}\sum_{r=0}^{s}\eta_m b_{m+s-2r}(\delta^{n-1}_{m+s-2r}+\delta^{n-3}_{m+s-2r}+\ldots)\Bigg]\Bigg\}. \quad (6.57)$$

Here we represent the deformation tensor e as:

$$e=\sum_k \overset{k}{e'}P_k\left(\frac{z}{h}\right)+\frac{1}{h}\sum_k \overset{k}{e''}P_k'\left(\frac{z}{h}\right).$$

These equations with system (6.9) and using the procedure of subsection 6.3 give a complete system of two-dimensional PDEs, representing the Vekua system in this case.

We remark that formula (6.57) constructed by Vashakmadze Tamara[1989].

7. On One New Model of Elastic Plates

The basic principles of section 6, evidently, hold for the case of anisotropic nonhomogeneous plates. At the same time, with increasing N — an approximation in Vekua's systems — the difficulties with the inversion of the corresponding operator L_N are considerably growing. It depends on the fact that matrix operators L_{v_α} are complete even in an isotropic homogeneous case, when d in conditions (6.2) is a zero matrix and inclusion (6.5) is fulfilled (see lemma 6.1).

On the other hand, the availability of the explicit expression for the remainder vector R characterizing the approximation of the three-dimensional problem (1.1) – (1.3) by refined theories, gave the possibility of error estimation *of transition unimprovable on the class of admissible functions (see section 5),* or, roughly speaking, the error may be arbitrarily large on the class of solutions.

In this respect, the finite model, free of defects, noted above, will be suggested below. The model is constructed on the basis of refined nonlocal representations which have been already set forth in sections 2–4. The reduced system, corresponding to this model, is reduced to the inversion of the operator of comparatively simple structure m-times, where m denotes the number of pseudo-layers and defines the exactness of approximation of the three-dimensional problem (1.1) – (1.3) by a two-dimensional one.

For convenience and for clearness, we shall consider below the case when

$$\Omega_H(x_1,x_2,x_3)=D(x_1,x_2)\times]0,H[,\quad H=\text{const},$$

is an isotropic homogeneous body , $D(x_1,x_2)$ is a two-dimensional connected domain.

In this connection the initial problem of finding the stress-strained state will take the form:

$$\begin{aligned}\sigma_{ij,j} &= f_i,\\ e_{i,j} &= \frac{1}{2\mu}\sigma_{ij} - \frac{\lambda\delta_{ij}}{2\mu(3\lambda+2\mu)}\sigma_{kk}\\ &= \frac{1}{2}(u_{i,j}+u_{j,i}), \quad x\in\Omega_H\end{aligned} \tag{7.1}$$

$$lu = g, \qquad S = \partial D\times]0,H[, \tag{7.2}$$

$$\sigma_3|_{x_3=0} = \sigma_3^- = g^-, \qquad S^- = \partial\Omega_H \times \{0\}, \tag{7.3$^-$}$$

$$\sigma_3|_{x_3=H} = \sigma_3^+ = g^+, \qquad S^+ = \partial\Omega_H \times \{H\}, \tag{7.3$^+$}$$

Let us consider, as was above, that the linear boundary operator l is such that conditions (7.2), (7.3) appear to be covering for (7.1), according to Lions, Magenes [1970].

7.1. Models corresponding to bending problems

Let h_i be the points of division of the interval [0,H], i.e.

$$h_i = ih, \quad mh = H, \quad i = 0,1,2,...,m, \quad m\geq 1$$

Let us introduce the averaged, with respect to pseudolayers (h_i, h_{i+1}), values according to Reissner, which are connected with the displacement vector and tensor stresses:

$$u_\alpha^i = \frac{12}{h^3}\int_{h_i}^{h_{i+1}}(t-h_{i+1/2})u_\alpha dt, \quad u_3^i = \frac{6}{h^3}\int_{h_i}^{h_{i+1}}\left[\frac{1}{4}h^2-(t-h_{i+1/2})^2\right]u_3 dt, \tag{7.4}$$

$$M_\alpha^i = \int_{h_i}^{h_{i+1}}(t-h_{i+1/2})\sigma_{\alpha\alpha}dt, \quad M_{12}^i = \int_{h_i}^{h_{i+1}}(t-h_{i+1/2})\sigma_{12}dt, \tag{7.5}$$

$$Q_\alpha^i = \int_{h_i}^{h_{i+1}}\sigma_{\alpha 3}dt, \quad R_\alpha^i = \frac{1}{2}\int_{h_i}^{h_{i+1}}\left(\frac{1}{4}h^2-(t-h_{i+1/2})^2\right)\sigma_{\alpha 3}dt, \tag{7.6}$$

Here $h_{i+1/2} = (i+\frac{1}{2})h, \quad i = 0,1,\ldots,m-1$.

Then basing on e.n.r. of sections 2, 3 for any layer $[h_i, h_{i+1}]$ and assuming for simplicity $\gamma = 0$ and $f_\alpha = g = 0$ we shall obtain:

$$u_\alpha^i = -u_{3,\alpha}^i + D_1 R_\alpha^i, \tag{7.7}$$

$$\begin{aligned}M_\alpha^i &= -D(\partial_\alpha^2+\nu\partial_{3-\alpha}^2)u_3^i + 2R_{\alpha,\alpha}^i\\ &-\frac{\nu}{1-\nu}\int_{h_i}^{h_{i+1}}\left[(t-h_{i+1/2})\sigma_{33}-\left(\frac{1}{4}h^2-(t-h_{i+1/2})^2\right)f_3\right]dt,\end{aligned} \tag{7.8}$$

$$M^i_{12} = -(1 - \nu) D u^i_{3,12} + R^i_{\alpha,3-\alpha}, \tag{7.9}$$

$$Q^i_\alpha - \Delta R^i_\alpha = -D\Delta u^i_{3,\alpha} - \frac{1}{1-\nu}\int\limits_{h_i}^{h_{i+1}} \Big[(t - h_{i+1/2})\sigma_{33} - \frac{1+\nu}{2}\left(\frac{1}{4}h^2 - (t-h_{i+1/2})^2\right)\Delta\Big] f_3 dt + \hat{Q}^i_\alpha, \tag{7.10}$$

$$D\Delta^2 u^i_3 = \sigma^{i+1}_{33} - \sigma^i_{33} - \frac{2-\nu}{1-\nu}\int\limits_{h_i}^{h_{i+1}} (t - h_{i+1/2})\Delta\sigma_{33} dt - \int\limits_{h_i}^{h_{i+1}} \Big[1 - \frac{\nu}{1-\nu}\left(\frac{1}{4}h^2 - (t-h_{i+1/2})^2\right)\Delta\Big] f_3 dt + \hat{Q}^i_{\alpha,\alpha}. \tag{7.11}$$

The following notation is introduced in these formulae:

$$\sigma^i_{k3} = \sigma_{k3}(x_1, x_2, ih), \ D_1 = \frac{1}{2(1-\nu)D}, \ D = \frac{Eh^3}{12(1-\nu^2)}, \ \hat{Q}^i_\alpha = \frac{h}{2}\left[\sigma^{i+1}_{\alpha 3} + \sigma^i_{\alpha 3}\right].$$

Equations (7.7) – (7.11) form a connected system, for availability of unknown values σ^i_{j3}. Below we shall give another form of this system that will allow to factorize the operator, corresponding to e.n.r. (7.7) – (7.11) in such a way that we should invert the same differential operator m-times with different right hand parts in order to find the solution of the system.

For this purpose, let us sum the system (7.7) – (7.11) with i, we obtain:

$$u^*_\alpha = -u^*_{3,\alpha} + D_1 R_\alpha, \tag{7.12}$$

$$M_\alpha = -D(\partial_{\alpha\alpha} + \nu\partial^2_{3-\alpha})u^*_3 + 2R_{\alpha,\alpha} - \frac{\nu}{1-\nu}\sum_{i=0}^{m-1}\int\limits_{h_i}^{h_{i+1}} \Big[(t-h_{i+1/2})\sigma_{33} - \left(\frac{h^2}{4} - (t-h_{i+1/2})^2\right) f_3\Big] dt, \tag{7.13}$$

$$M_{12} = -(1-\nu)Du_{3,12} + R_{\alpha,3-\alpha}, \tag{7.14}$$

$$Q_\alpha - \Delta R_\alpha = -D\Delta u^*_{3,\alpha} - \frac{1+\nu}{2(1-\nu)}\sum_{i=0}^{m-1}\int\limits_{h_i}^{h_{i+1}} \Big[\frac{2}{1+\nu}(t-h_{i+1/2})\sigma_{33,\alpha} - \left(\frac{h^2}{4} - (t-h_{i+1/2})^2\right) f_{3,\alpha}\Big] dt + \hat{Q}_\alpha, \tag{7.15}$$

$$D\Delta^2 u_3^* = \sigma_{33}^+ - \sigma_{33}^- - \sum_{i=0}^{m-1} \int_{h_i}^{h_{i+1}} \left[\frac{2-\nu}{1-\nu}(t - h_{i+1/2})\Delta\sigma_{33} + \left(1 - \frac{1}{1-\nu}\left(\frac{h^2}{4} - (t-h_{i+1/2})^2\right)\Delta\right) f_3 \right] dt + \hat{Q}_{\alpha,\alpha}, \quad (7.16)$$

In these formulae the following notation is introduced:

$$(u^*, Q, \hat{Q}, M, R) = \sum_{i=0}^{m-1} (u^i, Q^i, \hat{Q}^i, M^i, R^i)$$

In system (7.12) - (7.16) R_α and σ_{33} are also unknown. If we use the equalities

$$\sum_{i=0}^{m-1} \int_{h_i}^{h_{i+1}} (t - h_{i+1/2})\sigma_{33} dt = \frac{h^2}{12}(\sigma_{33}^+ - \sigma_{33}^-) + \sum_{i=0}^{m-1} \rho_{Sm}^i \left[(t - h_{i+1/2})\sigma_{33} \right], \quad (7.17)$$

$$R_\alpha = \frac{h^2}{12} Q_\alpha + \sum_{i=0}^{m-1} \rho_{Sm}^i \left[(t - h_{i+1/2}) \int^t \sigma_{\alpha 3} dt \right], \quad (7.18)$$

where ρ_{Sm}^i denotes the remainder term of Simpson's rule, then in the interval $[h_i, h_{i+1}]$ for M_α, Q_α and u_3^* we shall obtain:

$$\begin{aligned} M_\alpha &= -D(\partial_\alpha^2 + \nu\partial_{3-\alpha}^2)u_3^* + \frac{h^2}{12}Q_{\alpha,\alpha} - \frac{h^2\nu}{12(1-\nu)}(\sigma_{33}^+ - \sigma_{33}^-) \\ &+ \frac{\nu}{1-\nu} \sum_{i=0}^{m-1} \int_{h_i}^{h_{i+1}} \left(\frac{h^2}{4} - (t-h_{i+1/2})^2\right) f_3 dt + r_\alpha[M_\alpha], \end{aligned} \quad (7.19)$$

$$\begin{aligned} Q_\alpha &- \frac{h^2}{12}\Delta Q_\alpha \\ &= -D\Delta u_{3,\alpha}^* - \frac{h^2}{12(1-\nu)}(\sigma_{33,\alpha}^+ - \sigma_{33,\alpha}^-) \\ &+ \frac{1+\nu}{2(1-\nu)} \sum_{i=0}^{m-1} \int_{h_i}^{h_{i+1}} \left(\frac{h^2}{4} - (t-h_{i+1/2})^2\right) f_{3,\alpha} dt + \hat{Q}_\alpha + r_{2+\alpha}[Q_\alpha], \end{aligned} \quad (7.20)$$

$$\begin{aligned} D\Delta^2 u_3^* &= \left(1 - \frac{h^2(2-\nu)}{12(1-\nu)}\Delta\right)(\sigma_{33}^+ - \sigma_{33}^-) - \int_0^H f_3 dt \\ &+ \frac{1}{1-\nu} \sum_{i=0}^{m-1} \int_{h_i}^{h_{i+1}} \left(\frac{h^2}{4} - (t-h_{i+1/2})^2\right) \Delta f_3 dt + \hat{Q}_{\alpha,\alpha} + r_5[u_3^*], \end{aligned} \quad (7.21)$$

In these formulae:

$$r_\alpha[M_\alpha] = \sum_{i=0}^{m-1} \rho^i_{Sm} \left[(t - h_{i+1/2}) \left(2 \int^t \sigma_{\alpha 3,\alpha} dt - \frac{\nu}{1-\nu} \sigma_{33} \right) \right],$$

$$r_{2+\alpha}[Q_\alpha] = \sum_{i=0}^{m-1} \rho^i_{Sm} \left[(t - h_{i+1/2}) \left(\Delta \int^t \sigma_{\alpha 3} dt - \frac{1}{1-\nu} \sigma_{33,\alpha} \right) \right],$$

$$r_s[u_3^*] = -\frac{2-\nu}{1-\nu} \sum_{i=0}^{m-1} \rho^i_{Sm} \left[(t - h_{i+1/2}) \Delta \sigma_{33} \right],$$

We should notice that relations (7.20) and (7.21) constitute an independent system of differential equations with respect to Q_α and u_3^*, by means of which values $u_\alpha^*, M_{\alpha\beta}, R_\alpha$ are completely defined. Now let us introduce the other values connected with (7.4) – (7.6).

$$v^i = u^{i+1} - u^i, \quad P^i = Q^{i+1} - Q^i, \quad N^i = M^{i+1} - M^i, \quad T^i = R^{i+1} - R^i,$$

$$i = 0, 1, \ldots, m-2.$$

Evidently that e.n.r. (7.7) – (7.11) for these values give:

$$v^i_\alpha = -v^i_{3,\alpha} + D_1 P^i_\alpha, \tag{7.22}$$

$$N^i_\alpha = -D(\partial^2_\alpha + \nu \partial^2_{3-\alpha}) v^i_3 + 2T^i_{\alpha,\alpha} - \frac{\nu}{1-\nu} \left(\int_{h_i}^{h_{i+1}} (t - h_{i+1/2}) - \int_{h_{i-1}}^{h_i} (t - h_{i-1/2}) \right) \sigma_{33} dt$$
$$+ \frac{\nu}{1-\nu} \left(\int_{h_i}^{h_{i+1}} \left(\frac{h^2}{4} - (t - h_{i+1/2})^2 \right) - \int_{h_{i-1}}^{h_i} \left(\frac{h^2}{4} - (t - h_{i-1/2})^2 \right) \right) f_3 dt,$$

$$N^i_{12} = -(1-\nu) D v^i_{3,12} + T^i_{\alpha,3-\alpha}, \tag{7.23}$$

$$P^i_\alpha - \Delta T^i_\alpha = -D \Delta v^i_{3,\alpha} - \frac{1}{1-\nu} \left(\int_{h_i}^{h_{i+1}} (t - h_{i+1/2}) - \int_{h_{i-1}}^{h_i} (t - h_{i-1/2}) \right) \sigma_{33,\alpha} dt$$
$$+ \frac{1+\nu}{2(1-\nu)} \left(\int_{h_i}^{h_{i+1}} \left(\frac{h^2}{4} - (t - h_{i+1/2})^2 \right) - \int_{h_{i-1}}^{h_i} \left(\frac{h^2}{4} - (t - h_{i-1/2})^2 \right) \right) f_{3,\alpha} dt + \hat{P}^i_\alpha,$$

$$D\Delta^2 v_3^* = \sigma_{33}^{i+1} - 2\sigma_{33}^i + \sigma_{33}^{i-1} - \frac{2-\nu}{1-\nu}\left(\int\limits_{h_i}^{h_{i+1}} (t - h_{i+1/2}) - \int\limits_{h_{i-1}}^{h_i} (t - h_{i-1/2})\right) \Delta\sigma_{33} dt$$

$$- \left[\int\limits_{h_i}^{h_{i+1}} \left(1 - \frac{1}{1-\nu}\left(\frac{h^2}{4} - (t - h_{i+1/2})^2\right)\Delta\right) - \int\limits_{h_{i-1}}^{h_i} \left(1 - \frac{1}{1-\nu}\left(\frac{h^2}{4} - (t - h_{i-1/2})^2\right)\Delta\right)\right] f_3 dt + \hat{P}^i_{\alpha,\alpha},$$

The further simplification of the equations, corresponding to functions N, P, v gives:

$$N_\alpha^i = -D(\partial_{\alpha\alpha} + \nu\partial_{3-\alpha}^2)v_3^i + \frac{h^2}{6}\left(P^i_{\alpha,\alpha} + \frac{h\nu}{4(1-\nu)}P^i_{\beta,\alpha\beta}\right) - \frac{h^3\nu}{24(1-\nu)}\left[(f_3^{i+1} - f_3^{i-1}) - \left(f_3^{i+1/2} - f_3^{i-1/2}\right)\right] + r^i_{5+\alpha}[N^i_\alpha], \tag{7.24}$$

$$P^i_\alpha - \frac{h^2}{12}\left(\Delta P^i_\alpha + \frac{h}{2(1-\nu)}P^i_{\beta,\alpha\beta}\right) = -D\Delta v^i_{3,\alpha} - \frac{h^3}{24(1-\nu)}\left(f^{i+1}_{3,\alpha} - f^{i-1}_{3,\alpha}\right) - \frac{1+\nu}{2}\left(f^{i+1/2}_{3,\alpha} - f^{i-1/2}_{3,\alpha}\right) + \hat{P}^i_\alpha + r^i_{7+\alpha}[P^i_\alpha], \tag{7.25}$$

$$D\Delta^2 v_3^i = -\frac{h}{2}\left(1 - \frac{2-\nu}{1-\nu}\frac{h^2}{12}\Delta\right)P^i_{\alpha,\alpha} - \frac{h^3}{24(1-\nu)}\partial_{\alpha\alpha}\left[(2-\nu) \times \left(f_3^{i+1} - f_3^{i-1} - (f_3^{i+1/2} - f_3^{i-1/2})\right)\right] + \hat{P}^i_{\alpha,\alpha} + r^i_{10}[v^i_3] \tag{7.26}$$

Here

$$r^i_{5+\alpha}\left[N^i_\alpha\right] = \frac{\nu}{1-\nu}\left((\rho^i_{Sm} - \rho^{i-1}_{Sm})\left[\frac{2(1-\nu)}{\nu}(t-\hbar)\int^t \sigma_{\alpha 3,\alpha} dt - (t-\hbar)\sigma_{33} + \left(\frac{h^2}{4} - (t-\hbar)^2\right) f_3\right] - \frac{h^3}{24}(\rho^i_{tr} - \rho^{i-1}_{tr})\left[\sigma_{\alpha 3,\alpha}\right] - \frac{h}{8}\bar{\rho}^i_{Sm}\left[\sigma_{33} + (t-h_i)\sigma_{33,3}\right]\right);$$

$$r^i_{7+\alpha}\left[P^i_\alpha\right]=\frac{1}{1-\nu}\left((\rho^i_{Sm}-\rho^{i-1}_{Sm})\left[(t-\hbar)\left((1-\nu)\int^t\Delta\sigma_{\alpha3}dt-\sigma_{33,\alpha}\right)\right.\right.$$
$$\left.+\frac{1+\nu}{2}\left(\frac{h^2}{4}-(t-\hbar)^2\right)f_{3,\alpha}\right]-\frac{h^3}{24}(\rho^i_{tr}-\rho^{i-1}_{tr})\left[\sigma_{\beta3,\alpha\beta}\right]$$
$$\left.-\frac{h}{8}\bar{\rho}^i_{Sm,\alpha}\left[\sigma_{33}+(t-h_i)\sigma_{33,3}\right]\right);$$
$$r^i_{10}\left[v^i_3\right]=\frac{3}{2h}\bar{\rho}^i_{Sm}\left[\left(1-\frac{h^2(2-\nu)}{12(1-\nu)}\Delta\right)(\sigma_{33}+(t-h_i)\sigma_{33,3})\right]$$
$$+\frac{h}{2}(\rho^i_{tr}-\rho^{i-1}_{tr})\left[\left(1-\frac{h^2(2-\nu)}{12(1-\nu)}\Delta\right)\sigma_{\alpha3,\alpha}-\frac{2}{h}f_3\right]$$
$$-\frac{1}{1-\nu}(\rho^i_{Sm}-\rho^{i-1}_{Sm})\left[\Delta\left((2-\nu)(t-\hbar)\sigma_{33}-\left(\frac{h^2}{4}-(t-\hbar)^2\right)f_3\right)\right],$$

where, in its turn, the following reduced entry is made:

$$(\rho^i-\rho^{i-1})[\varphi(t;\hbar)]=\rho^i[\varphi(t;h_{i+1/2})]-\rho^{i-1}[\varphi(t;h_{i-1/2})].$$

In formulae (7.20) – (7.21) and (7.25) – (7.26) the values $Q, \hat{Q}, P$ and $\hat{P}$ are connected by means of the following relations:

$$\begin{aligned}
Q^i_\alpha &= \hat{Q}^i_\alpha+\rho^{i+1/2}_{tr}[\sigma_{\alpha3}];\\
Q_\alpha &= \hat{Q}_\alpha+\sum_{i=0}^{m-1}\rho^{i+1/2}_{tr}[\sigma_{\alpha3}]\\
&= \frac{h}{2}\left[\sigma^-_{\alpha3}+2(\sigma^1_{\alpha3}+\ldots+\sigma^{m-1}_{\alpha3})+\sigma^+_{\alpha3}\right]+\sum_{i=0}^{m-1}\rho^{i+1/2}_{tr}[\sigma_{\alpha3}],\\
P^i_\alpha &= \hat{P}^i_\alpha+\left(\rho^{i+3/2}_{tr}-\rho^{i+1/2}_{tr}\right)[\sigma_{\alpha3}].
\end{aligned}\tag{7.27}$$

Above, while obtaining formulae (7.24) – (7.26) which together with formulae (7.22) and (7.23) form a complete set of (7.12), (7.14), (7.19) – (7.21) type, we have used the sixth equation, average with respect to the thickness (according to the theory):

$$\int_{h_{i-1}}^{h_{i+1}}(t-h_i)\sigma_{3j,j}dt=\int_{h_{i-1}}^{h_{i+1}}(t-h_i)f_3dt.\tag{7.28}$$

As is known, the equation of type (7.28) is usually ignored while constructing the refined theories of plates or the theories of shallow shells, although the validity of this procedure is discussed, (e.g., see Goldenveizer [1976], p.74-78). Equation (7.28) in Vekua's theory is one of ordinary ones, and a number of consequences with respect to approximation $N=1$ are already obtained in Vekua [1985]. Here for the first time it was noted that the sixth equation in the moment theory appears to be algebraic.

From system (7.28), as is not difficult to see, a partial difference of second-order is defined.

$$\begin{aligned}\delta^2\sigma_{33}^i &= \sigma_{33}^{i+1} - 2\sigma_{33}^i + \sigma_{33}^{i-1}\\ &= \frac{h}{2}[f_3^{i+1} - f_3^{i-1} - P_{\alpha,\alpha}^i] + \frac{3}{2h}\rho_{Sm}^i[\sigma_{33} + (t-h_i)\sigma_{33,3}]\\ &+\frac{h}{2}(\rho_{tr}^i - \rho_{tr}^{i-1})[\sigma_{\alpha 3,\alpha}], \quad i=\overline{1,m-1}, \end{aligned} \tag{7.29}$$

formulae (7.29) were taken into account, while equations (7.24) – (7.26) were obtained from the preceding system with respect to N_α^i, P_α^i and v_3^i, where values $\delta^2\sigma_{33}^i$ played the role of complementary unknown quantities.

Let us return to systems (7.20) – (7.21) and (7.25) – (7.26). Taking into account equalities (7.27) and boundary conditions (7.3) we have:

$$\begin{aligned} &D\Delta u_{3,\alpha}^* - \frac{h^2}{12}\Delta Q_\alpha = F_\alpha(x_1,x_2) + r_{10+\alpha}[Q_\alpha],\\ &D\Delta^2 u_3^* - Q_{\alpha,\alpha} = F_3(x_1,x_2) + r_{13}[u_3^*], \end{aligned} \tag{7.30}$$

$$\begin{aligned} &D\Delta v_{3,\alpha}^i - \frac{h^2}{12}\left(\Delta P_\alpha^i + \frac{h}{2(1-\nu)}P_{\beta,\alpha\beta}^i\right) = F_{3+\alpha}^i + r_{13+\alpha}^i[P_\alpha^i],\\ &D\Delta^2 v_3^i - \left(1-\frac{h}{2}\left(1-\frac{2-\nu}{1-\nu}\frac{h^2}{12}\Delta\right)\right)P_{\alpha,\alpha}^i = F_6^i + r_{16}^i[v_3^i], \end{aligned} \tag{7.31}$$

where functions F_s, F_{s+3}^i, r_{10+s}, r_{13+s}^i $(s=1,2,3)$ are defined in the obvious way by means of right hand parts and remainder terms of the system (7.20) – (7.21), (7.25) – (7.26), (7.27) and functions $g_3^\pm$.

System (7.30) corresponds to the system of differential equations of plate bending according to refined theories (in the proper sense), when the parameter $\gamma = 0$ in the scheme (P_h). A certain difference between the form of the operator in system (7.30) and the usual model of Reissner's type is defined with the account of formula (7.27) in schemes (P_h) with respect to summands $h[g_\alpha^+ + g_\alpha^-]$ and $h[g_{\alpha,\alpha}^+ + g_{\alpha,\alpha}^-]$. It can be easily seen that each equation in systems (7.30) and (7.31) is linearly independent of the other. For this purpose let us differentiate (7.30) and (7.31) and add them. Thus the obtained equations will be linearly independent of (7.30) and (7.31) accordingly, as the symbolic determinant of system (7.30) coincides with the determinant of Reissner type refined theories. The determinant corresponding to system (7.30) is equal to

$$\Gamma_1(\partial_1,\partial_2) = \pm D\left(1-\frac{h}{2}\right)\left(I - \frac{h^2}{12}\Delta\right)\Delta^2.$$

The calculation of the symbolic determinant for refined theories, following from the scheme (P_h), gives:

$$\Gamma_2(\partial_1,\partial_2;\gamma) = \pm D\left(I - \frac{1+2\gamma}{12}h^2\Delta\right)\Delta^2,$$

that with $\gamma = 0$ coincides with the symbolic determinant of system (7.30):

$$\Gamma_2(\partial_1, \partial_2; 0) = \pm D\left(I - \frac{h^2}{12}\Delta\right)\Delta^2 = \frac{2}{2-h}\Gamma_1(\partial_1, \partial_2).$$

Now, we shall consider the equations corresponding to formulae (7.12), (7.14), (7.22) - (7.24), (7.30) and (7.31) without remainder terms. We shall obtain a two-dimensional system of partial differential equations. The boundary conditions should be added to them, too. For this purpose we shall use equation (7.2), the linear element in $[h_i, h_{i+1}]$ and define the corresponding conditions, acoording to Galiorkin's method. With the conditions covering the boundary operator and domain $D(x_1, x_2)$ from the constructed system we can define the approximate values for $\bar{u_3^*}, \bar{Q}_\alpha, v_3^i, \bar{P_\alpha^i}(i = 0, 1, \ldots, m-2)$ as the solution of two-dimensional boundary value problems, with operators, corresponding to systems (7.30) and (7.31). By means of these functions we can easily estimate the remaining values, introduced above. For this purpose we should take into account formulae of type:

$$u^i = \sum_{j=0}^{i-1} v^j + u^0, \qquad u^0 = \frac{1}{m}\left(u^* - \sum_{j=1}^{m}\sum_{i=0}^{j-1} v^i\right).$$

Evidently, analogous representations appear to be true for the other functions, too, which were defined by means of equalities (7.5) and (7.6).

In case of non-expansion of the averaged boudary conditions, generated from relation (7.2), evidently, the whole abovegiven system will be unsplit.

Thus the constructed scheme of finding the approximate solution for problem (7.1) - (7.3) gives us the opportunity of definition of values, which correspond to functions, characterizing the stress-strained state of plate Ω_H at bending.

7.2. Models corresponding to plane stress-strained problems

Let us now start with the question, concerning the definition of values, corresponding to the extension (compression) of the plate.

For this purpose, let us introduce the function averaged with respect to layers and connected with displacement vector:

$$u_\alpha^i = \frac{1}{2h}\int_{h_{i-1}}^{h_{i+1}} u_\alpha dt, \qquad (u_1^i, u_2^i)^T = u_+^i, \quad i = \overline{1, m-1} \tag{7.32}$$

The corresponding system of two averaged equations takes the form of (see section 5):

$$\mu\Delta u_+^i + (\lambda^* + \mu)\,\mathrm{grad\,div}\, u_+^i = \frac{\lambda}{2h(\lambda + 2\mu)}\int_{h_{i-1}}^{h_{i+1}} \mathrm{grad}\,\sigma_{33} dt \tag{7.33}$$

in system (7.33) σ_{33} is unknown, as well as u. Nevertheless, if we use system (7.29) without the remainder term, and take into account the fact that values $\bar{P_\alpha^i}$

are already approximately found together with boundary conditions (7.3), then we can define $\bar{\sigma}^i_{33}$ as follows:

$$\begin{aligned}
\bar{\sigma}^{i+1}_{33} &= a_{i+1}\bar{\sigma}^{i+2}_{33} + b_{i+1}, \quad \sigma^0_{33} = g^i_3, \quad \sigma^m_{33} = g^+_3, \\
a_{i+1} &= \frac{i+1}{i+2}, \quad b_{i+1} = a_{i+1}\left[b_i + \frac{h}{2}(f^{i+1}_3 - f^{i-1}_3 - \hat{P}^i_{\alpha,\alpha})\right]. \\
b_0 &= \sigma^0_{33} = g^-_3, \quad i = -1, 0, 1, \dots, m-2.
\end{aligned}$$

The line over a letter here and above denotes an approximate value.

After this procedure we shall return to system (7.33). Let us compute the right hand parts according to Simpson's formula. Now, we add averaged boundary conditions (7.2), with respect to the values, connected with u^i_+ to the system of different equations (7.33). Thus, the problem, from which we find u^i_+, is defined; subsequently, we define the values characterizing the stress–strain state compressed (expanded) lengthwise layer of plate Ω_H with transversal loads being also taken into account.

Conclusion 7.1: *in subsections 7.1, 7.2 the scheme for calculation of problem (7.1), (7.3) was constructed; the problem is reduced to the solution of split systems of differential equations. The estimation of interior stress–strain state of an isotropic homogeneous plate (shell) Ω_h is reduced to the definition of:*

a) values, characterizing the vertical deflection all over the thickness, intersecting forces, moments and angular turns in the horizontal direction, corresponding to the body bending as a whole;

b) values, characterizing the bending of each layer as an independent one;

c) values, characterizing extension (compression) of each layer, as an independent one.

In the case where conditions (7.2) admit the analogous splitting, then the problem of the definition of the stress–strain state of the body on the whole structure coincides with the problem, which circumscribes its inner state. From this remark it immediately follows that the problem of the investigation of unique solvability, finding solution and numerical realizations on computers, can be fulfilled for each layer separately; *the factorized scheme of the solution assumes complete parallelizability of the algorithm.*

With respect to the problems of substantiation of this kind of model, the estimation of error transition and convergence of the suggested process, the following reasoning will be true:

a) the proposed procedure arranges a closer connection between *the moment theory,* on the one hand, and *the method of finite elements (using broken lines as linear elements, the basis of space W^1_2 and Galiorkin's method),* on the other hand. When the number of layers is $m = 1$ the proposed procedure then appears to be the classic moment theory accurate up to the sixth equation (7.28). Further, the use of results of the theory of finite elements method (e.g., see Strang and Fix [1973]) allows the substantiation of validity of the above given procedure. We should note that, as the starting class of the solutions, the class of functions represented by integral averages according to Reissner and (7.32) was taken.

b) the investigation of estimation of error transition and convergence of the proposed process depends, as it is known, on problems of approximation and stability (the establishment of *a priori* energetic estimations) of the initial problem (7.1) - (7.3) by the proposed scheme (compare with results of Ovtchinnikov, Xantis [1997]).

In our case the error of the right hand parts appears to be the approximation, and, it is, indeed, the linear form of remainder terms of quadrature formulae of Simpson and trapezoid (see sections 2–3); the values must be obtained *a priori*, depend on D domain and boundary conditions (7.2). Further, the investigation of boundary value problems is reduced to the investigation of the operator, corresponding to $\Gamma_1(\partial_1, \partial_2), \Gamma_2(\partial_1, \partial_2, \gamma)$ as well as to the problems of definition of generalized plane stress state. These operators, as is known, are the subject of investigation of the plane theory of elasticity. Thus we have proved the validity of the richest heritage in the field of numerous investigations related to two-dimensional problems (with space coordinate), in particular, the theories of singular integral equations and generalized analytical functions Muskhelishvili [1968], Vekua [1959]. We have also proved the applicability of this heritage to the investigation of approximate solutions of the initial problem (see ch.III). At last, in connection with this investigation of problems within this section, we should note that if the boundary of domain ∂D and boundary conditions satisfy the conditions, known in literature, then estimation of error transition and the proof of convergence are carried out according to a standard scheme, based on Lax–Milgram's lemma.

7.3. Some generalizations

7.3.1. The statement of 7.1–7.2, evidently, admits the generalization in the case of nonhomogeneous plates of variable thickness, investigated in sections 2–5. It is not necessary to set forth the corresponding material as the principal aspects of the problems are the same as the considerations and the formulae necessary for this purpose have been given in the same section.

We should also note that in case of boundary conditions (6.2) with Dirichlet's or Newton's conditions on the surfaces, the proposed scheme of plate design requires essential changing. In subsection 12.4, while considering also one-dimensional problems, we shall return to this question again.

7.3.2. In conditions (6.2) the cases $d < +\infty$ and $d = +\infty$ appeared to be different in a certain sense. From subsections 6.2 - 6.4 and 7.1, 7.2 we can see that the diversity of these data appeared to be essential with the proof of Korn's type energy inequalities, in construction of resolving algorithms for corresponding cutting systems of Vekua, factorized schemes of section 6, etc..

As is known (e.g., see Rektorys [1980]), these conditions have one more essential distinction: (6.2) is stable in the case $d = +\infty$, and unstable with $d = 0$ and is connected with the notion of trace for the function from space $W_p^k(\Omega_h)$. According to Rektorys [1980]: the stability of boundary conditions is depended with the order of differential equation, and, if, let us say, for the equation of $2k$-order, boundary conditions are of $k-1$-order at most (i.e., with derivatives of $k-1$ order at most), then they are called stable ones. Conditions with derivatives of order higher than

k–1 are called unstable for the equation of order $2k$. Unstable boundary conditions are usually natural ones, and limit functions, defined by the process and connected with Vekua's method do not always satisfy these conditions. In the proposed variant of Vekua's method (in section 6) with $d \leq +\infty$, condition (6.2) on surfaces $S^{\pm}$ for $\forall N \leq +\infty$ is satisfied exactly.

Let us note then, that by virtue of the performed inclusion (6.5) components of stress tensor will be $\sigma_{ij} \in W_p^{1+\alpha}(\Omega_h)$. If we now use Beltrami–Mitchel's equations (e.g., see Koltunov, Kravtchuk, Maiboroda [1983] or Washizu [1982]), equations (2.60) – (2.61), then boundary conditions (7.3) will be main and stable with the initial problem being formulated in terms of a stress tensor. Thus the tensor σ obtained will be the weak solution, nevertheless, its trace on $\partial\Omega_h$ will be a continuous one. The last remark provides the numerical stability of Vekua's method within the class of solution, satisfying (6.2) with $d < +\infty$ and (6.5).

8. The Application of Vekua's Method. Extensions and Examples

8.1. Vekua–Kantorovich projective method for self-infinite and infinitive intervals

As it was mentioned in subsection 1.2, the principle of Saint-Venant [1855] in case of semi-infinite cylindrical domains was studied, i.e., in works of Love [1959], Muskhelishvili [1966].

8.1.1. The method developed in 6.2 – 6.4 can be immediatly applied to the problem (1.1) – (1.5) in case, when the initial domain has the form:

$$\Omega(x_1, x_2, z) = D(x_1, x_2) \times]0, +\infty[, \quad S = \partial D \times]0, \infty[, \quad S^{\pm} = \partial\Omega / S.$$

For definition, let us assume that S^+ is free, and the vector of exterior stress acts on S^-, i.e.,

$$\sigma_3|_{S^-} = g^-(x_1, x_2), \quad \sigma_3|_{S^+} = 0.$$

Let us assume that Fourier–Laguerre's incomplete series expansion is true (e.g., see Szegö [1959]):

$$\sigma_3 = e^{-\gamma z}\left[g^- + \sum_{k=0}^{\infty} \overset{k}{\sigma}_3(x_1, x_2)(L_k(z) - L_{k+1}(z))\right], \qquad \sigma_3 = (\sigma_{31}, \sigma_{32}, \sigma_{33})^T,$$

$$\sigma_{\alpha\beta} = e^{-\gamma z} \sum_{k=0}^{\infty} \overset{k}{\sigma}_{\alpha\beta} L_k(z), \quad u = e^{-\gamma z} \sum_{k=0}^{\infty} \overset{k}{u} L_k(z),$$

where γ is some parameter.

Then equilibrium equation (1.1) (in the linear case) in components of stress takes the form:

$$\begin{aligned} &\overset{n}{\sigma}_{\alpha\beta,\beta} + (1-\gamma)\overset{n}{\sigma}_{\alpha\beta} + \gamma \overset{n-1}{\sigma}_{\alpha 3} = \overset{n}{f}_{\alpha} + \gamma g_{\alpha}^- \delta_{n0}, \\ &\partial_\beta(\overset{n}{\sigma}_{\beta 3} - \overset{n-1}{\sigma}_{\beta 3}) + \gamma \overset{n-1}{\sigma}_{33} + (1-\gamma)\overset{n}{\sigma}_{33} = \overset{n}{f}_3 + (\gamma g_3^- - g_{\alpha,\alpha}^-)\delta_{n0}. \end{aligned} \tag{8.1}$$

Taking into account boundary conditions on S^- and Hooke's law for the isotropic case in system (8.1) we shall obtain the following system of differential equations:

$$\begin{aligned}
&l_2\overset{n}{u}_+ - \gamma\left[(\lambda+\mu)\operatorname{grad}\overset{n}{u}_3 - \gamma\mu\overset{n}{u}_+\right] \\
&\qquad + \sum_{k=n+1}^{\infty}\left[2\gamma\mu\overset{k}{u}_+ - (\lambda+\mu)\operatorname{grad}\overset{k}{u}_3 - \mu(k-n)\overset{k+1}{u}_+\right] \\
&\quad = \overset{n}{f}_\alpha + \gamma g_\alpha^-\delta_{n0} = \overset{n}{F}_+, \\
&\mu\Delta\overset{n}{u}_3 - \gamma\left[(\lambda+\mu)\operatorname{div}\overset{n}{u}_+ - \gamma(\lambda+2\mu)\overset{n}{u}_3\right] \\
&\qquad + \sum_{k=n+1}^{\infty}\left[2\gamma(\lambda+2\mu)\overset{k}{u}_3 - (\lambda+\mu)\operatorname{div}\overset{k}{u}_+ - (\lambda+2\mu)(k-n)\overset{k}{u}_3\right] \\
&\quad = \overset{n}{f}_3 + (\gamma g_3^- - g_{\alpha,\alpha}^-)\delta_{n0} = \overset{n}{F}_3, \\
&n = 0,1,2,\dots
\end{aligned}\tag{8.2}$$

Boundary conditions on the lateral surface are analogously reduced: if they are given in the sense of section 6, then $\forall n$ has three scalar conditions, and they are set up by means of formulae, similar to (8.1).

Let us now consider the reduced system (8.2) together with averaged conditions on ∂D.

The following theorem on approximation is valid:

Theorem 8.1. *Let the series $\sum k\overset{k}{u}_i$ and $\sum \operatorname{grad}\overset{k}{u}_i$ be absolutely and uniformly convergent in $D(x_1,x_2)$. Then for $\forall\varepsilon>0\ \exists N_0 = N_0(\varepsilon)$ such that the reduced system (8.2) approximates the initial boundary value problem with ε accuracy when $N > N_0$. The solution of scheme $(8.2)_N$ is reduced to the consequent inversion of two-dimensional differential problems in $D(x_1,x_2)$ for differential operators*

$$\begin{aligned}
&\Gamma_\alpha(\partial_1,\partial_2;\gamma) = \mu^\alpha(\lambda+2\mu)(\Delta+\gamma^2 I)^{\alpha+2} \quad (\alpha=1,2), \\
&l_2 + \mu\gamma^2 I_2 = -\mu\operatorname{rot}\operatorname{rot} + (\lambda+2\mu)\operatorname{grad}\operatorname{div} + \mu\gamma^2 I_2.
\end{aligned}\tag{8.3}$$

Proof. Let in system (8.2) $n = 1,2,\dots,N$. Then with $n = N$, for convenience lowering of index of the number of approximation, by virtue of $\overset{n}{u}_i = 0\,(n > N)$, we have:

$$\begin{aligned}
l_2 u_+ + \gamma^2\mu u_+ &= \gamma(\lambda+\mu)\operatorname{grad} u_3 + F_+, \\
\mu\Delta u_3 + \gamma^2(\lambda+2\mu)u_3 &= \gamma(\lambda+\mu)\operatorname{div} u_+ + F_3.
\end{aligned}\tag{8.4}$$

Hence, carrying out the ordinary procedure with respect to the first two equations of system (8.4), we shall obtain:

$$((\lambda+2\mu)\Delta + \gamma^2\mu I)\operatorname{div} u_+ = \gamma(\lambda+\mu)\Delta u_3 + \operatorname{div} F_+.$$

By virtue of (8.4), the last equation makes it possible to obtain an independent equation with respect to $u_3(x_1,x_2)$:

$$\Gamma_3 u_3 = \mu(\lambda+2\mu)(\Delta+\gamma^2 I)^2 u_3$$

$$= \gamma(\lambda + \mu)F_{\alpha,\alpha} + ((\lambda + 2\mu)\Delta + \gamma^2\mu)F_3 \,. \qquad (8.5)$$

As far as boundary conditions given on S are also split in the sense of section 6, then we shall define $\overset{N}{u}_3$ from (8.5). The finding of u_3 gives possibility to find u_+ from system (8.4). Further, by means of recurrent formulae originating from (8.2) we shall obtain the remaining components $\overset{n}{u}$ of displacement vector U_N.

On the other hand, the symbolic determinant of system (8.4), after a few simple computations, takes the following form:

$$\Gamma_4(\partial_1, \partial_2, \gamma) = \begin{vmatrix} l_2 + \gamma^2\mu I_2, -\gamma(\lambda+\mu)\,\mathrm{grad} \\ -\gamma(\lambda+\mu)\,\mathrm{div}, \mu\Delta + \gamma^2(\lambda+2\mu)I_1 \end{vmatrix}$$
$$= \mu^2(\lambda + 2\mu)(\Delta + \gamma_1^2 I)^3.$$

Here, as before, I_α is the unit operator of α dimension.

Theorem 8.1 is proved.

Let us return to system (8.2). Evidently, the known models with $N = 0$ and $\gamma = 0$ follow from this system: such as equations, corresponding to plane deformation, torsion and bending of homogeneous beams (Saint-Venant's problem) (e.g., see Muskhelishvili [1966], ch.II, 25, ch.VII §§129, 137). At the same time the parameter γ, characterizing the asymptotic behavior of the solution of the initial problem at infinity, plays an important role in operator $\Gamma_4(\partial_1, \partial_2, \gamma)$. The definition of admissible values for this parameter is obviously connected with substantiation of Saint-Venant's principle and theorems of Phragmen–Lindelöf type, investigated in works, such as Lax [1957]. The use of these results, as well as lemmas of type 6.1 – 6.3, in the case of expansion into incomplete series in terms of Laguerre's functions, the proof of which seems to be trivial for the known properties of system $\{L_n(z)\}$ makes it possible to substantiate the rightfulness of this method for semi-infinite cylindrical bodies (e.g., see Szegö [1959], ch.V, 5.1, IX, 9.2, and ch.V of Supplements from its russian version).

Let us dwell upon some problems, connected with investigation, methods of solution of a system like (8.4) and their numerical realization.

a) As can be easily seen, the differential equation corresponding to the operator $\Gamma_4(\partial_1, \partial_2, \gamma)$ and the system (8.4) are strongly elliptical, therefore, the common methods, developed for corresponding boundary value problems to be studied, are, certainly, applicable. At the same time the polyharmonic operator (8.3) with $\gamma = 0$ was the subject of investigation in Trabucho, Viagno [1988]. The basic properties of the solution of the polyharmonic equation together with Dirichlet's boundary conditions were studied there: the behaviour of the solution in the vicinity of conic (irregular) points of the boundary and at infinity, the estimation of the values for derivatives of generalized solutions. Extension of these results to the investigation of properties of some problems for system of type (8.7) ($\gamma \neq 0$) deserves a certain interest.

b) A method in the solution of two-dimensional problems, corresponding to operators $(\Delta + \gamma^2 I)^m$, is given in the monograph of Vekua [1948]; it is based on the representation of general solutions by means of complex potentials. So, the procedure set forth in this monograph can be properly applied to the system (8.7).

c) With numerical realization of the computational scheme of finding solution of the reduced system and with restoring the solution of three-dimensional initial problem as well as with calculating scalar products

$$\overset{n}{f} = \int_0^\infty e^{-\gamma z} L_n(z) f(x_1, x_2, z) dz,$$

the reliable computation of zeroes and weights of Laguerre's polynomials seems to be essential. In this connection we, together with Upor, *developed a computational scheme and standard programme for the calculation of orthogonal polynomials, Laguerre's polynomial inclusive, as well as scalar products of type* $\overset{n}{f}$ *in a wide range of occuring parameters with* $n \approx 300$ (see Vashakmadze [1982] or Remark 12.8. of section 12).

8.1.2. The projective method, developed above, evidently, is extended to anisotropic nonhomogeneous compound rod (bar) systems, and obtaining solvable equations, offers no difficulty. The consideration of non-infinite cylindrical body, in our view, appears to be more essential

$$\Omega(x_1, x_2, z) = D(x_1, x_2) \times]-\infty, +\infty[.$$

In this case the surface will be $S^\pm = \Omega \times \{\pm\infty\}$, the lateral surface is $S = \partial D \times]-\infty, +\infty[$.

Let Ω be an isotropic, homogeneous elastic body, then we consider the boundary value problem where the equation of the stress–strain state is given in the form of (1.7), the conditions at S are given in their classic form, i.e., surfaces $S^\pm$ are free:

$$\sigma_3|_{S^\pm} = 0. \tag{8.6}$$

(In the case where $S^\pm$ is under the action of forces, i.e., $\sigma_3 = g^\pm$, then by virtue of

$$\tau_3 = \sigma_3 - \frac{(1 + \operatorname{sign} z)g^+ + (1 - \operatorname{sign} z)g^-}{2} a(z)$$

functions with appropriate $a(z)$, for example, when it is a shearing function in the vicinity of infinitely distant point, we can obtain the realization of conditions (8.6); at the same time this substitution imposes additional restrictions of smoothness on function $g^\pm$).

Let us expand the desired field of displacement and stress, as in the previous case, into the incomplete series according to Hermite's functions

$$(\sigma, u) = e^{-\gamma z^2/2} \sum_{k=0}^\infty (\overset{k}{\sigma}, \overset{k}{u}) H_k(z).$$

Then formulae similar to expressions of (8.1), (8.2) have the following form:

$$\overset{n}{\sigma}_{\alpha\beta,\alpha} + (2-\gamma)n\overset{n+1}{\sigma}_{\beta3} - \frac{\gamma}{2}\overset{n-1}{\sigma}_{\beta3} = A_n^{-1}\overset{n}{f}_\beta,$$

$$\overset{n}{\sigma}_{3\alpha,\alpha} + (2-\gamma)n\overset{n+1}{\sigma}_{33} - \frac{\gamma}{2}\overset{n-1}{\sigma}_{33} = A_n^{-1}\overset{n}{f}_3,$$

$$l_2\overset{n}{u}_+ + (\lambda+\mu)\,\mathrm{grad}\left[(2-\gamma)n\overset{n+1}{u}_3 - \frac{\gamma}{2}\overset{n-1}{u}_3\right]$$
$$+\mu\left[(2-\gamma)^2 n(n+1)\overset{n+2}{u}_+ - \frac{\gamma}{2}(2-\gamma)(2n-1)\overset{n}{u}_+ + \frac{\gamma^2}{4}\overset{n+2}{u}_+\right]$$
$$= A_n^{-1}\overset{n}{f}_+ = \overset{n}{F}_+, \tag{8.7a}$$

$$\mu\Delta\overset{n}{u}_3 + (\lambda+\mu)\,\mathrm{div}\left[(2-\gamma)n\overset{n+1}{u}_+ - \frac{\gamma}{2}\overset{n-1}{u}_+\right]$$
$$+(\lambda+2\mu)\left[(2-\gamma)^2 n(n+1)\overset{n+2}{u}_3\frac{\gamma}{2}(2-\gamma)(2n-1)\overset{n}{u}_3 + \frac{\gamma^2}{4}\overset{n+2}{u}_3\right]$$
$$= A_n^{-1}\overset{n}{f}_+ = \overset{n}{F}_3, \tag{8.7b}$$

where

$$A_n = \int_{-\infty}^{+\infty} e^{-z^2/2}H_n^2(z)dz = \sqrt{\pi}2^n n!,$$

$$\overset{n}{f}(x_1,x_2) = \int_{-\infty}^{+\infty} e^{-(1-\gamma)z^2/2}H_n(z)f(x_1,x_2,z)dz,$$

$$H_0(z)=1,\ H_1(z)=2z,\ H_n(z)=2zH_{n-1}(z)-2(n-1)H_{n-2}(z),\quad n=2,3,\ldots$$

If we bound ourselves to the first N equations in system (8.7), then, evidently, it would be subdivided into two systems with respect to vector-functions $\overset{n}{v} = (\overset{n}{u}_+, \overset{n-1}{u}_3)^T$ for odd and even values of n which, in their turn, are linked together, as in the case of system $(6.13)_N$.

8.2. Generalization of the Vekua–Kantorovich method for elastic quasi-cylindrical bodies

Within this subsection the problem of construction of computational scheme for problems (1.1) – (1.5) is discussed in case of bodies with shell or quasi-cylindrical form, i.e., when the domain Ω being an elastic one, can be represented as the sum of two domains Ω_h and Ω_1 with

$$\Omega_h = D(x_1,x_2)\times]h_1(x_1,x_2),h_2(x_1,x_2)[$$

is cylindrical domain, considered above, $\Omega_1 \subseteq \Omega_2$, $\mathrm{meas}\,\Omega_2 = o(\varepsilon)$, where $\varepsilon > 0$ is a small parameter.

Let us consider two sequences of boundary value problems:

a) find the stress–strain state of an elastic body in the cylindrical domain Ω_h, the solution of which is denoted as $u_i^{[2n]}$ $(n = 0, 1, 2, \ldots)$;

b) find the sress–strain state of elastic body in domain Ω_2, the solution of which is denoted as $u_i^{[2n+1]}$ $(n = 0, 1, 2, \ldots)$.

We suppose that in the domain Ω_h the solution of the problem can be found by the procedure of sections 2–7.

As for the function $u_i^{[2n+1]}$ in the domain Ω_2, it is possible to use direct methods in order to find them: variational-difference Strang, Fix [1973], methods of volumetric formulae Sobolev [1975], the potential theory in case of smooth boundaries, investigated in the papers Kùpradze [1963,1967], algorithms, proposed by Cruse [1978], Parton, Perlin [1981], Veriuzski [1978] in the case where $\partial\Omega_2$ is piecewise smooth and based on hybrid methods. The smallness of the domain, evidently, brings to the linear system of algebraic equations with rather small number of equations.

The method of approximate solution of the problem is suggested below; it is based on a difference computational scheme for calculation of Cauchy's problem for evolution equations.

Let us consider the net domain $\omega_{2,\tau} \subset \Omega_2$, constructed in the usual way, and assume that the vector function $u^{[2n]}$ is known.

Let the domain $\omega_{2,\tau}$ be partitioned into layers, then, let us call the surface, situated as far as $i\tau$ $(i = 0, 1, 2, \ldots, m;\ m\tau = \varepsilon)$ from the surface $\partial\Omega_2^- \subset \bar{\Omega}$ in the direction of a small parameter, as the i-layer. Assuming $u^{[2n+1]} = u^{[2n]}$ at the zero and the first layers, we replace the equilibrium equation by its finite difference (or variational differentional) analogue. If we denote the approximate value of the vector function $u^{[2n+1]}$ as $\overset{i}{u}{}^{[2n+1]}$ on the i-layer, then, evidently, it will be defined by virtue of initial data, boundary conditions and difference equation, as is observed in the calculation of Cauchy's problems for non-stationary equations of the second order (e.g., see Godunov and Riabenki [1962]). For the incorrectness of the proposed scheme according to Hadamard, we should use the corresponding postulates of the theory of solving Cauchy's problem for calculation of elliptic equations (e.g., see Carleman [1926], Lavrentiev [1956], Tikhonov and Arsenin [1974]). Moreover, for smallness of ε the number of layers may be a few, too. It, for example, can be defined by virtue of *a priori* values of remainder vector in finite difference scheme, depending on the data of the initial problem. Further, the obtained results are compared with the data on some part of boundary $\partial\Omega_2 \subset \partial\Omega$. In the case of an unsatisfactory result we should perform the reverse calculation. Having corrected the data on the zero and the first layers for $u^{[2n+1]}$, we should find the solution $u^{[2n+2]}$, etc..

The proposed procedure coincides schematically with Schwartz's method (using the well known scheme of predictor–corrector in a domain of low dimension), the convergence of which originates from such works as Mikhlin [1951], Sobolev [1936] for any closed sub-domain Ω uniformly, assuming that the corresponding problems in Ω_h and Ω_2 to be solved exactly.

If we can represent the smaller domain Ω_2 as a cylindrical domain in a certain orthogonal curvilinear system of coordinates (α, β, γ) when γ coincides with the direction of generators of lateral surface for the shell Ω, then we may use the results of sections 2–7 instead of the abovegiven procedure.

8.3. Dynamic case

The method of sections 4–7 is immediately transferred to evolutionary equations and in particular to the dynamic problems of the theory of elasticity.

In this respect let us consider the system of differential equations for the movement of a three-dimensional elastic body in the domain $G_T = \Omega(x_1, x_2, x_3) \times (0, T)$.

$$L_t u = [L - B(\partial_t)]u = f. \tag{8.8}$$

The initial and lateral conditions will be written in the form:

$$b(\partial_t)u|_{t=0} = \varphi(x), \tag{8.9}$$

$$lu|_{\partial\Omega \times (0,T)} = g(x,t), \tag{8.10}$$

respectively.

It is assumed below that B is a differential operator of the first or second order, and conditions (8.9) are defined in a proper way.

Let G_T be the bi-cylindrical domain, $G_T = D(x_1, x_2) \times] - h, h[\times (0, T)$. For clearness, let us consider the case, when $l = I$ is a single operator with $(x, t) \in S_T^{\pm} = S^{\pm} \times (0, T)$.

Based on the procedure of section 6, especially of subsection 6.5, the system (8.8) – (8.10) will be the identity for the following class of operator equations:

$$Q_{s,t} \overset{s-2}{v} + P_{s,t} \overset{s}{v} + R_{s,t} \overset{s+2}{v} = \overset{s}{g} \quad (s = \overline{1, N+1}). \tag{8.11}$$

Here

$$P_{s,t} = P_s - h\{c_s, c_s, c_{s-1}\}B(\partial_t),$$

$$Q_{s,t} = Q_s - h\{a_{s-1}, a_{s-1}, a_{s-2}\}B(\partial_t),$$

$$R_{s,t} = R_s - h\{a_{s+1}, a_{s+1}, a_s\}B(\partial_t).$$

The solution of the operator equation (8.11), evidently, is completed in the inversion of operators

$$H_{s,t} = P_{s,t} - Q_{s,t}(H_{s-2,t})^{-1} R_{s-2,t},$$

on the class of three-component vector functions, satisfying the reduced initial and boundary conditions at S_T.

Besides the differential factorization according to subsections 6.5 we can use various approximate methods of calculation of evolutionary equations with (8.11) (e.g., see Janenko [1966], Samarski [1967]). The following theorem on equivalence is true in this respect.

Theorem 8.2. *The homogeneous system corresponding to (8.11) identifies the evolutionary equation:*

$$B(\partial_t) = Au. \tag{8.12}$$

Proof. The validity of this statement follows from

$$c_s \geq a_{s-1} + a_{s+1}, \qquad (s = \overline{1, N+1}).$$

(We should note that the strict inequality occurs when $s = 1, 2, N, N+1$) and further we have to use the properties of Jacobi's matrices.

By virtue of this theorem it is necessary to reduce system (8.11) to a normal form of (8.12).

Based on this procedure and subsection 12.1, in case of $\bar{\Omega} = [-1, 1]^3$, $l = I$ on $\partial\Omega \times (0, T)$, $b = \{I, \partial_t\}_{6\times 6}$, $B(\partial_t) = \{\partial_t^2, \partial_t^2, \partial_t^2\}$ for problem (8.8) – (8.10) a computational scheme was constructed together with the corresponding standard programs by J. Peradze; it was issued as a separate brochure.

In case, as one can easily see, when $d < +\infty$ in conditions (6.2), i.e., when tensions are given on $S_T^{\pm}$, the use of the procedure of subsection 6.3 brings us to a factorized scheme of (6.32) type.

At the same time, when $d = 0$, as follows from the article by Gordeziani, Samarski [1978], Vekua's reduced system has a normal form (with natural conditions on $S^{\pm}$) in this case, usual for hyperbolic systems of differential equations.

In some cases the investigation of stability of many-layered schemes adequately coincides with the investigation of a definite class of operator polynomials of one or several variables. In the case of the system (8.8) corresponding implicit schemes are two or three-layered. On the basis of Korn type inequalities, set in section 6 together with methods, developed in the paper of Vashakmadze [1978], where the usual polynomial system from one variable and Tchebishev's polynomial from two variables were used as operator polynomials, the stability of the considered schemes follows. The idea proposed in the work (Vashakmadze [1978]) on the use of basic functions and, in particular, orthogonal polynomials for the investigation of stability of many-layered difference schemes, corresponding to evolutionary equations was realized by J. Rogava in his thesis. He extended the theory of orthogonal polynomials of Tchebishev in a certain triangle to the case of two variables and used this system for the investigation of scheme with three layers.

8.4. Investigation of the boundary value problem for thermodynamical stress–strained state of isotropic bodies. Example 1

Let us consider the class of problems connected with the space theory of elasticity. The investigation of these problems, which for simplicity will be called examples, is of a proper interest, nevertheless, the given method of calculation, as has already stressed been, is connected with the previous material by its content.

Let us consider the problem of the thermo-elastic-dynamic state for an isotropic medium (e.g., see Kupradze, Gegelia, Basheleishvili, Burchuladze [1976], p.46–47):

$$\begin{aligned} &\sigma_{ij,j} - f_i = \rho\partial_{tt}u, \\ &\partial_{ii}\vartheta - \frac{1}{\kappa}\partial_t\vartheta - \eta\partial_t \operatorname{div} u = f_4, \\ &\sigma_{ij} = \delta_{ij}(\lambda \operatorname{div} u - \gamma\vartheta) + \mu(u_{i,j} + u_{j,i}), \end{aligned} \tag{8.13}$$

$$[-\mu \operatorname{rot}\operatorname{rot} + (\lambda + 2\mu)\operatorname{grad}\operatorname{div}]u - \gamma \operatorname{grad}\vartheta = \rho\partial_{tt}u + f \tag{8.14}$$

Relations (8.13) – (8.14) are represented by equations of thermo-elastic-dynamic state of the isotropic medium, characterized by $\lambda, \mu, \gamma, \eta, \kappa, \rho$ parameters, and corresponding to weight force f and heat source f_4. The initial boundary conditions have the form analogous to representation (8.9) – (8.10).

The following theorem is valid.

Theorem 8.3. *Let four-component solution of nonstationary problem u, ϑ be defined in $W_2^{2+\alpha}(G_T)$, where G_T is the star domain with piecewise-smooth boundary with respect to a certain sphere, and, besides, let the initial value corresponding to the displacement vector be differentiable.*

Then the process of finding the thermoelastic deformation of medium is defined as a consequently solutions for Cauchy's problems with evolutionary equations of the first order and the second order with respect to $\vartheta(x,t)$ and $u(x,t)$ accordingly. Schemes of semi-discretization are stable ones.

Proof. Let us use Kirchhoff's formula (e.g., see Vladimirov [1961] p.233, formula (19)) for expression $\operatorname{div} u$, defined from (8.14) in a certain way. Taking into account the initial boundary conditions and including the obtained expression into (8.13), we define $\vartheta(x,t)$, as the solution for Cauchy's problem for evolutionary equations of the first order with retarded potential. After that, u will be defined uniquely, as the solution of the dynamic problem of the theory of elasticity.

Let us assume now that nonstationary problems are solved by means of semi-discretization. Then for the approximate solution of the thermodynamic problem and investigations of problems of stability and convergence the reasoning of the previous subsection can be completely applied.

The theorem is proved.

The following examples will be stationary problems of definition of the stress–strain state of orthotropic elastic plates and isotropic circular cylinder of finite length.

Let us assume that the displacement vector on surfaces $S^{\pm}$ is equal to zero (faces are fixed), i.e.,

$$u(x_1, x_2, x_3) = 0, \quad x \in S^{\pm},$$

On lateral surfaces S we shall write down the conditions in the usual form (see, e.g., (1.3) for the definition) :

$$l(\partial_1, \partial_2, \partial_3) = g, \quad x \in S.$$

Using procedure of subsection 6.4, we obtain the following equilibrium equations (Gotsuliac, Guliaev, Tchibiriakov [1973])

$$ha_k\left[\partial_\alpha\sqrt{g}\,\overset{k}{\sigma}_\alpha - \sqrt{g}\,\overset{k}{f}\right] - ha_{k-2}\left[\partial_\alpha\sqrt{g}\,\overset{k-2}{\sigma}_\alpha - \sqrt{g}\,\overset{k-2}{f}\right] - 2\sqrt{g}\,\overset{k-1}{\sigma}_3 = 0 \quad (8.15)$$

Here, as before, $a_k = \dfrac{2}{2k+1}$, g a determinant of metric form, $\sqrt{g} = 2h\sqrt{a}$, $\sqrt{a}$ is a square element of an average surface.

8.5. The problems of definition of the stress–strained state of an orthotropical elastic plates. Example 2

Let D be a simply connected domain. On lateral surface S the components of the stress tensor are given, i.e.

$$\sigma_\nu|_S = q \tag{8.16}$$

The finite difference method was used in the solution of the two-dimensional system of differential equations. The following scheme was chosen as a model

$$\begin{aligned} &0.5[\sigma_1 + \sigma_2 + 0.5(\sigma_{3,3} - f)]_{i+0.5,j+0.5} \\ &\quad +0.5[\sigma_1 - \sigma_2 + 0.5(\sigma_{3,3} - f)]_{i+0.5,j-0.5} \\ &\quad -0.5[-\sigma_1 + \sigma_2 + 0.5(\sigma_{3,3} - f)]_{i-0.5,j+0.5} \\ &\quad +0.5[-\sigma_1 - \sigma_2 + 0.5(n\sigma_{3,3} - f)]_{i-0.5,j-0.5} = 0 \end{aligned} \tag{8.17}$$

Here the representation $[\ ]_{i,j}$ denotes that the value of the function is taken at the point $(x_1 = ih_1,\ x_2 = jh_2) \in \omega_n \subset D$ of net domain, h_α is a step to x_α-directions. Further (8.17) is used in equations (8.15) – (8.16) together with Cauchy's relations and Hooke's law in the linear case.

We obtain the system of linear algebraic equations with a matrix of diagonal-block form if the system of differential equations is substituted by a finite difference scheme. This matrix form makes it possible to perform inversion with respect to block and to use iterative process for solving algebraic equations according to Saidel's method. .

Convergence takes place by virtue of symmetry and positive definiteness of the corresponding matrix, following from subsection 6.4.

On the basis of composing algorithms the design of stress–strain state for the following bodies are fulfilled(Vashakmadze [1982]:

a) the domain D is an orthotropic thick plate of rectangular section, at the same time two lateral faces are free, on the other rectangular faces the normal efforts are acting;

b) thickwall plate with rectangular notch, exposuring into a lateral surface.

We denote the lateral surface $S = \bigcup S_i$ $(i = 1,\ 2,\ 3,\ 4)$ at the same time let $S_1 = \bigcup S_{1i}$ $(i = 1,\ 2,\ 3,\ 4,\ 5)$ be a face with notch.

The problem is solved when S_i $(i = 2,3,4)$ are free, and normal forces act at S_{1i}.

c) a quarter of circular hollow cylinder (cylindrical thick panel) when only its internally part is loaded.

While calculating the problem considered in example 2, we paid attention to the possibility of increment accuracy of solution. A numerical experiment was carried out for this purpose. The net thickening and the increasing of N (the number of terms in expansion into Legendre's polynomials) appear to be necessary mainly for more exact description of stress distribution at zone of exterior forces, i.e., at the points of boundary layer where there is a high concentration of stress. At the same time far from the applied forces inside the plate the stabilization of influence by N is observed, and the corresponding data of designs is practically the same.

In the corresponding programme N is restricted, being in dependence on h_α steps of net and it is connected with the volume of on-line storage. For example, the use of 5 polynomials makes it possible to choose the difference net with dimension 30×30. Nevertheless, by virtue of the abovegiven analysis, the developed programme may be used without any changes (as a standard one) for calculation of a problem in the zone with larger variability of stresses, which makes it possible whether to increase N to a great extent (that's particularly important in the case of thick plates), or diminish the net step.

8.6. Isotropic circular cylinder of finite length. Example 3

Let D be a circular two-layered ring. If the direct or iterative method for the solution of system (8.15) according to the procedure of subsection 6.4 is used, then the solution of two-dimensional problems for the system of elliptic differential equations of the plane theory of elasticity and Helmholtz's equation will be one of the basic elements.

In this respect we use the representation of the basic characteristics of stress–strain state of isotropic circular ring in polar system of coordinates and consider the following boudary–contact problem for homogeneous isotropic two-layered ring $\bar{D} = [r_0, r_2] \times [0, 2\pi]$

$$\begin{gathered}\left(\partial_{rr} - 1 + \frac{1-\nu_\alpha}{2}\partial_{\varphi\varphi}\right) u_\alpha + \partial_\varphi \left(\frac{1-\nu_\alpha}{2}\partial_r - \frac{3-\nu_\alpha}{2}\right) v_\alpha = \frac{1-\nu_\alpha^2}{E_\alpha} e^{2r} f_{1\alpha},\\ \partial_\varphi \left(\frac{1+\nu_i}{1-\nu_i}\partial_r + \frac{3-\nu_i}{1-\nu_i}\right) u_\alpha + \left(\partial_{rr} - 1 + \frac{2}{1-\nu_i^2}\partial_{\varphi\varphi}\right) v_\alpha = \frac{2(1+\nu_\alpha)}{E_\alpha} e^{2r} f_{2\alpha},\\ (r,\varphi) \in D_\alpha, \quad D = D_1 \bigcup D_2,\end{gathered} \tag{8.18}$$

with free interior boundary of ring ∂D_1, i.e.:

$$\begin{aligned}\partial_r u_1 + \nu_1 u_1 + \nu_1 \partial_\varphi v_1 = 0,\\ \partial_r v_1 - v_1 + \partial_\varphi u_1 = 0,\end{aligned} \tag{8.19}$$

Contact conditions are given on the circle $r = r_1$:

$$u_1 = u_2, \qquad v_1 = v_2,$$

$$\begin{aligned}\frac{E_1 e^{-r_1}}{1-\nu_1^2}(\partial_r u_1 + \nu_1 u_1 + \nu_1 \partial_\varphi v_1) = \frac{E_2 e^{-r_1}}{1-\nu_2^2}(\partial_r u_2 + \nu_2 u_2 + \nu_2 \partial_\varphi v_2);\\ \frac{E_1 e^{-r_1}}{2(1+\nu_1)}(\partial_r v_1 - v_1 + \partial_\varphi u_1) = \frac{E_2 e^{-r_1}}{2(1+\nu_2)}(\partial_r v_2 - v_2 + \partial_\varphi v_2);\end{aligned} \tag{8.20}$$

Mixed conditions of the following form are given at exterior boundary:

$$\begin{aligned}(1 - a(\varphi))\frac{E_2 e^{-r_2}}{1-\nu_2^2}(\partial_r u_2 + \nu_2 u_2 + \nu_2 \partial_\varphi v_2) + u_2 = 0,\\ (1 - a(\varphi))\frac{E_2 e^{-r_2}}{2(1+\nu_2)}(\partial_r v_2 - v_2 + \partial_\varphi v_2) + v_2 = 0,\end{aligned} \tag{8.21}$$

where $a(\varphi)$ is a characteristic function of $[-\varphi_0, \varphi_0] \subseteq [-\pi, \pi]$ the interval.

Let us write problem (8.18) - (8.21) schematically in the form

$$L(\partial_r, \partial_\varphi)u(r,\varphi) = f$$

We shall represent the desired vector function in the following way:

$$u(r,\varphi) = \frac{\overset{0}{u}_1(r)}{2} + \sum_{k=1}^{n} \left(\overset{k}{u}_1(r) \cos k\varphi + \overset{k}{u}_2 \sin k\varphi \right),$$

and use the method of reduction to ordinary differential equations of Kantorovich [1933]. In order to find the unknown coefficient we shall obtain the system which can be divided into separate subsystems of the following form:

$$\begin{aligned} &\overset{k}{u}{}''_\alpha - \left(1 + \frac{1-\nu_\alpha}{2}k^2\right)\overset{k}{u}_\alpha + k\frac{1+\nu_\alpha}{2}\overset{k}{v}{}^+ - k\frac{3-\nu_\alpha}{2}\overset{k}{v}_\alpha = \overset{k}{g}_{\alpha_1}, \\ &\overset{k}{v}{}''_\alpha - \left(1 + \frac{2}{1-\nu_\alpha}k^2\right)\overset{k}{v}_\alpha - k\frac{1+\nu_\alpha}{1-\nu_\alpha}\overset{k}{u}{}'_\alpha - k\frac{3-\nu_\alpha}{1-\nu_\alpha}\overset{k}{u}_\alpha = \overset{k}{g}_{\alpha_2}, \\ &(k = 1, 2, \ldots, n). \end{aligned} \tag{8.22}$$

The general solution of the last system, evidently, has the form:

$$\begin{aligned} \overset{k}{u}_\alpha &= c_{1k}e^{(k+1)(r-r_0)} + c_{2k}e^{-(k-1)(r-r_0)} + c_{3k}e^{-(k+1)(r-r_0)} + c_{4k}e^{(k-1)(r-r_0)} \\ &\quad + \overset{k}{G}_{1\alpha}, \\ \overset{k}{v}_\alpha &= \mu_{k1}c_{1k}e^{(k+1)(r-r_0)} + \mu_{k2}c_{2k}e^{-(k-1)(r-r_0)} + c_{3k}e^{-(k+1)(r-r_0)} - c_{4k}e^{(k-1)(r-r_0)} \\ &\quad + \overset{k}{G}_{2\alpha}. \end{aligned}$$

Here G is a particular solution, c_{ik} $(i = \overline{1,4})$ are arbitrary constants, which are defined from boundary and contact conditions (8.20), (8.21), $\mu_{k\alpha}$ are fixed numbers, providing the compatibility of homogeneous systems. Consequently we obtain the system of linear algebraic equations with matrix, being the linear form from Toeplitz and Hankel matrices (see, e.g., Grenander, Szegö [1958]).

The proposed approach reveals a deep connection between the classic moment problem (see, e.g., Akhiezer [1961]) and representation of function by virtue of Fourier's integral, on the one hand, and method of the potential theory in the solution of strongly elliptical systems with constant coefficients in the strip region, on the other hand.

Convenience in using Toeplitz and Hankel matrices in the solution of algebraic systems makes it possible to solve the problem (originated from (8.22)) by direct methods on computers for $n \leq 100$, while iterative methods are used with $n > 100$. The calculation of two-dimensional integrals, occuring in the right hand parts, represents the main computational difficulty.

Numerical experiments are carried out on examples, when $\bar{f}_i$– right hand parts depend on single variable, or can be represented by the trigonometric polynomials.

With investigation of problems on error estimation and the process of convergence, connected with the method of reduction to ordinary differential equations

Kantorovich [1933] we should take into account the behaviour of the solution at the point of partition of boundary conditions on $r = r_2$ with $\varphi = \pm\varphi_0 \neq 0$ (see Muskhelishvili [1968]).

8.7. Some generalizations

Alongside the abovesaid, there are some ways which make it possible to extend the procedure of investigation and methods of calculation to more general problems of elastic theory. To such ones we can ascribe:

a) problems connected with design of elastic shells when the main Vekua's supposition is violated. In this case, as follows from paper Guliaev, Bazgenov, Lizunov [1978], Vekua's reduced systems corresponding to the case of $d < +\infty$ in condition (6.2) have a structure similar to the system with Jacobi matrix with respect to the higher derivatives. In case the first degree in z is preserved it will be three-diagonal, when in z^2 it will be penta-diagonal, etc. (see subsection 11.2).

The investigation of such multipoint systems of partial differential equations in principle is not different from that considered in section 6, nevertheless, difficulties connected with their numerical realization are growing to a great extent, sometimes they are inseparable;

b) problems, connected with the use of more perfect models, such as the following, in particular:

1) containing nonhomogeneity towards Oz and anisotropy of more general character;

2) consisting of composite material in case of bodies in a certain sense approaching semi-infinite rod systems;

3) taking into account the influence of electromagnetic field.

Within the investigation of problem on the use of the given procedure the essential matter is the possibility of application and justification of the perturbation method.

The application of method of small parameter for all enumerated classes of problems (with possibility of error estimation and process of convergence) may be realized with the account of the following remarks:

1) In the monograph Lomakin [1976] (see bibliography there, too) nonhomogeneous respect z problems of space theory of elasticity were studied. The problem of plane theory of elasticity for anisotropic case is reduced to the orthotropic one by means of the method of small parameter, which depends on variability of physical model (e.g., see Signorini's perturbation method Ciarlet [1988], Oleinik, Losifan, Shamaev [1990], Vorovich [1965] etc., and also subsection 10.2).

Expansion of coefficients of boundary value problems by degrees, corresponding to small parameter brings us to uniform problems with constant coefficients or to models, containing less general anisotropy. Evidently, radius of convergence of such expansions depends on given values of problem and it is to be effectively estimated.

2) Further, in papers by Guz, Nemish [1984], Khatiashvili [1983, 1985] processes of the solution of problems of Saint-Venant, Almansi–Mitchel's type by means of method of perturbation were given for the case of compound bodies with quasi-cylindrical form. Application of these methods and the approaches of subsection

8.1 without any assumption of carrying out Saint-Venant's hypotheses make it possible to reduce the initial three-dimensional problem to successive solution of two-dimensional problem in the unperturbed domain.

3) In the monograph Ambartsumian, Bagdasarian, Belubekian [1977] the problems on definition of stress–strain state of plates in electromagnetic fields were studied. Kirchhoff's theory was taken for the initial model. As it follows from developments of section 2, application of influence of such a kind of fields to any of the refined theories offers no difficulty.

4) Application of e.n.r. of sections 2–4 to multi-layer plates leads to the connected system of differential equations. Application of some results of Bolotin [1963], Donnel [1976], Panč [1975] are also reduced to different two-dimensional models for plates and beams .

In particular, for isotropic many-layered plates, lying on elastic foundation, in case Winkler's hypotheses (see, e.g., Panč [1975]) are performed, the equation for deflection has the form:

$$D\Delta^2 w - k_2 \Delta w + k_1 w = g, \tag{8.23}$$

where

$$\begin{aligned} g &= g^+ - g^- - f_3 - \Delta\left[\frac{2-\nu_{m-1}}{1-\nu_{m-1}}\frac{1+2\gamma_{m-1}}{3}\hbar_{m-1}^2 g^+ - \frac{2-\nu_0}{3}\frac{1+2\gamma_0}{3}\hbar_0^2 g^-\right], \\ k_2 &= \frac{2-\nu_0}{1-\nu_0}\frac{1+2\gamma_0}{3}\hbar_0^2 k_1, \quad \hbar_0 = \frac{(h_1+h_0)}{2}, \quad \hbar_{m-1} = \frac{(h_m+h_{m-1})}{2}, \\ \Omega_H &= D\times]0,H[, \quad 0 = h_0 < h_1 < h_2 < \cdots < h_m = H, \end{aligned}$$

k_1 is Winkler's foundation.

By choosing γ. from (8.23) we can obtain different models, including the known ones, of isotropic plates and beams (comp. Ambartsumian [1967], Donnel [1976]). At last, we should note that in the theory of cracks (e.g., see Morozov [1984]) two dimensional models, corresponding to the plane theory of elasticity, to problems of bending in homogeneous isotropic plates according to Kirchhoff, von Karman, Reissner, are investigated. Statement of sections 2–4, 7 and subsection 8.1 make it possible to extend this procedure to the remaining two-dimensional models,. originating from scheme (P_h) and to systems, given in subsections 7.1–7.3 together with differential operators, defined by process of (8.2).

5) In the work Gordeziani, Evseev [1977] on the basis of Vekua theory for equations of cylindric shells of variable thickness a model corresponding to the mixed BVP, as well as at its geometric and analytic characteristics are built up. By means of the obtained finite element scheme a model of the arched dam is computed in case $N = 1$. The results of the computation are also given.

9. Refined Theories for Piezoelectric and Electrically Conductive Elastic Plates

We developed methods of section 2 for the piezoelectrically and electrically conductive elastic plates.

For the sake of clearity and simplicity we restrict our consideration here to the linear boundary value problem with constant thickness (see, e.g., Parton, Kudriavtsev [1988]):

$$\sigma_{ij,j} = f_i, \quad \operatorname{div} D = 0, \quad \varepsilon_{ij} = 0.5(u_{i,j} + u_{j,i}), \quad E = \operatorname{grad} \varphi, \quad x \in \Omega_h,$$

$$\sigma_{3i}|_{S^\pm} = g_i^\pm, \quad D_3|_{S_1^\pm} = d_3^\pm, \quad \varphi|_{S_2^\pm} = v_0^\pm, \quad S_1^\pm \cup S_2^\pm = S^\pm, \quad l[u, \varphi] = g \quad x \in S.$$

Here σ, ε, u stand for the stress and strain tensors and for displacement vector respectively, D is an electrical induction vector, E means a tension of electric field, φ stands for the electrical potential.

Hooke's law generalized for transversally isotropic piezoceramics material is written as follows Parton, Kudriavtsev [1988]:

$$\sigma_{\alpha\beta} = b_{\alpha\beta}\varepsilon_{\beta\beta} + b_{13}\varepsilon_{33} - e_{31}E_3, \quad \sigma_{33} = b_{13}\varepsilon_{\alpha\alpha} + b_{33}\varepsilon_{33} - e_{33}E_3,$$

$$\sigma_{\alpha 3} = 2b_{44}\varepsilon_{\alpha 3} - e_{15}E_\alpha, \quad \sigma_{12} = 2b_{66}\varepsilon_{12},$$

$$D_\alpha = \epsilon_{11}E_\alpha + 2e_{15}\varepsilon_{\alpha 3}, \quad D_3 = \epsilon_{33}E_3 + \epsilon_{31}\varepsilon_{\alpha\alpha} + e_{33}\varepsilon_{33}.$$

Here b are elasticity moduli, the quantities e are piezoelectric constants and ϵ are dielectric constants.

The use of some results of section 2 for this case will be set forth below.

Equations corresponding to the components of the rotation of normals to the middle plane have the following form

$$u_\alpha^* = -u_{3,\alpha}^* + (2hb_{44})^{-1}(1 + 2\gamma)(Q_\alpha + e_{15}\Phi_{1,\alpha}) + R_\alpha(u_\alpha^*, \gamma),$$

where

$$\Phi_1 = \int_{-h}^{h} \varphi dt = (\varphi, 1),$$

$$R_\alpha(u_\alpha^*, \gamma) = \frac{3}{2b_{44}h^3}[(1-\gamma)\rho_{Sm} + \gamma\rho_{tr}]\left[t\int_0^t (\sigma_{\alpha 3} + e_{15}E_\alpha)dt\right],$$

γ is an arbitrary parameter.

The scheme constructing e.n.r. of this case is similar to considerations of subsection 2.2 for linear case and taking into account the boundary conditions on $S^\pm$ with respect to D_3 or φ.

In the first case when D_3 is given on $S_1^\pm$, we have:

$$\varepsilon_{33} = p_{31}\varepsilon_{\alpha\alpha} + p_{32}\sigma_{33} + p_{33}D_3,$$

$$E_3 = q_{31}\varepsilon_{\alpha\alpha} + q_{32}\sigma_{33} + q_{33}D_3.$$

Here

$$p_{31} = -\frac{1}{\delta}(b_{13}\epsilon_{33} + e_{31}e_{33}), \quad p_{32} = \frac{1}{\delta}\epsilon_{33}, \quad p_{33} = \frac{1}{\delta}e_{33}, \quad \delta = b_{33}\epsilon_{33} + e_{33}^2$$

$$q_{31} = \frac{1}{\delta}(b_{13}e_{33} - b_{33}e_{31}), \quad q_{32} = -\frac{1}{\delta}e_{33}, \quad q_{33} = \frac{1}{\delta}b_{33},$$

From Hooke's law for $\sigma_{\alpha\alpha}$ by these formulae we have:

$$\sigma_{\alpha\alpha} = p_{11}\varepsilon_{\alpha\alpha} + p_{12}\varepsilon_{3-\alpha 3-\alpha} + p_{13}\sigma_{33} + p_{14}D_3,$$

where for coefficients p_{1i} $(i = \overline{1,4})$ we have:

$$p_{11} = b_{11} + b_{13}p_{31} - e_{31}q_{31}, \quad p_{12} = b_{12} + b_{13}p_{31} - e_{31}q_{32},$$
$$p_{13} = b_{13}p_{32} - e_{31}q_{32}, \qquad p_{14} = b_{13}p_{33} - e_{31}q_{33}.$$

For bending and twisting moments we have:

$$M_\alpha = -\frac{2h^3}{3}(p_{11}\partial_\alpha^2 + p_{12}\partial_{3-\alpha}^2)u_3^* + (3b_{44})^{-1}h^2(1+2\gamma)\Bigg[e_{15}(p_{11}\partial_\alpha^2 + p_{12}\partial_{3-\alpha}^2)\Phi_1 + (p_{11} - p_{12})Q_{\alpha,\alpha}. + (b_{44}p_{13} - p_{12})(g_3^+ - g_3^-) + b_{44}p_{14}(d_3^+ - d_3^-) + p_{12}\int_{-h}^{h} f_3 dt\Bigg] + R_{\alpha+2}(M_\alpha;\gamma),$$

$$M_{12} = -\frac{4h^3}{3}b_{66}u_{3,12}^* + \frac{2h^2}{3b_{44}}b_{66}(1+2\gamma)(e_{15}\Phi_{1,12} + Q_{\alpha,3-\alpha}) + R_5(M_{12};\gamma).$$

Equations corresponding to the shearing forces have the following form:

$$Q_\alpha = h(g_\alpha^+ + g_\alpha^-) - \int_{-h}^{h} tf_\alpha dt + (p_{12}+b_{66})\int_{-h}^{h} f_{3,\alpha}dt - \frac{2h^2}{3}(p_{11}\partial_\alpha^2 + (p_{12}+2b_{66})\partial_{3-\alpha}^2)[hu_{3,\alpha}^* - \frac{1}{2b_{44}}(1+2\gamma)e_{15}\Phi_{1,\alpha}] + \frac{h^2}{3b_{44}}(1+2\gamma)[((p_{11}-p_{12}-b_{66})\partial_\alpha^2 + b_{66}\partial_{3-\alpha}^2)Q_\alpha + (b_{44}p_{13} - p_{12} - b_{66})(g_{3,\alpha}^+ - g_{3,\alpha}^-) + b_{44}^{-1}p_{14}(d_{3,\alpha}^+ - d_{3,\alpha}^-)] + R_{5+\alpha}(Q_\alpha;\gamma).$$

The equation for Reissner's averaged deflection is written out as follows:

$$\frac{2h^3}{3}p_{11}\Delta^2 u_3^* = \frac{e_{15}h^2}{3b_{44}}(1+2\gamma)p_{11}\Delta^2\Phi_1 + g_3^+ - g_3^- - \int_{-h}^{h} f_3 dt + h(g_{\alpha,\alpha}^+ + g_{\alpha,\alpha}^-) - \int_{-h}^{h} tf_{\alpha,\alpha}dt + (1+2\gamma)\frac{h^2}{3b_{44}}\Bigg[(b_{44}p_{13} - p_{11})\Delta(g_3^+ - g_3^-) + b_{44}^{-1}p_{14}\Delta(d_3^+ - d_3^-) + p_{11}\int_{-h}^{h}\Delta f_3 dt\Bigg] + R_8(u_3^*,\gamma).$$

From the equation $\operatorname{div} D = 0$, *we deduce*

$$(\epsilon_{11} + b_{44}^{-1} e_{15}^2)\Delta\Phi_1 = b_{44}^{-1} e_{15} \left(g_3^+ - g_3^- - \int_{-h}^{h} f_3 dt \right) - (d_3^+ - d_3^-) + R_9(\Phi_1).$$

In another case, when φ is given on $S_2^{\pm}$ surfaces, the equations for u_α^* remain unchanged, whereas others take some different forms. The equations with respect to M_{12}, Q_α, u_3^* and Φ_1 must be replaced by the following ones:

$$\begin{aligned} M_\alpha = &-\frac{2h^3}{3}(p'_{11}\partial_\alpha^2 + p'_{12}\partial_{3-\alpha}^2)u_3^* - p'_{13}\Phi_1 + p'_{14}h(\varphi^+ + \varphi^-) \\ &+\frac{h^2}{3b_{44}}(1+2\gamma)\left[(p'_{11}\partial_\alpha^2 + p'_{12}\partial_{3-\alpha}^2)e_{15}\Phi_1 + p'_{11}Q_{\alpha,\alpha}\right. \\ &\left. + p'_{12}Q_{3-\alpha,3-\alpha} + \frac{b_{13}b_{44}}{b_{33}}(g_3^+ - g_3^-)\right] + R'_{6+\alpha}[M_\alpha;\gamma], \\ M_{12} = &-\frac{4}{3}b_{66}h^3 u_{3,12}^* + \frac{2b_{66}h^2}{3b_{44}}(1+2\gamma)\left[e_{15}\Phi_{1,12} + \frac{1}{2}(Q_{1,2} + Q_{2,1})\right] + R'_8[M_{12};\gamma], \\ Q_\alpha = &-\frac{2h^3}{3}p'_{11}\Delta u_{3,\alpha}^* - p'_{13}\Phi_{1,\alpha} \\ &+\frac{h^2}{3b_{44}}(1+2\gamma)\left[(e_{15}p'_{11}\Delta\Phi_{1,\alpha} + b_{66}\Delta Q_\alpha) + (b_{33}^{-1}b_{13}b_{44} - p'_{11} + b_{66})\right. \\ &\left. \times(g_{3,\alpha}^+ + g_{3,\alpha}^-) + (p'_{11} - b_{66})\int_{-h}^{h} t f_{3,\alpha} dt\right] \\ &+h(g_\alpha^+ + g_\alpha^-) + p'_{14}\partial_\alpha h(\varphi^+ + \varphi^-) - \int_{-h}^{h} t f_\alpha dt + R_{\alpha+1}[Q_\alpha;\gamma], \end{aligned}$$

$$\begin{aligned} \frac{2h^3}{3}p'_{11}\Delta^2 u_3^* = &(3b_{44})^{-1}e_{15}h^2 p'_{11}(1+2\gamma)\Delta^2\Phi_1 - p'_{13}\Delta\Phi_1 + g_3^+ - g_3^- \\ &+h(g_{\alpha,\alpha}^+ + g_{\alpha,\alpha}^-) - \int_{-h}^{h} f_3 dt - \int_{-h}^{h} t f_{\alpha,\alpha} dt \\ &+(3b_{44})^{-1}h^2(1+2\gamma)\Delta\left[(b_{13}b_{33}^{-1}b_{44} - p'_{11} - b_{66})(g_3^+ - g_3^-) + p'_{14}h\right. \\ &\left. \times(\varphi^+ + \varphi^-) + (p'_{11} - b_{66})\int_{-h}^{h}(h^2 - t^2) f_3 dt\right] \\ &+R'_8(u_3^*;\gamma), \end{aligned}$$

$$(\epsilon_{11}+b_{44}^{-1}e_{15}^2)\Delta\Phi_1 = b_{44}^{-1}e_{15}\left(g_3^+ - g_3^- - \int_{-h}^{h} f_3 dt\right) - \frac{1}{2h}\Big[b_{44}^{-1}e_{15}(g_{\alpha,\alpha}^+ + g_{\alpha,\alpha}^-)$$
$$+(\epsilon_{11}+b_{44}^{-1}e_{15}^2)\Delta(v_0^+ + v_0^-)\Big] + R_9'(\Phi_1).$$

In these e.n.r. $\sigma_{\alpha\alpha}$ and coefficients p_{1i}' have the following form:

$$\sigma_{\alpha\alpha} = p_{11}'\varepsilon_{\alpha\alpha} + p_{12}'\varepsilon_{3-\alpha 3-\alpha} + p_{13}'\sigma_{33} + p_{14}'E_3.$$

$$p_{11}' = b_{11} - b_{13}^2 b_{33}^{-1},\quad p_{12}' = b_{12} - b_{13}^2 b_{33}^{-1},\quad p_{13}' = b_{13}^2 b_{33}^{-1} - b_{13},\quad p_{14}' = p_{13}' - e_{31}.$$

We also remind that in this paragraph ε_{ij} are components of small deformations tensor.

The above system without the remainder terms R_i gives the system of differential equations with respect to u_3^h, Q_α^h, Φ_1^h with arbitrary parameters γ_i. Choosing γ respectively we get refined theories in a wide sense (see section 1), which correspond to plate bending equations for the elastic case. For $\gamma = -0.5$ from e.n.r. follow models of Parton, Kudriavtsev [1988].

Other systems of differential equations, appropriate for extension (compression) of a plate for piezoceramic bodies are the following:
1) when D_3 is given on $S_1^\pm$, we have:

$$b_{66}\Delta\bar{u}_+ + (p_{11} - b_{66})\,\mathrm{grad}\,\mathrm{div}\,\bar{u}_+$$
$$= (2h)^{-1}\left[\int_{-h}^{h} f_+ dt - (g_+^+ - g_+^-) - \int_{-h}^{h} \mathrm{grad}(p_{11}\sigma_{33} - p_{14}D_3)dt\right];$$

2) when φ is known on $S_2^\pm$, we have:

$$b_{66}\Delta\bar{u}_+ + (p_{11}' - b_{66})\,\mathrm{grad}\,\mathrm{div}\,\bar{u}_+$$
$$= (2h)^{-1}\left[\int_{-h}^{h} f_+ dt - b_{13}b_{33}^{-1}\int_{-h}^{h} \mathrm{grad}\,\sigma_{33} dt - p_{31}'\,\mathrm{grad}(v_0^+ - v_0^-)\right].$$

Here, as before,

$$\bar{u}_\alpha = (2h)^{-1}\int_{-h}^{h} u_\alpha dt,\quad \bar{u}_+ = (\bar{u}_1, \bar{u}_2)^T.$$

Boundary conditions which follow from the data given on S, must be in agreement with the order of the system of two-dimensional differential equations to avoid several paradoxes mentioned in section 3.

Now we calculate surface efforts, defined by formulae:

$$T_{\alpha\beta} = \int_{-h}^{h} \sigma_{\alpha\beta} dt, \quad T_{\alpha\alpha} = T_\alpha.$$

For the first case we have:

$$T_\alpha = 2h(p_{11}\bar{u}_{\alpha,\alpha} + p_{12}\bar{u}_{3-\alpha,3-\alpha}) + p_{14}(d_3^+ - d_3^-) + p_{13}(\sigma_{33}^+ - \sigma_{33}^-) + R_{9+\alpha}[T_\alpha],$$

$$T_{12} = 2hb_{66}(\bar{u}_{1,2} + \bar{u}_{2,1}) + R_{12}[T_{12}].$$

For the second case when $(x_1, x_2) \in S_2^{\pm}$, we have:

$$T_\alpha = 2h(p'_{11}\bar{u}_{\alpha,\alpha} + p'_{12}\bar{u}_{3-\alpha,3-\alpha} + p'_{31}v_0) + \frac{b_{13}}{b_{33}} \int_{-h}^{h} \sigma_{33} dt + R'_{9+\alpha}[T_\alpha],$$

$$T_{12} = 2hb_{66}(\bar{u}_{1,2} + \bar{u}_{2,1}) + R'_{12}[T_{12}].$$

Here as above R denotes the remainder terms.

It should be noted that the coefficients involved in the above adduced equations are formed by means of the constants b, as well as the numbers e and ϵ.

In order to obtain a complete formulation (compare, e.g., Parton, Kudriavtsev [1988]), ch. III, §§13, 14), let us introduce a moment of the first order for $\Phi_2 = (\varphi, t)$.

Then we get for the first case $(x_1, x_2) \in S_1^{\pm}$

$$(\epsilon_{11} + b_{44}^{-1}e_{15}^2)\Delta\Phi_2 + (3b_{44})^{-1}e_{15}h^2(1+2\gamma)(g_{\alpha,\alpha}^+ - g_{\alpha,\alpha}^-) = R_{13}(\Phi_2; \gamma).$$

In the second case we have for Φ_2 the same expression:

$$(\epsilon_{11}+b_{44}^{-1}e_{15}^2)\Delta\Phi_2+\frac{h^2}{3}(1+2\gamma)\left[b_{44}^{-1}e_{15}(g_{\alpha,\alpha}^+ - g_{\alpha,\alpha}^-) + D_{3,3}^+ - D_{3,3}^-\right] = R'_{13}(\Phi_2; \gamma),$$

where

$$D_{3,3}^{\pm} \approx (\epsilon_{11} + b_{44}^{-1}e_{15}^2)\Delta v_0^{\pm} + b_{44}^{-1}e_{15}g_{\alpha,\alpha}^{\pm}.$$

10. Some new mathematical problems of the theory of nonlinear elasticity

In the report on STAMM-94 Vashakmadze [1995] formulated some mathematical problems (**P.**·)) of nonlinear solid mechanics when piezoelectrically and electric conductive creeping thermodynamic elastic beams, plates and shallow shells with small or finite deformations subjected to electromagnetic fields are anisotropic and non-homogeneous. Below we continue to adduce some new results which substantiate possibility of research and solving of the corresponding mathematical problems.

10.1. On the problems of thermoelasticity

Let us consider for clearness a static state of isotropic homogeneous media when on the surfaces $S^{\pm}$ the boundary conditions have the classical form:

$$l^{\pm}[u] = \sigma_{i3} + \sigma_{j3}u_{i,j} = g_i^{\pm}, \quad \vartheta = T - T_0 = g_4^{\pm}, \quad x \in S^{\pm}.$$

In this case the corresponding system of differential equations disintegration (see, e.g., Kupradze, Gegelia, Basheleishvili, Burchuladze [1976]) and it is necessary to add to the system of subsection 2.1, the averaged equation with respect to the function $\vartheta(x_1, x_2, x_3)$.

Before to constructing a finding equation, we have to recall the remark (Kovalenko [1975], p.74): "thus, in a simple connected freely medium, being in conditions of a plane deformation or generalized plane stress state, a stationary temperature field without thermal sources doesn't provoke stresses $\sigma_1, \sigma_2, \sigma_{12}$. This result in case of plane deformation for the first time was established by Muskhelishvili [1916]".

Below, *basing ourselves on some conclusions of section 2 and Green's functions, we construct two-dimensional Helmholz differential equation with a small parameter, the presence of which proves that in case of an elastic plate even without thermal sources in Ω_h there may arise important redistributions of a temperature, that denote to quantitive changes of the right hand-side of an equilibrium equation (1.1).*

Thus, we consider Poisson equation, corresponding to the function $\vartheta = \vartheta(x_1, x_2, x_3)$:

$$\nabla^2\vartheta = f_4, \ \vartheta|_S = g_4. \tag{10.1}$$

From this equation immediately follows:

$$\vartheta(x_1, x_2, z) = \frac{h+z}{2h} g_4^+ + \frac{h-z}{2h} g_4^- - \frac{1}{2h}\int_{-h}^{h} k(z,t)\frac{\partial^2\vartheta}{\partial t^2}dt, \tag{10.2}$$

$$k(z,t) = \begin{cases} (h-z)(h+t), & t \le z, \\ (h+z)(h-t), & t \ge z, \end{cases} \quad k(z,t) = k(t,z). \tag{10.3}$$

In its turn, the last equation gives:

$$\bar{\vartheta}(x_1, x_2) = \frac{1}{2h}\int_{-h}^{h} \vartheta dt = \frac{1}{2}(g_4^+ + g_4^-) - \frac{1}{4h^2}\int_{-h}^{h} dz \int_{-h}^{h} k(z,t)\frac{\partial^2\vartheta}{\partial t^2}dt. \tag{10.4}$$

If we change the integrate order in the last member, we shall then have:

$$I = \int_{-h}^{h} dz \int_{-h}^{h} k(z,t)\frac{\partial^2}{\partial t^2}\vartheta(x_1, x_2, t)dt = \int_{-h}^{h} \frac{\partial^2}{\partial t^2}\vartheta(x_1, x_2, t)dt \int_{-h}^{h} k(z,t)dz$$

$$= \int_{-h}^{h} \frac{\partial^2}{\partial t^2}\vartheta(x_1, x_2, t)dt \left[\int_{-h}^{t} k(t,z)dz + \int_{t}^{h} k(t,z)dz\right].$$

$$\int_{-h}^{t} k(t,z)dz + \int_{t}^{h} k(t,z)dz = h(h^2 - t^2).$$

If we now use (10.1), taking into acount the last equality, from (10.4) follows

$$\bar{\vartheta}(x_1,x_2) = \frac{1}{2}(g_4^+ + g_4^-) + \frac{1}{4h}\int_{-h}^{h}(h^2-t^2)[\Delta\vartheta(x_1,x_2,t) - f_4(x_1,x_2,t)]dt.$$

Using the parametrical representation of integrals with weight $h^2 - t^2$ from section 2 (see (2.16)), we will have:

$$\bar{\vartheta}(x_1,x_2) = \frac{1}{2}(g_4^+ + g_4^-) + \frac{(1+2\gamma)h^2}{3}\frac{1}{2h}\int_{-h}^{h}\Delta\vartheta(x_1,x_2,t) - \frac{h^2}{3}f_4^* +$$

$$+\frac{1}{2h}[(1-\gamma)\rho_{Sm} + \gamma\rho_{tr}][t\int_{0}^{t}\Delta\vartheta dt],$$

or

$$\frac{(1+2\gamma)h^2}{3}\bar{\vartheta} - \bar{\vartheta} = \frac{h^2}{3}f_4^* - \frac{1}{2}(g_4^+ + g_4^-) + R_1[\vartheta;\gamma], \tag{10.5}$$

where $\bar{\vartheta}$ is averaged value in the integral sense of $\vartheta = \vartheta(x_1,x_2,z)$,

$$R_1(\vartheta;\gamma) = -\frac{1}{2h}[(1-\gamma)\rho_{S_m} + \gamma\rho_{tr}][t\int_{0}^{t}\Delta\vartheta(x_1,x_2,t)]dt].$$

It is easy to show that *exact nonlocal representation* type (10.5) is true also for the deflection averaged by Reissner :

$$\vartheta^* = \frac{3}{4h^3}\int_{-h}^{h}(h^2-t^2)\vartheta(x_1,x_2,t)dt.$$

For this purpose, if we multiply (10.2) on $3(4h^3)^{-1}(h^2-t^2)$ and integrate such expression, we get:

$$2\vartheta^* = \frac{3}{4h^3}\frac{1}{2h}\int_{-h}^{h}(h^2-t^2)[(h+t)g_4^+ + (h-t)g_4^-]dt$$

$$+\frac{3}{4h^3}\frac{1}{2h}\int_{-h}^{h}(h^2-t^2)dt\int_{-h}^{h}k(t,z)\frac{\partial^2\vartheta}{\partial z^2}dz.$$

This expression and the procedure considered above finally give such an exact nonlocal representation for ϑ^*:

$$h^2\frac{1+2\gamma}{3}\Delta\vartheta^* - \vartheta^* = \frac{h^2}{4}f_4^* - \frac{g_4^+ + g_4^-}{2} + R_2(\vartheta;p_m(x_3)).$$

Here

$$\vartheta^*(x_1,x_2)=\int_{-h}^{h}p_m(t)\vartheta(x_1,x_2,t)dt,\quad \int_{-h}^{h}p_m(t)dt=1,\quad \int_{-h}^{h}p'_m(t)dt=0$$

$p_m(t)$ are polynomials of m order, f_4^* is an averaged function of fourth component of the right hand–side of initial three-dimentional system. For the remainder term R the following estimation holds

$$\|R_2(\vartheta;p_m)\|_{L_2(D)}\leq ch^4\|\partial_{33}\vartheta\|_{L_2(\Omega_h)},$$

Here $c>0$ is constant independent of h and ϑ.

Remark 10.1. The cases where the density of thermal flux is given or it is lacking on the $S^{\pm}$ are investigated analogously by scheme of section 3 because the problem of construction of an averaged differential equation with respect to ϑ^* is the same.

Remark 10.2. The thermodynamical case are investigated immediately if we apply also the method of subsection 3.2.

10.2. On homogeneity of Vekua theory of plates and shallow shells

The method of reducibility of the initial problem to the investigation of the operational equation of the comparatively simple structure is given in the paper Vashakmadze Tamaz, Vashakmadze Tamara [1993] on the example of one class of Vekua differential equations. In recent times the domain of application of the Vekua theory of elastic plates and shells has been sufficiently expanded. Among the publications of this kind we shall mention Meunargia [1991], Vashakmadze Tamara [1989], Zgenti [1988]. They, together with the monograph Oleynik, Iosifian, Shamaev [1990], dealt with the problems of homogeneity of strongly non-homogeneous mediums.

Let Ω_h be a three-dimensional domain, corresponding to an anisotropic and nonhomogeneous with respect to thickness, elastic plate or shell. Then, if we denote by the symbol $L(\partial_1,\partial_2)$ corresponding to the works Meunargia [1991], Vashakmadze Tamara [1989], Zgenti [1988], Vekua's differential operator, and by η the parameter measure of non-homogeneity and anisotropy, also if we denote by the symbol $M(\partial_1,\partial_2)$ the disturbance operator, then the problem of the elasticity theory of shells and plates can be written in the form ($\forall N\leq+\infty$):

$$(L(\partial_1,\partial_2)-\eta M(\partial_1,\partial_2))u(x_1,x_2)=0,\quad (x_1,x_2)\in\bar{D}\qquad(10.6)$$

Below, depending on the context, we shall also assume that the averaged by Vekua conditions given on $S=\partial D\times]-h,h[$ are included in the formula (10.6). We shall also recall that on account of the representation of the type (6.4b) the boundary conditions on face surfaces $S^{\pm}$ are fulfilled.

As we have already mentioned in the monograph Oleinik, Iosifian, Shamaev [1990] on the basis of the asymptotic method the solution of the initial problem is reduced to the successive solution of more simple problems. In our opinion, the method stated in Oleinik, Iosifian, Shamaev [1990] requires further justification to prove the correctness of the mentioned process.

The way stated below differs from the methodology of the works of the kind Oleinik, Iosifian, Shamaev [1990]. Instead of asymptotic representation or Fourier–Legendre incomplete series we use the expression

$$u = \gamma \sum_{k=0}^{\infty} u_k(x_1, x_2) P_k(\eta/\hbar) + (1-\gamma) \sum_{k=0}^{\infty} v_k(x_1, x_2)(\eta/\hbar)^k. \tag{10.7}$$

Below we consider the case when $\gamma = 1$ (see Vashakmadze Tamaz, Vashakmadze Tamara [1993]). Evidently, at $\gamma = 0$ we may apply the well known asymptotic procedure (see, e.g., Oleinik, Iosifian, Shamaev [1990]) to an operator equation (10.6).

If we denote $\eta = \tau\hbar$, then (10.6) taking into account (10.7), by the method of Galiorkin–Vekua we have:

$$\begin{aligned} &L(\partial_1, \partial_2)u_0 - \frac{1}{3}\hbar M(\partial_1, \partial_2)u_1 = 0 \\ &L(\partial_1, \partial_2)u_i - \hbar M\left[\frac{i}{2i-1}u_{i-1} + \frac{i+1}{2i+3}u_{i+1}\right] = 0 \quad (i = 1, 2, \ldots) \end{aligned} \tag{10.8}$$

The following theorem holds true.

Theorem 10.1. *Let the number of equations in system (10.8) be finite* $(n < \infty)$.

Then find the solution (if (10.6) are BVP) or general solution (if (10.6) are system of PDE) closed in the inversion of the operator L n-times and the application operator $\hbar M$ to the solution φ of an equation $L\varphi = f$, where f are known functions formulated by recursive processes.

Proof. Denote $Lu_k = b_k$, $-\hbar M u_k = t_k$ and confine ourselves to the cutting system, we shall have:

$$\begin{aligned} b_0 &= \frac{1}{3}t_1, \ldots, \\ b_{2(n-k)} &= \frac{2(n-k)}{4(n-k)-1}t_{2(n-k)-1} + \frac{2(n-k)+1}{4(n-k)+3}t_{2(n-k)+1}, \\ b_{2n} &= \frac{2n}{4n-1}t_{2n-1}; \end{aligned} \tag{10.9}$$

and

$$\begin{aligned} b_1 &= t_0 + \frac{2}{5}t_2, \ldots, \\ b_{2(n-k)-1} &= \frac{2(n-k)-1}{4(n-k)-3}t_{2(n-k)-2} + \frac{2(n-k)}{4(n-k)+1}t_{2(n-k)}, \\ b_{2n-1} &= \frac{2n-1}{4n-3}t_{2n-2}; \end{aligned} \tag{10.10}$$

If we express values t from the system (10.9) by b we shall have

$$\frac{2n}{4n-1}t_{2n-1} = b_{2n}, \quad \frac{2n-2}{4n-5}t_{2n-3} = b_{2n-2} - \frac{2n-1}{2n}b_{2n}, \ldots,$$

$$\frac{2(n-k)}{4(n-k)-1}t_{2(n-k)-1}$$

$$= b_{2(n-k)} - \frac{2(n-k)+1}{4(n-k)-3}t_{2(n-k)+1}$$

$$= b_{2(n-k)} - \frac{2(n-k)+1}{2(n-k)+2}b_{2(n-k)+2} + \ldots$$

$$+(-1)^k \frac{(2(n-k)+1)!! \ldots (2n-1)}{2(n-k)+2) \ldots 2n}b_{2n}, \ldots,$$

$$\frac{1}{3}t_1 = \frac{1}{2}b_2 - \frac{1\cdot 3}{2\cdot 4}b_4 + \frac{1\cdot 3\cdot 5}{2\cdot 4\cdot 5}b_6 + \ldots + (-1)^{(n-1)}\frac{(2n-1)!!}{(2n)!!}b_{2n}, \quad \frac{1}{3}t_1 = b_0.$$

From this system there follows the equality:

$$b_0 - \frac{1}{2}b_2 - \frac{1\cdot 3}{2\cdot 4}b_4 + \ldots + (-1)^n \frac{(2n-1)!!}{(2n)!!}b_{2n} = 0. \tag{10.11}$$

If we remember, that $b_k = Lu_k$ and assume, that the operator L is invertible, i.e., from the equation $Lu = 0$ if follows $u = \psi$, where ψ is either the solution of the boundary value problem , or the general solution, then from (10.11) it follows that

$$\sum_{i=0}^{n}(-1)^i \frac{(2i-1)!!}{(2i)!!}u_{2i} = \psi_1. \tag{10.12}$$

Now, if we apply the operator $\hbar M$ to the last expression, we shall have

$$\sum_{i=0}^{n-1}(-1)^i \frac{(2i-1)!!}{(2i)!!}t_{21} = \psi_2, \quad \hbar M\psi_1 = -\psi_2. \tag{10.13}$$

Analogously, for the system (10.10) we shall have:

$$\frac{2n-1}{2n-3}t_{2n-2} = b_{2n-1}, \quad \frac{2n-3}{4n-7}t_{2n-4} = b_{2n-3} - \frac{2n-2}{2n-1}b_{2n-1}, \ldots,$$

$$\frac{2(n-k)-1}{4(n-k)-3}t_{2(n-k)-2} = \sum_{i=n-k-1}^{n-1} (-1)^{i-(n-k)+1} \frac{(2(n-k)) \ldots (2i)}{(2(n-k)+1) \ldots (2i+1)}b_{2i+1},$$

or

$$\frac{2k+1}{4k+1}t_{2k} = \sum_{i=k}^{n-1}(-1)^{i-k} \frac{(2k+2) \ldots (2i)}{(2k+3) \ldots (2i+1)}b_{2i+1}.$$

If we denote

$$\alpha_j = \sum_{i=0}^{j} \frac{(2i-1)!!}{(2i)!!}\frac{4i+1}{2i+1}\frac{(2i+2) \ldots (2j)}{(2i+3) \ldots (2j+1)},$$

then from the last formulae taking into account (10.12) and (10.13) it follows

$$\sum_{j=0}^{n-1}(-1)^j \alpha_j b_{2j+1} = \psi_2.$$

Analogously, as above, from the last expression we have:

$$\sum_{j=0}^{n-1}(-1)^j \alpha_j u_{2j+1} = \psi_3. \tag{10.14}$$

Here ψ_3 is the result of the inversion of the operator L.
From the formula (10.14), by applying the operator $\hbar M$ we shall have:

$$t_{2n-1} = (-1)^{n-1}\sum_{i=0}^{n-1}(-1)^i \frac{\alpha_i}{\alpha_{2n-1}} t_{2i+1} + \psi_4. \tag{10.15}$$

Here $\alpha_{2n-1}\psi_4 = \hbar M \psi_3$.

If we insert the expression (10.15) in the formulae (10.9), for the first n equations b will be expressed with the help of $t_{2k-1} (k < n)$. In this way we find all the functions u_k $(k = 0, 1, 2, \dots, 2n)$ with the help of the successive multiple inversion of the operator L and applying the operator $\hbar M$ to the functions u and ψ.

10.3. Research design and unsolved problems

P.1) Creation and justification of a new class of two-dimensional mathematical models without simplifying assumptions for the initial spatial nonlinear BVP. Transition error estimation of some subclass of linear problems.

P.2) Creation and justification of dimensional reduction method of Vekua–Kantorovich type for anisotropic nonhomogeneous piezo-thermodynamic creeping plates and shallow shells with small deflection. Investigation of problems connected with error estimation, convergence, solvability by using energy inequalities, statical and dynamical cases. Numerical realizations.

P.3) Development of spectral theory of linear singular integral operators when the initial systems of integro-differential equations are two-dimensional models kind of refined theories with Volterra integral members for transisotropic creeping plates with small deformations.

P.4) Study of two-dimensional non-linear systems of Kadomtsev–Petviashvili equations type by methods of the theory of solitons. Such systems arise in macrostructure solid mechanics when the initial BVP of thermoelasticity with large deflection are substituted by methods of section 2. These models make the corresponding systems of von Karman and August Fëpl precise.

P.5) Creation of refined theories for shells in the sense of section 2. Korn type inequalities for two dimensional models for linearly elastic shells (in curvilinear coordinates).

P.6) Construction of general solution for Vekua system with variable coefficients of the plate theory by the method of homogeneity.

P.7) Numerical experiment and comparison of any two-dimensional models by which the data will be collected, analyzed and interpreted.

P.8) Justification of the introduction of the product of deflection onto intersecting forces in the BVPs with local loading (see point 2.3.12 or (2.24)).

P.9) Introduction and justification of the correction member characterizing a new edge effect (described in subsection 3.3) in plates and shells for multi-layer composite medium.

P.10) Construction of refined theories (in wide sense) by procedure of section 2 when boundary conditions on S have nonlocal character; investigation correspondently BVPs arise in dynamical case. Numerical realizations.

P.11) Construction of refined theories (P_h) section 2 control parameters $\gamma = \gamma(x, y)$ for BVPs of thermoelasticity, investigation problems of exact controllability (Lagnese, Lions [1988]).

11. A Brief Mathematical Review: Some Justifications of the Vekua Theory for Cusped, Non-shallow and Nonhomogeneous Shells

Below we describe some typical results belonging to Vekua's collaborators and developing some parts of section 6 for case of elastic shells.

11.1. Elastic bodies with non-smooth boundaries—cusped plates and shells

This part gives an up to date overview of theory of cusped plates and shells, related problems of partial differential equations with order degeneration, and their application to problems of elastisity theory for bodies with non-smooth boundaries.

In order to establish the mechanical sense of the effect of Keldish for degenerate elliptic equations, in 1955 Vekua (see f.e.Vekua [1985], and the references cited therein) raised the problem of investigation of cusped plates, i.e., such ones whose thickness on the part of boundary or on the whole one vanishes. Since 1972 Jaiani has been systematically studying the above problem, investigating also the tension-compression problem of cusped plates on the base of Vekua theory of plates and shallow shells. Summing up his results in mechanical sense one can make the following conclusion.

Conclusion 11.1 *By tension–compression on the cusped edge of plate where*

$$0 \leq \frac{\partial h}{\partial \nu} < +\infty$$

(such an edge is called a sharp cusp) one can not give displacements but where

$$\frac{\partial h}{\partial \nu} = +\infty$$

(such an edge is called a blunt cusp) the displacements can be given. By bending the cusped edge, where

$$\frac{\partial h}{\partial \nu} = O(d^{\kappa-1}), \quad d \to 0, \quad \kappa = const > 0,$$

(d is the distance from inside to cusped edge), can not be fixed if $\kappa \geq \frac{1}{3}$, and it can be fixed if $0 < \kappa < \frac{1}{3}$; it can not be free supported if $\kappa \geq \frac{2}{3}$, and it can be free supported if $0 < \kappa < \frac{2}{3}$; it can be free or arbitrary loaded with intersecting force and bending moment if $\kappa > 0$.

The same result remains true also in the case of cusped shells. Recently Jaiani has shown that it does not depend on anisotropy of material of elastic body. The problem under consideration mathematicaly leads to the question of setting and solving of BVPs for even order equations and systems of elliptic type with order degeneration. Following the functional-theoretical method developed by Fichera with some modifications, the degenerate system containing the Vekua system for elastic plates in the general case of cusped plate has been investigated. The classical bending of the plate with the stiffness $D(x_1, x_2)$:

$$D_1 x_2^\kappa \leq D(x_1, x_2) \leq D_2 x_2^\kappa, \quad D_1, D_2, \kappa = \text{const} > 0,$$

in energetic and weighted Sobolev spaces has been studied. In the case

$$2h = h_0 x_2^\kappa, \quad h_0, \kappa = const > 0, \quad x_2 \geq 0,$$

for investigation and solving in explicit form of corresponding BVPs the method of fundamental solutions depending on polar angle (in contrast to the method of fundamental solutions depending on radius) and the operator factorization method when the solutions of BVPs for higher order equations can be constructed by means of solutions of simple BVPs (e.g., Dirichlet's problem) for equations of less order have been used. Further, it had found to be important to explore cusped bodies (i.e., 3-D bodies with non-smooth boundaries, when the edges and vertices of body are locally lying on such surfaces that the orthogonal to the latter dimension of the body is also locally sufficiently small than at least one of two others. One can consider such parts of body as shell (plate) or beam with above surface as "middle surface (plane)" as mentally composite bodies constituted from shells, plates, beams, and essentially 3-D parts. Taking the interfaces normal to "the middle surfaces" so that to have mentally "cut out" shells, plates, and beams from body, one can consider for the whole body transmission problem with corresponding equations in each part, and, e.g., if one uses N-th approximation of Vekua theory of shallow shells with conditions on interfaces Γ_l, $l = \overline{1,m}$, as follows:

$$\begin{cases} \underset{s}{\overset{(r)}{u_i}}|_{\Gamma_l} = \underset{b}{\overset{(r)}{u_i}}|_{\Gamma_l}, \quad \underset{s}{\overset{(r)}{\sigma_{ij}}}\nu_j|_{\Gamma_l} = \underset{b}{\overset{(r)}{\sigma_{ij}}}\nu_j|_{\Gamma_l}, \quad i = 1,2,3; \quad r = \overline{1,N} \\ \left\| \sum\limits_{r=N+1}^{\infty} \underset{b}{\overset{(r)}{u_i}} P_r(\frac{x_3}{h}) \right\|_{\Gamma_l} < \epsilon, \quad \left\| \sum\limits_{r=N+1}^{\infty} \underset{b}{\overset{(r)}{\sigma_{ij}}}\nu_j P_r(\frac{x_3}{h}) \right\|_{\Gamma_l} < \epsilon, \quad l = \overline{1,m}; \end{cases}$$

or

$$\left| \underset{b}{u_i} - \sum_{r=0}^{N} \underset{s}{\overset{(r)}{u_i}} P_r\left(\tfrac{x_3}{h}\right) \right|\Bigg|_{\Gamma_l} < \epsilon,$$

$$\left| \underset{b}{\sigma_{ij}} \nu_j - \sum_{r=0}^{N} \underset{s}{\overset{(r)}{\sigma_{ij}}} \nu_j P_r\left(\tfrac{x_3}{h}\right) \right|\Bigg|_{\Gamma_l} < \epsilon, \quad i = 1,2,3; \; l = \overline{1,m},$$

where ε is exactness of approximation $\underset{s}{\overset{(r)}{u_i}}, \underset{s}{\overset{(r)}{\sigma_{ij}}}$ are r-th moments of displacements u_i and of stress tensor σ_{ij} of shell part, and $\underset{b}{\overset{(r)}{u_i}}, \underset{b}{\overset{(r)}{\sigma_{ij}}}$ the same quantities computed from the essentially 3-dimensional part. Details and bibliography you can see in Jaiani [1982,1996].

11.2. On nonlinear nonshallow isotropic shells

In the present subsection the three-dimensional problems of the nonlinear theory of elasticity are reduced to the two dimensional problems of non-shallow shells by means of Vekua method.

Below, by constructions under non-shallow shells we shall mean elastic bodies when

$$a_\alpha^\beta - x_3 b_\alpha^\beta \neq a_\alpha^\beta, \quad |x_3 b_\alpha^\beta| < 1.$$

Complete system of equations of the three dimensional non-linear theory of elasticity may be written as:

$$\partial_i(\sqrt{g}\boldsymbol{\sigma}^i) + \sqrt{g}\boldsymbol{\phi} = 0, \quad \boldsymbol{\sigma}^i = \sigma^{ij}(\boldsymbol{R}_j + \partial_j \boldsymbol{u}), \tag{11.1}$$

Hooke's law for isotropic bodies has the form

$$\boldsymbol{\sigma}^i = \sigma^{ij}(\boldsymbol{R}_j + \partial_j \boldsymbol{u}) = E^{ijpq}\varepsilon_{pq}(\boldsymbol{R}_j + \partial_j \boldsymbol{u}). \tag{11.2}$$

Here E^{ijpq} and ε_{pq} are contravariant and covariant components of tensors of elasticity and strain respectively,

$$E^{ijpq} = \lambda g^{ij} g^{pq} + \mu(g^{ip}g^{jq} + g^{iq}g^{jp}), \quad (g^{ij} = \boldsymbol{R}^i \boldsymbol{R}^j) \tag{11.3}$$

$$\varepsilon_{pq} = \frac{1}{2}(\boldsymbol{R}_p \partial_q \boldsymbol{u} + \boldsymbol{R}_q \partial_p \boldsymbol{u} + \partial_p \boldsymbol{u} \partial_q \boldsymbol{u}), \tag{11.4}$$

The stress vector, acting on the area with the normal $\hat{\boldsymbol{l}}$, has the form

$$\boldsymbol{\sigma}_{(\hat{l})} = \boldsymbol{\sigma}^i \hat{l}_i, \quad (\hat{l}_i = \hat{\boldsymbol{l}}\boldsymbol{R}_i). \tag{11.5}$$

For reducing three-dimensional problems to two-dimensional ones it is necessary to write (11.1)–(11.5) in bases of the middle surface, therefore it is expedient to consider the coordinate system, normally connected with the middle surface, i.e.,

$$\boldsymbol{R}(x_1, x_2, x_3) = \boldsymbol{r}(x_1, x_2) + x_3 \boldsymbol{n}(x_1, x_2).$$

where $\boldsymbol{R}$ and $\boldsymbol{r}$ are radius vectors of the space and surface, and $\boldsymbol{n}$ is a normal of this surface. The dependence between covariant and contravariant base vectors of the space and the surface are expressed as Vekua [1985]:

$$\boldsymbol{R}_i = A_{ij}\boldsymbol{r}^j = A^{j}_{.i}\boldsymbol{r}_j, \quad \boldsymbol{R}^i = A^{ij}\boldsymbol{r}_j = A^{i}_{.j}\boldsymbol{r}^j, \boldsymbol{R}_3 = \boldsymbol{R}^3 = \boldsymbol{n} \quad (A^{\alpha.}_{.3}A^{.\alpha}_{3.} = 0),$$

$$A_{ij} = a_{ij} - x_3 b_{ij}, \sqrt{g}A^{ij} = \sqrt{a}[a^{ij} + x_3(b^{ij} - 2Ha^{ij}) + x_3^2 K a^i_3 a^j_3],$$
$$A^{j}_{i.} = a^j_i - x_3 b^j_i, \sqrt{g}A^{i}_{.j} = \sqrt{a}[a^i_j + x_3(b^i_j - 2Ha^i_j) + x_3^2 K a^i_3 a^3_j],$$

$$a_{i3} = a^i_3 = a^{i3} = \delta^i_3, \quad b_{i3} = b^i_3 = b^{i3} = 0, \quad \sqrt{g} = \sqrt{a}[1 - 2Hx_3 + Kx_3^2].$$

Hooke's law now takes the form

$$\boldsymbol{\sigma}^i = M^{klns}A^{i}_{.k}A^{p}_{.n}(\boldsymbol{r}_s\partial_p\boldsymbol{u} + \frac{1}{2}A^q_s\partial_p\boldsymbol{u}\partial_q\boldsymbol{u})(\boldsymbol{r}_l + A^{j}_{.l}\partial_j\boldsymbol{u}), \tag{11.6}$$

where

$$M^{klns} = \lambda a^{kl}a^{ns} + \mu(a^{kn}a^{ls} + a^{ks}a^{ln}). \tag{11.7}$$

The stress tensor on face surface $x_3 = \pm h$, with the unit normal $\overset{(\pm)}{\boldsymbol{l}}$, where

$$\overset{(\pm)}{\boldsymbol{l}} = \frac{\partial_1\boldsymbol{R}(x_1, x_2, \pm h) \times \partial_2\boldsymbol{R}(x_1, x_2, \pm h)}{|\partial_1\boldsymbol{R} \times \partial_2\boldsymbol{R}|} = \frac{\boldsymbol{n} - \partial_\alpha h\boldsymbol{R}^\alpha(x_1, x_2, \pm h)}{\sqrt{1 + g^{\alpha\beta}(x_1, x_2, \pm h)\partial_\alpha h\partial_\beta h}},$$

$$\left(g^{\alpha\beta}(x_1, x_2, \pm h) = (\boldsymbol{R}^\alpha\boldsymbol{R}^\beta)_{x_3=\pm h}\right),$$

$$\sqrt{1 + g^{\alpha\beta}(x_1, x_2, \pm h)\partial_\alpha h\partial_\beta h}\,\boldsymbol{\sigma}_{\overset{\pm}{(l)}} = \boldsymbol{\sigma}^3(x_1, x_2, \pm h) \mp \boldsymbol{\sigma}^\alpha(x_1, x_2, \pm h)\partial_\alpha h. \tag{11.8}$$

The stress tensor of the lateral surface $dS = d\hat{s}dx_3$ with the normal $\hat{\boldsymbol{l}}$

$$\hat{\boldsymbol{l}} = [(1 - x_3 k_s)\boldsymbol{l} - x_3\tau_s\boldsymbol{s}]\frac{ds}{d\hat{s}}, \left(\hat{\boldsymbol{s}} = \frac{d\boldsymbol{R}}{d\hat{s}} = [(1 - x_3 k_s)\boldsymbol{s} + x_3\tau_s\boldsymbol{l}]\frac{ds}{d\hat{s}}\right),$$

has the form

$$\boldsymbol{\sigma}_{(\hat{l})} = \boldsymbol{\sigma}^\alpha(\hat{\boldsymbol{l}}\boldsymbol{R}_\alpha) = \sqrt{\frac{g}{a}}\boldsymbol{\sigma}^\alpha(\boldsymbol{l}\boldsymbol{r}_\alpha)\frac{ds}{d\hat{s}}, (\hat{\boldsymbol{l}} \times \hat{\boldsymbol{s}} = \boldsymbol{n}), \tag{11.9}$$

where $\boldsymbol{l}$ and $\boldsymbol{s}$ are unit vectors of the tangential normal and tangent lateral curve of middle surface ($\boldsymbol{l} \times \boldsymbol{s} = \boldsymbol{n}$), k_s and τ_s is the normal curvature and geodesic torsion of this curve, $d\hat{s}$ and ds are linear element of the surface $x_3 =$ const and middle surface, where

$$d\hat{s} = \sqrt{1 - 2x_3 k_s + x_3^2(k_s^2 + \tau_s^2)}\, ds,$$

The formula (11.9) is proved with dyadal representation of the differential $d\boldsymbol{R}$ of the surface $x_3 =$ const:

$$d\boldsymbol{R} = d\boldsymbol{r}(\nabla \otimes \boldsymbol{R}), \quad (\nabla = \boldsymbol{r}^\alpha\partial_\alpha)$$

and determination of the tangential normal

$$
\begin{aligned}
\hat{\boldsymbol{l}} &= \frac{d\boldsymbol{R}}{d\hat{s}} \times \boldsymbol{n} = \frac{d\boldsymbol{r}(\nabla \otimes \boldsymbol{R})}{ds} \times \boldsymbol{n}\frac{ds}{d\hat{s}} \\
&= s^{\alpha}\boldsymbol{R}_{\alpha} \times \boldsymbol{n}\frac{ds}{d\hat{s}} = s^{\alpha}\mathbb{C}_{\alpha 3\beta}\boldsymbol{R}^{\beta}\frac{ds}{d\hat{s}} = s^{\alpha}\sqrt{\frac{g}{a}}\sqrt{a}\epsilon_{\alpha 3\beta}\boldsymbol{R}^{\beta}\frac{ds}{d\hat{s}} \\
&= s^{\alpha}\sqrt{\frac{g}{a}}(\boldsymbol{r}_{\alpha} \times \boldsymbol{n})\boldsymbol{r}_{\beta}\boldsymbol{R}^{\beta}\frac{ds}{d\hat{s}} = \sqrt{\frac{g}{a}}l_{\beta}\boldsymbol{R}^{\beta}\frac{ds}{d\hat{s}} \Rightarrow \hat{\boldsymbol{l}}\hat{\boldsymbol{R}}_{\beta} = \sqrt{\frac{g}{a}}(\boldsymbol{l}\boldsymbol{r}_{\beta})\frac{ds}{d\hat{s}},
\end{aligned}
$$

where $\epsilon_{\alpha 3\beta}$ is Levi-Civita symbol, $\mathbb{C}_{\alpha 3\beta} = \sqrt{g}\epsilon_{\alpha 3\beta}$.

Let the expansion

$$
\left(\sqrt{\frac{g}{a}}\boldsymbol{\sigma}^{i}, \boldsymbol{u}, \sqrt{\frac{g}{a}}\boldsymbol{\phi}\right) = \sum_{m=0}^{\infty}(\overset{(m)}{\boldsymbol{\sigma}}{}^{i}, \overset{(m)}{\boldsymbol{u}}, \overset{(m)}{\boldsymbol{\phi}})P_m\left(\frac{x_3}{h}\right), \tag{11.10}
$$

hold true,where

$$
\left(\overset{(m)}{\boldsymbol{\sigma}}{}^{i}, \overset{(m)}{\boldsymbol{u}}, \overset{(m)}{\boldsymbol{\phi}}\right) = \frac{2m+1}{2h}\int\limits_{-h(x_1,x_2)}^{h(x_1,x_2)}\left(\sqrt{\frac{g}{a}}\boldsymbol{\sigma}^{i}, \boldsymbol{u}, \sqrt{\frac{g}{a}}\boldsymbol{\phi}\right)P_m\left(\frac{x_3}{h}\right)dx_3.
$$

By substituting these expansions into (11.1), (11.6), (11.9), satisfying beforehand the boundary conditions on face surface, *we get the following infinite system of two-dimensional equations:*

for equilibrium equations:

$$
\nabla_{\alpha}\overset{(m)}{\boldsymbol{\sigma}}{}^{\alpha} + \partial_{\alpha}\ln h\overset{(m)}{\underline{\underline{\boldsymbol{\sigma}}}}{}^{\alpha} - \frac{1}{h}\overset{(m)}{\underline{\boldsymbol{\sigma}}}{}^{3} + \overset{(m)}{\boldsymbol{F}} = 0, \tag{11.11}
$$

for Hooke's law:

$$
\begin{aligned}
\overset{(m)}{\boldsymbol{\sigma}}{}^{i} = M^{klns} \\
\times \sum_{m_1=0}^{\infty}\Bigg\{ & \overset{(m)}{\underset{(m_1)}{A}}{}^{i.p.}_{k.n.}\left(\boldsymbol{r}_s D_p\overset{(m_1)}{\boldsymbol{u}}\right)\boldsymbol{r}_l \\
& + \sum_{m_2=0}^{\infty}\Bigg[\overset{(m)}{\underset{(m_1m_2)}{A}}{}^{ijp.}_{kln.}\left(\boldsymbol{r}_s D_p\overset{(m_1)}{\boldsymbol{u}}\right)D_j\overset{(m_2)}{\boldsymbol{u}} \\
& + \frac{1}{2}\overset{(m)}{\underset{(m_1m_2)}{A}}{}^{i.pq}_{k.ns}\left(D_p\overset{(m_1)}{\boldsymbol{u}}D_q\overset{(m_2)}{\boldsymbol{u}}\right)\boldsymbol{r}_l \\
& + \frac{1}{2}\sum_{m_3=0}^{\infty}\overset{(m)}{\underset{(m_1m_2m_3)}{A}}{}^{ijpq}_{klns}\left(D_p\overset{(m_1)}{\boldsymbol{u}}D_q\overset{(m_2)}{\boldsymbol{u}}\right)D_j\overset{(m_3)}{\boldsymbol{u}}\Bigg]\Bigg\},
\end{aligned} \tag{11.12}
$$

for boundary conditions:

$$
\overset{(m)}{\boldsymbol{\sigma}}{}_{(l)} = \overset{(m)}{\sigma}{}_{(ll)}\boldsymbol{l} + \overset{(m)}{\sigma}{}_{(ls)}\boldsymbol{s} + \overset{(m)}{\sigma}{}_{(ln)}\boldsymbol{n} = \frac{2m+1}{2h}\int\limits_{-h}^{h}\boldsymbol{\sigma}_{(\hat{l})}\frac{d\hat{s}}{ds}P_m\left(\frac{x_3}{h}\right)dx_3, \tag{11.13}
$$

where

$$\underset{\alpha}{\underline{\underline{\overset{(m)}{\boldsymbol{\sigma}}}}} = m \overset{(m)}{\boldsymbol{\sigma}}{}^{\alpha} + (2m+1)\left[\overset{(m-2)}{\boldsymbol{\sigma}^{\alpha}} + \overset{(m-4)}{\boldsymbol{\sigma}^{\alpha}} + \ldots\right],$$

$$\underline{\overset{(m)}{\boldsymbol{\sigma}}}{}^3 = (2m+1)\left[\overset{(m-1)}{\boldsymbol{\sigma}^3} + \overset{(m-3)}{\boldsymbol{\sigma}^3} \ldots\right]$$

$$\overset{(m)}{\boldsymbol{F}} = \overset{(m)}{\boldsymbol{\Phi}} + \frac{2m+1}{2h}\sqrt{1+g^{\alpha\beta}(x_1,x_2,\pm h)\partial_\alpha h \partial_\beta h}\left(\sqrt{\frac{g_+}{a}}\boldsymbol{\sigma}_{\underset{(l)}{+}} - (-1)^m\sqrt{\frac{g_-}{a}}\boldsymbol{\sigma}_{(\bar{l})}\right],$$

$$\sqrt{g_\pm} = \sqrt{a}(1 \mp 2Hh + kh^2),$$

$$\underset{(m_1m_2m_3)}{\overset{(m)}{A}}{}^{ijpq}_{klns} = \frac{2m+1}{2h}\int_{-h}^{h}\sqrt{\frac{g}{a}}A^{i\cdot}_{\cdot k}A^{j\cdot}_{\cdot l}A^{p\cdot}_{\cdot n}A^{q\cdot}_{\cdot s}\prod_{i=0}^{3}P_{m_i}\left(\frac{x_3}{h}\right)dx_3\,, \tag{11.14}$$

$$D_p \overset{(m)}{\boldsymbol{u}} = \delta^\alpha_p \partial_\alpha \overset{(m)}{\boldsymbol{u}} + \delta_{3p}\overset{(m)}{\boldsymbol{u}'} - \partial_\alpha \ln h \cdot \overset{(m)}{\boldsymbol{u}''}\delta^\alpha_p$$

$$\overset{(m)}{\boldsymbol{u}'} = \frac{2m+1}{n}\left(\overset{(m+1)}{\boldsymbol{u}} + \overset{(m+3)}{\boldsymbol{u}} + \ldots\right),$$

$$\overset{(m)}{\boldsymbol{u}''} = m\overset{(m)}{\boldsymbol{u}'_3} + (2m+1)\left(\overset{(m+2)}{\boldsymbol{u}} + \overset{(m+4)}{\boldsymbol{u}} + \ldots\right).$$

The integrals of the form (11.14) proved to yield to explicit calculations, for example,

$$\begin{aligned}\underset{(m_1)}{\overset{(m)}{A}}{}^{\alpha 3\gamma 3}_{\beta 3\delta 3} &= \frac{2m+1}{2h}\int_{-h}^{h}\frac{B^\alpha_\beta(\frac{x_3}{h})B^\gamma_\delta(\frac{x_3}{h})P_{m_1}(\frac{x_3}{h})P_m(\frac{x_3}{h})}{1-2Hx_3+Kx_3^2}dx_3 \\ &= \frac{(b^\alpha_\beta - 2Ha^\alpha_\beta)(b^\gamma_\delta - 2Ha^\gamma_\delta)}{K}\delta^{m_1}_m \\ &\quad + \frac{2m+1}{2h\sqrt{E}}\left[B^\alpha_\beta(y)B^\gamma_\delta(y)\left\{\begin{array}{ll} P_{m_1}(y)Q_m(y), & m_1 \le m \\ Q_{m_1}(y)P_m(y), & m_1 \ge m\end{array}\right\}\right]^{y_2}_{y_1},\end{aligned}$$

where Q_m are *Legendre functions of the second order,* when

$B^\alpha_\beta(y) = (1-2Hhy)a^\alpha_\beta + hb^\alpha_\beta y$, $y_{1,2} = [(H \mp \sqrt{E})h]^{-1}$, $E = H^2 - K$,
$[f(y)]^{y_2}_{y_1} = f(y_2) - f(y_1)$, $(\alpha, \beta, \gamma, \delta = 1, 2)$.

Further, let is write out the integral, containing the multiplier of four Legendre polynomials

$$\underset{(m_1m_2m_3)}{\overset{(m)}{A}}{}^{\alpha\gamma\nu\varkappa}_{\beta\delta\eta\sigma}$$

$$= \frac{2m+1}{2h} \int\limits_{-h}^{h} \frac{B^{\alpha}_{\beta}(\frac{x_3}{h}) B^{\gamma}_{\delta}(\frac{x_3}{h}) B^{\nu}_{\eta}(\frac{x_3}{h}) B^{\varkappa}_{\sigma}(\frac{x_3}{h})}{(1-2Hx_3+Kx_3^2)^3} \prod_{i=0}^{3} P_{m_i}\left(\frac{x_3}{h}\right) dx_3$$

$$= \frac{2m+1}{4K^3h^6} \sum_{r_1=0}^{m_4} \alpha_{m_1m_2r_1} \sum_{r_2=0}^{m_5} \alpha_{mm_3r_2} \frac{\partial^4}{\partial y_1^2 \partial y_2^2}$$

$$\times \left(\frac{1}{y_1 - y_2} \left[B^{\alpha}_{\beta}(y)\, B^{\gamma}_{\delta}(y)\, B^{\alpha}_{\beta}(y)\, B^{\varkappa}_{\sigma}(y)\, \Gamma_{n_1,n_2}(y) \right]_{y_1}^{y_2} \right).$$

Here

$$\Gamma_{n_1,n_2}(y) = \begin{cases} P_{n_1}(y) Q_{n_2}(y), & \text{if } n_1 \le n_2 \\ Q_{n_1}(y) P_{n_2}(y), & \text{if } n_1 \ge n_2 \end{cases},$$

$m_4 = \min(m_1, m_2)$, $m_5 = \min(m, m_3)$, $n_1 = m_1 + m_2 - 2r_1$, $n_2 = m + m_3 - 2r_2$, $m_0 = m$.

In particular, if $m_1 + m_2 + m_3 \le m$ we will have

$$\underset{(m_1m_2m_3)}{\overset{(m)}{A}}{}^{\alpha\gamma\nu\varkappa}_{\beta\delta\eta\sigma} = \frac{2m+1}{4K^3h^6} \frac{\partial^4}{\partial y_1^2 \partial y_2^2} \left\{ \frac{1}{y_1 - y_2} \left[B^{\alpha}_{\beta}(y) B^{\gamma}_{\delta}(y) B^{\nu}_{\eta}(y) B^{\varkappa}_{\sigma}(y) \right.\right.$$

$$\left.\left. \times P_{m_1}(y) P_{m_2}(y) P_{m_3}(y) Q_m(y) \right]_{y_1}^{y_2} \right\}.$$

Here we have used the well known Neumann–Adams formula:

$$\int\limits_{-1}^{1} \frac{P_n(t)dt}{x-t} = 2Q_n(x), (|x|>1), \; P_m(x)P_n(x) = \sum_{r=0}^{\min(m,n)} \alpha_{mnr} P_{m+n-2r}(x),$$

$$\alpha_{mnr} = \frac{A_{m-r}A_rA_{n-r}}{A_{m+n-r}} \frac{2(m+n)-4r+1}{2(m+n)-2r+1}, \; A_m = \frac{(2m-1)!!}{m!},$$

as well as the representations

$$\frac{1}{1-2Hx_3+Kx_3^2} = \begin{cases} \frac{1}{2\sqrt{E}} \left(\frac{1}{\frac{1}{H+\sqrt{E}} - x_3} - \frac{1}{\frac{1}{H-\sqrt{E}} - x_3} \right), & E \ne 0 \\ \frac{1}{(1-Hx_3)^2}, & E = 0. \end{cases}$$

The transform to a finite system is carried out by considering finite segments in the expansion (11.10), where $m = 0, 1, ..., N$.

In conclusion finite system two-dimensional equations of nonlinear and isotropic non-shallow shells looks as follows:

equilibrium equations of the components tensor of stress (11.11):

$$\nabla_\alpha \overset{(m)}{\sigma}{}^{\alpha}_{\beta} - b_{\alpha\beta} \overset{(m)}{\sigma}{}^{\alpha}_{3} + \partial_\alpha lnh \, \overset{(m)}{\underline{\underline{\sigma}}}{}^{\alpha}_{\beta} - \frac{1}{h} \overset{(m)}{\sigma}{}^{3}_{\beta} + \overset{(m)}{F}_{\beta} = 0,$$

$$\nabla_\alpha \overset{(m)}{\sigma_3}{}^{\alpha} + b^\beta_\alpha \overset{(m)}{\sigma_\beta}{}^{\alpha} + \partial_\alpha lnh \overset{(m)}{\underline{\underline{\sigma}}}{}^{\alpha}_{3} - \frac{1}{h}\overset{(m)}{\underline{\underline{\sigma}}}{}^{3}_{3} + \overset{(m)}{F_3} = 0,$$

$$(m = 0, 1, ..., N)$$

where ∇_α *is the symbol of covariant derivation on the surface;*

for Hooke's law:

$$\overset{(m)}{\sigma}{}^{it} = M^{klns}\sum_{m_1=0}^{N}\left\{\underset{(m_1)}{\overset{(m)}{A}}{}^{i3p3}_{k3n3}(\boldsymbol{r}_s D_p \overset{(m_1)}{\boldsymbol{u}})a^t_l\right.$$

$$+\sum_{m_2=0}^{N}\left[\underset{(m_1m_2)}{\overset{(m)}{A}}{}^{ijp3}_{kln3}(\boldsymbol{r}_s D_p \overset{(m_1)}{\boldsymbol{u}})(D_j \overset{(m_2)}{\boldsymbol{u}}\boldsymbol{r}^t) + \right.$$

$$+\frac{1}{2}\underset{(m_1m_2)}{\overset{(m)}{A}}{}^{i3pq}_{k3ns}(D_p \overset{(m_1)}{\boldsymbol{u}} D_q \overset{(m_2)}{\boldsymbol{u}})a^t_l$$

$$\left.\left.+\frac{1}{2}\sum_{m_3=0}^{N}\underset{(m_1m_2m_3)}{\overset{(m)}{A}}{}^{ijpq}_{klns}(D_p \overset{(m_1)}{\boldsymbol{u}} D_q \overset{(m_2)}{\boldsymbol{u}})(\boldsymbol{r}^t D_j \overset{(m_3)}{\boldsymbol{u}})\right]\right\},$$

for boundary conditions on the contours Γ*:*

$$\overset{(m)}{\sigma}_{(ll)} = \overset{(m)}{f_1}, \overset{(m)}{\sigma}_{(ls)} = \overset{(m)}{f_2}, \overset{(m)}{\sigma}_{(ln)} = \overset{(m)}{f_3}.$$

$$(m = 0, 1, ...N)$$

Let's consider shallow shells, i.e.,

$$a^\beta_\alpha - x_3 b^\beta_\alpha \cong a^\beta_\alpha, \quad \boldsymbol{R}_\alpha \cong \boldsymbol{r}_\alpha, \quad \boldsymbol{R}^\alpha \cong \boldsymbol{r}^\alpha, \quad \sqrt{g} \cong \sqrt{a};$$

then the integrals (11.14) greatly simplify and system equations of nonlinear theory shell takes the form:

for equilibium equations does not change appearance;

for Hooke's law

$$\overset{(m)}{\sigma}_{is} = M^{ijpq}\Big\{(\boldsymbol{r}_q D_p \overset{(m)}{\boldsymbol{u}})a_j^s + \sum_{m_1,m_2=0}^{N}\Big[\frac{1}{2}(D_p \overset{(m_1)}{\boldsymbol{u}} D_q \overset{(m_2)}{\boldsymbol{u}})a_j^s + (\boldsymbol{r}_q D_p \overset{(m_1)}{\boldsymbol{u}})(\boldsymbol{r}^s D_j \overset{(m_2)}{\boldsymbol{u}})\Big]\sum_{r=0}^{m_4}\alpha_{m_1m_2r}\delta^m_{m_1+m_2-2r},$$

$$+\frac{1}{2}\sum_{m_1,m_2,m_3=0}^{N}(D_p \overset{(m_1)}{\boldsymbol{u}} D_q \overset{(m_2)}{\boldsymbol{u}})(\boldsymbol{r}^s D_j \overset{(m_3)}{\boldsymbol{u}}) \times\sum_{r_1=0}^{m_4}\alpha_{m_1m_2r_1}\sum_{r_2=0}^{m_5}\alpha_{mm_3r_2}\frac{2m+1}{2n_1+1}\delta^{n_2}_{n_1}\Big\}$$

for boundary conditions on the contours Γ *does not change appearance.*

Appearance of these equations are more simplified for the plates of variable thickness:

for equilibrium equation takes the form

$$\partial_\alpha \overset{(m)}{\sigma^\alpha} + \partial_\alpha \ln h \, \overset{(m)}{\underline{\underline{\sigma}}^{is}} - \frac{1}{h}\overset{(m)}{\underline{\sigma}^3} + \overset{(m)}{\boldsymbol{F}} = 0,$$

for Hooke's law has the same form as the shallow shells, only the formula for M^{klns} *takes the form*

$$M^{klns} = \lambda\delta^{kl}\delta^{ns} + \mu(\delta^{kn}\delta^{ls} + \delta^{ks}\delta^{ln})$$

These results from the present subsection 11.2 belong to Meunargua.

11.3. To investigation of Vekua theory for isotropic thickwalled shells of non-homogeneous structure

These results belong to Zgenti [1991].

Let on S the coordinate lines be lines of curvature:

$$R_\alpha = (1 - k_\alpha x^3)r_\alpha, \; R_3 = n - R_3^3, \; R^\alpha = (1 - k_\alpha x^3)r^\alpha.$$

Then the equilibrium equations of shells have such form (Vekua [1985]:

$$\frac{1}{\sqrt{a}}\partial_\alpha(\sqrt{a}P^\alpha) - \partial_3 P^3 + \vartheta F = 0, \quad \vartheta = (1 - k_1 x^3)(1 - k_1). \tag{11.15}$$

Using method of Vekua in our case we have:

$$\overset{(k)}{P^{\alpha\beta}} = \sum_{m=0}^{\infty}\Big\{\lambda^{km}_{3-\alpha,\gamma}a^{\alpha\beta}(\nabla_\gamma \overset{(m)}{u^\gamma} - b^\gamma_\gamma \overset{(m)}{u_3}) + \frac{1}{h}\lambda^{km}_{3-\alpha}a^{\alpha\beta}\overset{(m)}{u_3'} + \mu^{km}_{3-\alpha,\gamma}[a^{\alpha\gamma}(\nabla_\gamma \overset{(m)}{u^\beta} - -b^\beta_\gamma \overset{(m)}{u_3}) + a^{\beta\gamma}(\nabla_\gamma \overset{(m)}{u^\alpha} - b^\alpha_\gamma \overset{(m)}{u_3})]\Big\},$$

$$\overset{(k)}{P}{}^{\alpha 3} = \sum_{m=0}^{\infty} \left[\mu_{3-\alpha,\gamma}^{km} a^{\alpha\gamma} (\nabla_\gamma \overset{(m)}{u_3} + b_{\gamma\beta} \overset{(m)}{u^\beta}) + \frac{1}{h} \mu_{3-\alpha}^{km} \overset{(m)}{u'^\alpha} \right], \tag{11.16}$$

$$\overset{(k)}{P}{}^{3\alpha} = \sum_{m=0}^{\infty} \left[\mu_{3-\alpha,\gamma}^{km} a^{\alpha\gamma} (\nabla_\gamma \overset{(m)}{u_3} + b_{\gamma\beta} \overset{(m)}{u^\beta}) + \frac{1}{h} \mu_*^{km} \overset{(m)}{u'^\alpha} \right],$$

$$\overset{(k)}{P}_{33} = \sum_{m=0}^{\infty} \left[\lambda_{3-\gamma}^{km} (\nabla_\gamma \overset{(m)}{u^\gamma} - b_\gamma^\gamma \overset{(m)}{u_3}) + \frac{1}{h} (\lambda_*^{km} + 2\mu_*^{km}) \overset{(m)}{u'_3} \right],$$

where

$$\lambda_{\alpha\beta}^{km} = \left(k + \frac{1}{2}\right) \frac{1}{h} \int_{-h}^{h} \lambda A_\alpha^\beta P_k(x^3/h) P_m(x^3/h) dx^3,$$

$$\mu_{\alpha\beta}^{(km)} = \left(k + \frac{1}{2}\right) \frac{1}{h} \int_{-h}^{h} \mu A_\beta^\alpha P_h(x^3/h) P_m(x^3/h) dx^3,$$

$$\lambda_\alpha^{km} = \left(h + \frac{1}{2}\right) \frac{1}{h} \int_{-h}^{h} \lambda A_\alpha P_k(x^3/h) P_m(x^3/h) dx^3,$$

$$\mu_\alpha^{km} = \left(k + \frac{1}{2}\right) \frac{1}{h} \int_{-h}^{h} \mu A_\alpha P_k(x^3/h) P_m(x^3/h) dx^3,$$

$$\lambda_*^{km} = \left(k + \frac{1}{2}\right) \frac{1}{h} \int_{-h}^{h} \lambda \vartheta P_k(x_3/h) P_m(x^3/h) dx^3,$$

$$\mu_*^{km} = \left(k + \frac{1}{2}\right) \frac{1}{h} \int_{-h}^{h} \mu \vartheta P_k(x^3/h) P_m(x^3/h), \quad A_\alpha = 1 - k_\alpha x^3, \; A_\beta^\alpha = \frac{A_\alpha}{A_\beta}.$$

Let $g^\pm$ are defined by 1.2. Then we have

$$P^{3(+)} = \vartheta^{(+)} g^+, \quad P^{3(-)} = \vartheta^{(-)} g^-$$
$$\vartheta^{(+)} = (1 - k_1 h)(1 - k_2 h), \quad \vartheta^{(-)} = (1 + k_1 h)(1 + k_2 h). \tag{11.17}$$

The solution of (11.15), following to Vashakmadze [1981c], has the form

$$P^\alpha = \sum_{k=0}^{\infty} \overset{(k)}{P}{}^\alpha P_k(x^3/h);$$
$$P^3 = \frac{1}{2} \left[\left(1 + \frac{x^3}{h}\right) \vartheta^{(+)} g^+ + \left(1 - \frac{x^3}{h}\right) \vartheta^{(-)} g^- \right]$$
$$+ \sum_{k=0}^{\infty} P_*^{3(k)} [P_k(x^3/h) - P_{k+2}(x^3/h)]; \tag{11.18}$$

If we use method of Vekua [1985] we have:

$$\frac{1}{\sqrt{a}}\partial_\alpha(\sqrt{a}P^{\alpha(k)}) - \frac{2k+1}{h}\sum_{S=0}^{[\frac{k-1}{2}]} P^{3(k-2S-1)} + F^{(k)} = 0\,, \quad (k=0,1,\ldots), \qquad (11.19)$$

where

$$F^{(k)} = \frac{2k+1}{2}(g^+ g_3^+ - (-1)^k g^- g_3^- \,.$$

Cutting the infinite system (11.19) we have:

$$\begin{aligned}
&\nabla_\alpha \overset{(k)}{P^{\alpha\beta}} - b_\alpha^\beta \overset{(k)}{P^{\alpha 3}} - \frac{2k+1}{h}\sum_{k=0}^{[\frac{k-1}{2}]} \overset{(k-2s-1)}{P^{3\beta}} + \overset{(k)}{F^\beta} = 0 \quad (k=0,1,\ldots,N)\\
&\nabla_\alpha \overset{(k)}{P^{\alpha\beta 3}} + b_{\alpha\beta} \overset{(k)}{P^{\alpha\beta}} - \frac{2k+1}{h}\sum_{k=0}^{[\frac{k-1}{2}]} \overset{(k-2s-1)}{P_{33}} + \overset{(k)}{F^3} = 0 \quad (k=0,1,\ldots,N).
\end{aligned} \qquad (11.20)$$

or

$$\begin{aligned}
\overset{(k)}{P^{\alpha\beta}} &= \sum_{m=0}^{N}\Big\{\lambda^{km}_{3-\alpha,\gamma} a^{\alpha\beta}(\nabla_\gamma \overset{(m)}{u^\gamma} - b_\gamma^\gamma \overset{(m)}{u_3}) + \frac{2m+1}{h}\lambda^{km}_{3-\alpha} a^{\alpha\beta}\sum_{S=0}^{[\frac{N-m-1}{2}]} \overset{(m+2S+1)}{u_3}\\
&\quad + \mu^{km}_{3-\alpha,\gamma}[a^{\alpha\gamma}(\nabla_\gamma \overset{(m)}{u^\beta} - b_\gamma^\beta \overset{(m)}{u_3}) + a^{\beta\gamma}(\nabla_\gamma \overset{(m)}{u^\alpha} - b_\gamma^\alpha \overset{(m)}{u_3})]\Big\};\\
\overset{(k)}{P^{\alpha 3}} &= \sum_{m=0}^{N}\Big[\mu^{km}_{3-\alpha,\gamma} a^{\alpha\gamma}(\nabla_\gamma \overset{(m)}{u_3} + b_{\alpha\beta}\overset{(m)}{u^\beta})\\
&\quad + \frac{2m+1}{h}\mu^{km}_{3-\alpha}\sum_{s=0}^{[\frac{N-m-1}{2}]} \overset{(m+2s+1)}{u^\alpha}\Big],\\
\overset{(k)}{P^{3\alpha}} &= \sum_{m=0}^{N}\Big[\mu^{km}_{3-\gamma} a^{\alpha\gamma}(\nabla_\gamma \overset{(m)}{u_3} + b_{\gamma\beta}\overset{(m)}{u^\beta}) + \frac{2m+1}{h}\mu^{km}_{*}\sum_{s=0}^{[\frac{N-m-1}{2}]} \overset{(m+2s+1)}{u^\alpha}\Big],\\
\overset{(k)}{P^{33}} &= \sum_{m=0}^{N}\Big[\lambda^{km}_{3-alpha,\gamma}(\nabla_\gamma \overset{(m)}{u^\gamma} - b_\gamma^\gamma \overset{(m)}{u_3}) + \frac{2m+1}{h}(\lambda^{km}_{*} + 2\mu^{km}_{*})\sum_{s=0}^{[\frac{N-m-1}{2}]} \overset{(m+2s+1)}{u_3}\Big].
\end{aligned} \qquad (11.21)$$

These equations (11.20), (11.21) represent closed system of differential equations respect to $\overset{(k)}{u^j}$ functions.

Consider now when Ω_h is the spherical shell with constant thickness $2h$, R denotes the radius. Let $x^1 = x$, $x^2 = y$, $x^3 = x_3$. In our case we evidently have:

$$a_{11} = a_{22} = \Lambda; \quad a_{12} = a_{21} = 0.$$

If ν, φ are geographical coordinates of a sphere, then as the isometrical coordinate system may be have

$$x = \tan\frac{\nu}{2}\cos\varphi; \quad y = \tan\frac{\nu}{2}\sin\varphi$$

and

$$\Lambda = R^2\Lambda_0^2; \quad \Lambda_0 = \frac{2}{1+x^2+y^2} = 2\cos^2\frac{\nu}{2}.$$

The system of equations (11.20), (11.21) following to Vekua [1985] has such complex form:

$$\frac{1}{\Lambda}\frac{\partial}{\partial z}(\overset{(k)}{P}_{11} - \overset{(k)}{P}_{22} + 2i\overset{(k)}{P}_{12}) + \frac{\partial}{\partial \bar z}(\overset{(k)}{P}{}_1^1 + \overset{(k)}{P}{}_2^2) + \frac{1}{R}(\overset{(k)}{P}_{13} + i\overset{(k)}{P}_{23})$$

$$-\frac{2h+1}{h}\sum_{s=0}^{\left[\frac{k-1}{2}\right]}(\overset{(k-2s-1)}{P}_{31} + i\overset{(k-2s-1)}{P}_{32}) + \overset{(k)}{F}_1 + i\overset{(k)}{F}_2 = 0$$

$$(k = 0, 1, \ldots, N), \tag{11.22}$$

$$2Re\left[\frac{1}{\Lambda}\frac{\partial}{\partial z}(\overset{(k)}{P}_{13} + i\overset{(k)}{P}_{23})\right] - \frac{1}{R}(\overset{(k)}{P}{}_1^1 + \overset{(k)}{P}{}_2^2)$$

$$-\frac{2k+1}{h}\sum_{s=0}^{\left[\frac{k-1}{2}\right]}\overset{(k-2s-1)}{P}_{33} + \overset{(k)}{F}_3 = 0 \quad (k = 0, 1, \ldots, N),$$

$$\overset{(k)}{P}_{11} - \overset{(k)}{P}_{22} + 2i\overset{(k)}{P}_{12} = 4\sum_{m=0}^{N}\mu_{km}\Lambda\frac{\partial}{\partial \bar z}\frac{1}{\Lambda}(\overset{(m)}{u}_1 + \overset{(m)}{u}_2);$$

$$\overset{(k)}{P}{}_1^1 - \overset{(k)}{P}{}_2^2 = 2\sum_{s=0}^{N}\left\{(\lambda_{km} + \mu_{km})\left(\overset{(m)}{\theta} + \frac{2}{R}\overset{(m)}{u}_3\right)\right.$$

$$\left. + \frac{2m+1}{h}\lambda_{km}^{(1)}\sum_{s=0}^{\left[\frac{N-m-1}{2}\right]}\overset{(m+2s+1)}{u}_3\right\}, \tag{11.23a}$$

$$\overset{(k)}{P}_{13} + i\overset{(k)}{P}_{23} = \sum_{m=0}^{N}\left\{\mu_{km}\left[2\frac{\partial u_3}{\partial \bar z} - \frac{1}{R}(\overset{(m)}{u}_1 + i\overset{(m)}{u}_2)\right]\right.$$

$$\left. + \frac{2m+1}{h}\mu_{km}^{(1)}\sum_{s=0}^{\left[\frac{N-m-1}{2}\right]}(\overset{(m_2s+1)}{u}_1 + i\overset{(m+2s+1)}{u}_2)\right\},$$

$$\overset{(k)}{P}_{31}+i\overset{(k)}{P}_{32}=\sum_{m=0}^{N}\left\{\mu_{km}^{(1)}\left[2\frac{\partial u_3}{\partial\overline{z}}-\frac{1}{R}(\overset{(m)}{u_1}+i\overset{(m)}{u_2}\right]\right.$$

$$\left.+\frac{2m+1}{h}\mu_{km}^{(2)}\times\sum_{s=0}^{\left[\frac{N-m-1}{2}\right]}(\overset{(m_2s+1)}{u_1}+i\overset{(m+2s+1)}{u_2})\right\},$$

$$\overset{(k)}{P}_{33}=\sum_{m=0}^{N}\left\{\lambda_{km}^{(1)}\left(\overset{(m)}{\theta}+\frac{2}{R}\overset{(m)}{u_3}\right)\right. \tag{11.23b}$$

$$\left.+\frac{2m+1}{1}(\lambda_{km}^{(2)}+2\mu_{km}^{(2)})\sum_{s=0}^{\left[\frac{N-m-2}{2}\right]}\overset{(m+2s+1)}{u_3}\right\},$$

where

$$\overset{(m)}{\theta}=\nabla_\gamma\overset{(m)}{u^\gamma}=2Re\left[\frac{1}{\Lambda}\frac{\partial}{\partial z}(\overset{(m)}{u_1}+i\overset{(m)}{u_2})\right],$$

$$z=x+iy;\quad \overline{z}=x-iy;\quad \frac{\partial}{\partial z}=\frac{1}{2}\left(\frac{\partial}{\partial x}-i\frac{\partial}{\partial y}\right);$$

$$\frac{\partial}{\partial\overline{z}}=\frac{1}{2}\left(\frac{\partial}{\partial x}+i\frac{\partial}{\partial y}\right);$$

$$\lambda_{km}=\lambda_{11}^{km};\quad \lambda_{km}^{(\alpha)}=\left(k+\frac{1}{2}\right)\frac{1}{h}\int_{-h}^{h}\left(1+\frac{x_3}{R}\right)^\alpha P_k(x_3/h)P_m(x_3/h)dx_3;$$

$$\mu_{km}=\mu_{11}^{km};\quad \mu_{km}^{(\alpha)}=\left(k+\frac{1}{2}\right)\frac{1}{h}\int_{-h}^{h}\left(1+\frac{x_3}{R}\right)^\alpha P_k(x_3/h)P_m(x_3/h)dx_3;$$

Assume that λ_{km}^α; μ_{km}^α are constants and

$$\overset{(k)}{F_1}=\overset{(k)}{F_2}=\overset{(k)}{F_3}=0.$$

Below we use the equality:

$$\frac{1}{\Lambda}\frac{\partial}{\partial z}\Lambda\frac{\partial}{\partial\overline{z}}\frac{1}{\Lambda}\frac{\partial}{\partial\overline{z}}(\cdot)=\frac{1}{4R^2}(\nabla^2+2)(\cdot),$$

where ∇^2 is Laplace's operator on sphere with unit radius:

$$\nabla^2=\frac{1}{\Lambda_0^2}\left(\frac{\partial^2}{\partial x^2}+\frac{\partial^2}{\partial y^2}\right)=\frac{4}{\Lambda_0^2}\frac{\partial^2}{\partial z\partial\overline{z}}.$$

The solution of equations (11.22)-(11.29) we find in such form:

$$\overset{(k)}{u_1}+i\overset{(k)}{u_2}=R^2\frac{\partial}{\partial\overline{z}}(W_{k+1}+i\Omega_{k+1})\quad(k=0,1,\ldots,N), \tag{11.24}$$

where W_{k+1}, Ω_{k+1} are arbitrary real functions. Let introduce definitions:

$$\overset{(k)}{u_3} = hW_{N+2+k} \quad (k = 0, 1, \ldots, N). \tag{11.25}$$

Using these representations we finally have such system:

$$\sum_{j=1}^{2N+1} D_{kj} \nabla^2 W_j - \sum_{j=1}^{2N+2} L_{kj} W_j = 0 \quad (k = 1, 2, \ldots, 2N + 2); \tag{11.26}$$

$$\sum_{j=1}^{N+1} d_{kj} \nabla^2 \Omega_j - \sum_{j=1}^{N+1} l_{kj} \Omega_j = 0 \quad (k = 1, 2, \ldots, N + 1). \tag{11.27}$$

or in the matrix form:

$$D \nabla^2 W - LW = 0; \tag{11.28}$$

$$d \nabla^2 \Omega - l\Omega = 0, \tag{11.29}$$

where W, Ω are colomn vectors. Matrices equations (11.23) and (11.24) we can reduce to a form:

$$\nabla^2 W - AW = 0, \quad A = D^{-1}L;$$

$$\nabla^2 \Omega - B\Omega = 0, \quad B = d^{-1}l.$$

We remark that for finding of the general solution of system it is possible to apply method Vekua [1988]. If we assume that matrices A and B have simple eigenvalue numbers (see Zgenti, Khvoles [1982]) and vectors $\alpha_1, \ldots, \alpha_{2N+2}$; $\beta_1, \ldots, \beta_{N+1}$; $X^{(1)}, \ldots, X^{(2N+2)}$; $Y^{(1)}, \ldots, Y^{(N+1)}$ respectively, then general solutions of these equations have such form:

$$W = \sum_{m=1}^{2N+2} X^{(m)} \psi_m; \quad \Omega = \sum_{m=1}^{N+1} Y^{(m)} \chi_m, \tag{11.30}$$

where ψ_m, χ_m are arbitrary solutions of the following scalar equations:

$$\nabla^2 \psi_m - \alpha_m \psi_m = 0 \quad (m = 1, \ldots, 2N + 2); \tag{11.31}$$

$$\nabla^2 \chi_m - \beta_m \chi_m = 0 \quad (m = 1, \ldots, N + 1). \tag{11.32}$$

At last for $\overset{(k)}{u_j}$ we have:

$$\overset{(k)}{u_1} + i \overset{(k)}{u_2} = R^2 \frac{\partial}{\partial \bar{z}} \left(\sum_{m=1}^{2N+2} X_{k+1}^{(m)} \psi_m + i \sum_{m=1}^{N+1} Y^{(m)} \chi_m \right),$$

$$\overset{(k)}{u_3} = h \sum_{m=1}^{2N+2} X_{N+2+k}^{(m)} \Psi_m, \quad (k = 0, 1, \ldots, N).$$

11.4. Shell theory using two basic surfaces

In this subsection are investigated problems of constructing shallow shell theory of the first order using two basic surfaces $S^{\pm}$ instead of one basic surface S_h as in the classical case. There is found the following kinematic hypothesis:

Let the initial three-dimensional BVP be linear, while the shell configuration before deformation is the following:

$$r(x^1, x^2, x^3) = r^-(x^1, x^2) + x^3(r^+(x^1, x^2) - r^-(x^1, x^2)).$$

The deformed configuration is:

$$R(x^1, x^2, x^3) = R^-(x^1, x^2) + x^3(R^+(x^1, x^2) - R^-(x^1, x^2)),$$

satisfying the condition that $R^+ - R^-$ vector is orthogonal to S^-.

Then for the unknown values of displacement vector u on the $S^{\pm}$:

$$u^- = R^- - r^-, \quad u^+ = R^+ - r^-,$$

the closed strongly elliptical system of PDEs of twelfth order with corresponding boundary conditions on ∂S^+ and ∂S^- is constructed.

Further, these equations were applied to the construction of two-dimensional models of multi-layer shallow shells and had solved approximately some BVPs of practical charachters. Details and bibliography are in Tvalchrelidze [1984].

Chapter III. Some Approximate Methods and Numerical Realizations

In this chapter we consider methods of approximate solution of two-dimensional boundary value problems for the system of differential equations, which arose also as a necessary step for solving the initial problem (1.1)–(1.5).

Computer tecnique introduced a change of principal characters in favour of numerical processes for solving BVP of continious medium, although it is evident that the importance of analytical methods remains fundamental.

There have appeared many monographs and practical handbooks for solving some problems of the theory of elasticity by computers; It consideration that the especially popular, at the same time, universal and quite effective methods, are finite elements method and a method of boundary integral equations (Brebbia [1978], Ciarlet [1978], Kravchuk [1974], Cruse [1978], Oliveira [1970], Perlin [1982], Strang, Fix [1973],...).

As known, these methods were treated as engineering technical methods of construction of algebraic analogues, corresponding to these on those problems and also to elasticity theory. However, due to investigations which Courant [1943] had begun a deep connection was detected between an engineering technical approach and variation-projective methods.

As was remarked, at the solution Vekua's reduced models or schemes, offered in section 7, the permitting algorithm comprises the solution of two-dimensional differential problems in the area $D(x_1, x_2)$, as one of the elements. Alongside with known methods, depending on the form of the area and boundary conditions in a number of cases the more preferable variant of a variation-discrete method is advanced.

Though the structure of a method stated here can be interpeted as a particular realization of general *variation-projective methods;* however, it is *an optimum* party of account on subclass of two-dimensional problems.

The method is based on use of linear combinations from orthogonal polynomials as coordinate functions; questions of stability of computing processes and methods of the solution of offered systems of linear algebraic equations are investigated.

In subsection 12.3 is advanced a way of constructing discrete-difference schemes on the basis of parallel applications of finite-difference and variation-discrete schemes. Such schemes have properties of finite-difference schemes and do not require the curvilinear part of a boundary approximations of boundary conditions.

In a class of methods of the solution of two-dimensional differential problems is advanced a continual analogue of the alternating direction method of Peaceman, Rachford [1955]. Thus we have to decide systems of one-dimensional differential problems. In this connection in sections 13—15 is stated a method of the solution of boundary value problems for ordinary equations of second order with an arbitrary (fixed) order of accuracy, which may be used immediately also for methods developing in subsections 12.3 and 12.4.

12. Methods of Solving Two-Dimensional Boundary Value Problems

12.1. Variant of a variation-discrete method

I. In this subsection we consider the variant of a variation-discrete method of solution of two-dimensional problems for equations of thin elastic shells and planar theory of elasticity in the rectangular area.

As is known, in a variation-discrete method the main thing is a selection of coordinate functions (see, e.g., Courant [1943]). The class of coordinate functions usually uses values of unknown vector functions (or its derivative) in points of a net as unknown parameters. We will have some deviation from such approach, namely: we select coordinate functions so that moments of unknown vector would satisfy an algebraic equations system which is received by a small perturbation of variation–discrete scheme, appropriate to a two-dimensional BVP.

For this purpose we introduce the *product of differences from Legendre's polynomials:*

$$\varphi_{ij}^{\alpha,\beta}(x,y) = \chi^{\alpha} P_i(x) \chi^{\beta} P_j(y), \tag{12.1}$$

where α, β stand nonnegative integers and

$$\chi^{\alpha} P_i(x) = \sqrt{\frac{2i+1}{2}} \sum_{s=0}^{\alpha} (-1)^s \binom{s}{\alpha} \prod_{\substack{j=-s \\ j\neq\alpha-2s}}^{\alpha-s} (2(i+j)+1)^{-1} P_{i+\alpha-2s}(x). \tag{12.2}$$

$$\chi^0 P_i(x) \equiv P_i(x), \quad \chi^1 P_i(x) = \frac{1}{\sqrt{2(2i+1)}} (P_{i+1}(x) - P_{i-1}(x)), \quad \chi^{\alpha} P_i = \chi^{\alpha-1} \chi P_i.$$

We shall find approximately the function $u(x,y)$ in the form

$$\overset{n}{u}(x,y) = \sum_{i,j=0}^{n} \overset{n}{u}{}^{ij} \varphi_{ij}^{\alpha,\beta}(x,y). \tag{12.3}$$

Then the components $\overset{n}{u}{}^{ij}$ will satisfy the following system of variational-difference equations

$$R_n \overset{n}{u} = \overset{n}{g}, \; \overset{n}{g}_{ij} = \iint_D f(x,y) \psi_{ij}^{\alpha\beta} \, d\omega. \tag{12.4}$$

for Dirichlet's problem for the system of differential equations in D. The system follows from the initial problem immediatly if we use Galorkin's method. ψ is *a complete system in L_2 enlarging of φ.*

The numerical process includes itself: the inversion of matrix R_n, calculation of right parts $\overset{n}{g}$, coordinate functions $\chi^\alpha P_i(x)$ and unknown vector in a point $(x,y) \in D$ by the formula (12.3). Together with questions of stability, the convergence and valuation of an error, should also be paid attention for *optimization of numerical process,* determined by means of (12.1)-(12.2).

We consider the possibility of constructing a matrix R_n of a relatively simple structure as a criterion for the volume of the necessary computation work to be minimal. The specified criterion of optimum limits the class of two-dimentional BVP.

Below the variant of variation-difference method is applied to a boundary value problem for equations of thin elastic spherical shell Vekua [1968], though the *distribution of this method* for more *general strongly elliptical systems with constant coefficients does not present difficulties.*

We shall consider Dirichlet's problem for a following system of differential equations in the area $\bar{D} = [-1,1]^2$

$$L(\partial_1, \partial_2)u = f, \qquad u = (u_1, u_2, ... u_5)^T|_{\partial D} = 0 \tag{12.5}$$

Or in a extensive form:

$$L_1[u_+] = \mu \Delta u_+ + (\lambda^* + \mu)\,\mathrm{grad}\,\mathrm{div}\, u_+ = f_+, \qquad u_+ = (u_1, u_2)^T, \tag{12.5a}$$

$$L_2[u_3, \overset{1}{u}_+] = \mu \Delta u_3 + \mu\,\mathrm{div}\,\overset{1}{u}_+ = f_3, \qquad \overset{1}{u}_+ = (u_4, u_5)^T, \tag{12.5b}$$

$$L_3[u_3, \overset{1}{u}_+] = \frac{h^2\mu}{2}\Delta \overset{1}{u}_+ + \frac{h^2}{2}(\lambda^* + \mu)\,\mathrm{grad}\,\mathrm{div}\,\overset{1}{u}_+ - \mu(\mathrm{grad}\, u_3 + \overset{1}{u}_+) = \overset{1}{f}_+. \tag{12.5c}$$

We shall put in a system of functions (12.1) $\alpha = \beta = 1$. Then, obviously, the solution of a system (12.5) may be represented in a form (12.3) and the corresponding variation-discrete scheme for the problem (12.5) will have the form:

$$\begin{aligned} L_{11}^{ij}[u_+] &= -[(\lambda^* + 2\mu)c_i + \mu c_j]u_1^{ij} + a_{i-1}(\lambda^* + 2\mu)u_1^{i-2,j} + a_{j-1}\mu u_1^{i,j-2} \\ &\quad + a_{i+1}(\lambda^* + 2\mu)u_1^{i+2,j} + \mu a_{j+1}u_1^{i,j+2} \\ &\quad - (\lambda^* + \mu)[b_{i+1,j}u_2^{i+1,j-1} + b_{i,j+1}u_2^{i-1,j+1} \\ &\quad - b_{i+1,j+1}u_2^{i+1,j+1} - b_{i,j}u_2^{i-1,j-1}] \\ &= g_1^{ij}, \end{aligned} \tag{12.6}$$

$$\begin{aligned} L_{12}^{ij}[u_+] &= -(\mu c_i + (\lambda^* + 2\mu)c_j)u_2^{ij} + a_{i-1}\mu u_2^{i-2,j} + a_{j-1}(\lambda^* + 2\mu)u_2^{i,j-2} \\ &\quad + a_{i+1}\mu u_2^{i+2,j} + (\lambda^* + 2\mu)a_{j+1}u_2^{i,j+2} - (\lambda^* + \mu)[b_{i,j+1}u_1^{i+1,j-1} \\ &\quad + b_{i,j+1}u_1^{i-1,j+1} - b_{i+j,j+1}u_1^{i+1,j+1} - b_{ij}u_1^{i-1,j-1}] \\ &= g_2^{ij}, \end{aligned}$$

$$\begin{aligned} L_2^{is}[u_3, \overset{1}{u}_+] &= \mu \Delta^{ij}[u_3] + \mu I_1^{ij}[u_4] + \mu I_2^{ij}[u_5] \\ &= -\mu[(c_i + c_j)u_3^{ij} - a_{i-1}u_3^{i-2,j} - a_{j-1}u_3^{i,j-2} - a_{i+1}u_3^{i+2,j} - a_{j+1}u_3^{i,j+2}] \end{aligned}$$

$$
\begin{aligned}
&+\mu[a_{j-1}(-e_{i+1}u_4^{i+1,j-2}+e_iu_4^{i-1,j-2})+c_i(e_{i+1}u_4^{i+1,j}-e_iu_4^{i-1,j})\\
&+a_{j+1}(-e_{i+1}u_4^{i+1,j+2}+e_iu_4^{i-1,j+2})]\\
&+\mu[a_{i-1}(-e_{j+1}u_5^{i-2,j+1}+e_ju_5^{i-2,j-1})+a_i(e_{j+1}u_5^{i,j+1}-e_ju_5^{i,j-1})\\
&+a_{i+1}(-e_{j+1}u_5^{i+2,j+1}+e_ju_5^{i+2,j-1})]\\
&=g_3^{ij}, \qquad (12.7)
\end{aligned}
$$

$$
\begin{aligned}
L_{31}^{ij}[\overset{1}{u}_+,u_3] &= \frac{h^2}{2}(L_{11}^{ij}[\overset{1}{u}_+])-\mu I_1^{ij}[u_3]-\mu I^{ij}[u_4]\\
&= \frac{h^2}{2}L_{11}^{ij}[\overset{1}{u}_+]-\mu I_1^{ij}[u_3]-\mu(c_ic_ju_4^{ij}-c_ia_{j-1}u_4^{i-2,j}-c_ja_{i-1}u_4^{i,j-2}\\
&\quad -c_ia_{j+1}u_4^{i+2,j}-c_ja_{i+1}u_4^{i,j+2}+\sum_{\alpha=\beta=1}a_{i\pm\alpha}a_{j\pm\beta}u_4^{i\pm2\alpha,j\pm2\beta})\\
&= g_4^{ij}, \qquad (12.8)
\end{aligned}
$$

$$
L_{32}^{ij}[\overset{1}{u}_+,u_3] = \frac{h^2}{2}L_{12}^{ij}[\overset{1}{u}_+]-\mu I_2^{ij}[u_3]-\mu I^{ij}[u_5]=g_5^{ij},
$$

where

$$
\begin{aligned}
a_i &= \frac{1}{(2i+1)\sqrt{(2i-1)(2i+3)}},\\
b_{ij} &= e_ie_j = \frac{1}{\sqrt{(2i-1)(2i+1)(2j-1)(2j+1)}},\\
c_i &= \frac{2}{(2i-1)(2i+3)}\,.
\end{aligned}
$$

The scheme (12.6)–(12.8) is recorded so that it can be used for constructing algebraic analogues for other, more general systems of differential equations: in it templates, appropriate to operations of differentiation of the second (direct and mixed), first and zero orders, as well as differential operators $l_2=-\mu\,\mathrm{rot}\,\mathrm{rot}+(\lambda+2\mu)\,\mathrm{grad}\,\mathrm{div}$ and $\Delta=\partial_{\alpha\alpha}$ are allocated.

Questions of convergence of the approximate solution and stability of numerical process, stipulated by scheme (12.6)–(12.8) are studied below.

We shall prove first the following lemma.

Lemma 12.1. *The process of finding u^{ij} is Ritz's process.*

Proof. For validity of lemma we shall prove, that the operator L on classes of function, vanished in a zero on ∂D, is self-adjointed and negatively definited.

Self-adjointness is obvious.

The negative definiteness follows from a following inequality:

$$
-(L_1u,u)\le-(Lu,u)\le-(L_2u,u),^1 \qquad (12.9)
$$

[1] In practice, this inequality is a Korn type of inequality for (P_h) schemes, when $\gamma\ge-0.5$ with Dirichlet conditions on ∂D.

where

$$L_1 = \{\mu\Delta, \mu\Delta, \mu(1-\varepsilon)\Delta, \quad \frac{(1-\alpha)h^2\mu}{2}\Delta - \mu(1-\frac{1}{\varepsilon}+2h^2\alpha)I,$$

$$\frac{(1-\alpha)h^2\mu}{2}\Delta - \mu(1-\frac{1}{\varepsilon}+2h^2\alpha)I\},$$

$$L_2 = \{(2\lambda^* + 3\mu)\Delta, \quad (2\lambda^* + 3\mu)\Delta, \quad 2\mu\Delta, \frac{h^2}{2}(2\lambda^* + 3\mu)\Delta - 2\mu I,$$

$$\frac{h^2}{2}(2\lambda^* + 3\mu)\Delta - 2\mu I\},$$

Here $0 < \alpha < 1$ is any number, and ε satisfies to a inequality

$$(1 + 2h^2\alpha)^{-1} < \varepsilon < 1.$$

The right part of an inequality (12.9) is obvious. To prove the left hand part one should take advantage of the $\varepsilon-$inequality

$$2ab \leq \frac{1}{\varepsilon}a^2 + \varepsilon b^2, \quad (\varepsilon > 0),$$

and then by the inequality of Friedrichs [1928].

We shall prove now the theorem on convergence of the approximate solution $\overline{u} = u_n(x,y)$ to the exact solution $u(x,y)$ of a problem (12.5), as well as we shall evaluate the error

$$\| u(x,y) - \bar{u}(x,y) \|_{C(D)} = \max_D | u(x,y) - \bar{u}(x,y) | .$$

Theorem 12.1. *a) In the energetic space H_{-L} the sequence $\bar{u} = \overset{n}{u}(x,y)$ is minimizing; b) on classes of functions, having square integrable derivatives of order $p+1$, there holds such error estimation:*

$$\max_D | u(x,y) - \overset{n}{u}(x,y) | < c\frac{\ln n}{n^p}. \tag{12.10}$$

Proof. Let the energetic space H_{-L} consists of vectors - functions, second derivatives of which are square integrable and satisfy the condition $u|_{\partial D} = 0$.

In H_{-L} we shall introduce scalar product by energy of operator L :

$$(u,v)_{-L} = \int\int_D -Luv\, d\omega.$$

For the convergence of Ritz's process, the coordinate functions should satisfy to following conditions (Mikhlin [1970]):

1. $\varphi_{ij}(x,y) \in H_{-L}$, [1]
2. For $\forall n$ $\{\varphi_{ij}(x,y)\}$ – systems φ_{ij} should be linearly independent.

[1] For simplicity of notation in $\varphi_{ij}^{1,1}$ omit the upper indexes.

3. This system must be complete in H_{-L}.

First two conditions are obvious. The third condition follows from the well known theorem (see, e.g.; de la Vallée-Poussin [1933] p. 120).

First part of theorem 12.1 is proved.

From a inequality (12.9) follows

$$\begin{aligned} \varepsilon_n &= -(L(u - \overset{n}{u}), u - \overset{n}{u}) \\ &\leq -(L_2(u - \overset{n}{u}), u - \overset{n}{u}) \leq \sum_{i=1}^{5} \varepsilon_n^i, \end{aligned} \tag{12.11}$$

where, $\varepsilon_n^i = -(2\lambda^* + 3\mu)(\Delta(u_i - \overset{n}{u}_i),\ u_i - \overset{n}{u}_i)$.

Using now results Kantorovich, Krilov [1962] (Ch. IV, 4) in case of a coordinate system (12.1), for each components $u_i - \overset{n}{u}_i$ follows

$$| u_i(x, y) - \overset{n}{u}_i(x, y) | < c\frac{\ln n}{n^p}, \qquad c > 0.$$

From the last inequality and (12.11) the validity of valuation (12.10) follows immediately.

We shall proceed to a question of stability of numerical process. The stability is understood in the following sense (see, e.g., Mikhlin [1966]): *the numerical process is stability, if there exist constants* a, b_α, c_α *independent from* n, *such, that at* $||\delta R_n|| \leq a$ *and for any* $\delta \overset{n}{g}$, *the system (12.4) has unique solution and following inequalities are fulfiled:*

$$||\overset{nij}{u} - \overset{nij}{v}|| \leq b_1||\delta R_n|| + c_1||\delta \overset{n}{g}||,$$

$$||\overset{n}{u} - \overset{n}{v}|| \leq b_2||\delta R_n|| + c_2||\delta \overset{n}{g}||,$$

where $\overset{nij}{v}$ *is a solution of a system*

$$(R_n + \delta R_n)\overset{n}{v} = \overset{n}{g} + \delta \overset{n}{g},$$

and it is represented in a following form:

$$\overset{n}{v} = \sum_{ij} \overset{nij}{v} \chi P_i(x) \chi P_j(y).$$

The following theorem is true also:

Theorem 12.2. *The system of coordinate functions* $\{\chi P_i(x)\chi P_j(y)\}$ *is strongly minimum in the energetic space* H_{-L} *and the numerical process is stability.*

Proof. Let through $\bar{\Delta}$ and Δ_n are designated accordingly finite-difference and variation-discrete in the sense (12.4) of Laplace's operators.

Then, obviously, the equality is true:

$$\Delta_n = \bar{\Delta} + E_\Delta, \qquad ||E_\Delta||_1 \leq 0.1.$$

We shall estimate now scalar product $(R_n \overset{n}{u}, \overset{n}{u})$ from below. With the help of (12.9) we have:

$$\begin{aligned}
-(R_n \overset{n}{u}, \overset{n}{u}) &\geq -\mu[(\Delta_n \overset{n}{u}_1, \overset{n}{u}_1) + (\Delta_n \overset{n}{u}_2, \overset{n}{u}_2) + (1-\varepsilon)(\Delta_n \overset{n}{u}_3, \overset{n}{u}_3) \\
&\quad + \frac{(1-\alpha)h^2}{2}((\Delta_n \overset{n}{u}_4, \overset{n}{u}_4) + (\Delta_n \overset{n}{u}_5, \overset{n}{u}_5))] \\
&\geq -\mu \min\{\frac{(1-\alpha)h^2}{2}, 1-\varepsilon\} \sum_{i=1}^{5}((\bar{\Delta} - 0.1)\overset{n}{u}_i, \overset{n}{u}_i) \\
&\geq \mu \min\{\frac{(1-\alpha)h^2}{2}, 1-\varepsilon\}(\lambda_1 - 0.1)\|\overset{n}{u}\|^2.
\end{aligned} \tag{12.12}$$

From here it follows that the system $\{\chi P_i(x)\chi P_j(x)\}$ is strong minimal in H_{-L} as $h \neq 0$ and $\lambda_1 > 4$. From this immediately follows the stability of the numerical process (see Mikhlin [1966]).

Now we shall consider ways of the solution of a variation-difference scheme and problems of realization of the numerical process on computers.

The variation-difference scheme (12.6)–(12.8) split up into independent subsystems. To each subsystem it is possible to apply, for example, Saidel's method. The convergence is provided from (12.9) and symmetry L (see, e.g., Fadeev, Fadeeva [1960]).

We shall give iterative process of the solution of a system (12.6)–(12.8) (the index in square brackets designates the number of iteration). We shall consider only the last three equations of this system. We have:

$$\mu\Delta^{ij}[u_3^{[k]}] = g_3^{ij} - \mu I_1^{ij}[u_4^{[k-1]}] - \mu I_2^{ij}[u_5^{[k-1]}], \tag{12.13a}$$

$$L_{11}^{ij}[u_4^{[k]}] = g_4^{ij} + 2h^{-2}\mu(I^{ij}[u_4^{[k-1]}] + I_1^{ij}[u_3^{[k-1]}], \tag{12.13b}$$

$$L_{12}[u_5^{[k]}] = g_5^{ij} + 2h^{-2}\mu(I^{ij}[u_5^{[k-1]}] + I_2^{ij}[u_3^{[k-1]}]), \tag{12.13c}$$

$$u_i^{[k]} = u_i^{[k-1]} - \xi_k v_i^{[k]}, \quad (i = 3, 4, 5).$$

The parameter ξ_k is selected so as to provide convergence of processes (12.13). From negative definiteness and symmetry of the operator, appropriate to equations (12.13), it is possible always (see, e.g., Godunov, Prokopov [1968]). Evidently, the solution of systems (12.6) and (12.8) presents the same type of problem.

The following statement to the solvability of a system is also true.

Theorem 12.3. *For the system (12.6)–(12.8) the alternating direction method converges.*

Proof. The validity of this result follows if we use the method of Konovalov [1964] for systems with band matrixes.

When realizing the abovestated process on computer it is important to calculate the right parts and coordinate functions .

For simplicity let $f(x, y)$ be written as a polynomial:

$$f(x, y) = \sum_{mn} a_{mn} x^m y^n,$$

then it is evident, that

$$g^{ij} = \sum_{mn} a_{mn}(J[m, i+1] - J[m, i-1])(J[n, j+1] - J[n, j-1]),$$

where

$$J[m, i] = \int_{-1}^{+1} x^m P_i(x)dx = \frac{i+1}{2i+1} J[m-1, i+1] + \frac{i}{2i+1} J[m-1, i-1].$$

By virtue of the values $J[i, j] = 0 (i < j)$ the process of calculation of other values of $J[i, j]$ are stable with respect to initial data. In the case where $f \in L_1(D)$ is a function with fixed singularities, the calculation is reduced (if we use well known results of representation by Bernstein's polynomials) to calculate values of $f^*(\pm m/M, \pm n/N)$, where f^* is regular (continuous) part of f. The coordinate functions, determined by means of (12.2), satisfy three-point recurrence relations:

$$\chi^\alpha P_{i+1}(x) = \frac{1}{i+\alpha}(\sqrt{(2i+1(2i+3} x\chi^\alpha P_i(x)$$
$$-(i-2)\sqrt{\frac{2i+3}{2i-1}}\chi^\alpha P_{i-1}(x)),$$

which is proved, for example, by an induction method.

If L is a biharmonic differential operator and we have a problem (ν is an external normal):

$$Lu \equiv \Delta^2 u = f,$$
$$u\Big|_{\partial D} = \frac{\partial u}{\partial \nu}\Big|_{\partial D} = 0, \tag{12.14}$$

then in (12.1) $\alpha = \beta = 2$ and a solution of the problem (12.14) is found in the following way:

$$u = \sum_{i,j=0}^{n} u^{ij}\varphi_{ij}^{22},$$

where $\chi^2 P_i(x)$, $\chi^2 P_j(y)$ are second order divided deferences from Legender's polynomials:

$$\chi^2 P_i = \sqrt{\frac{2i+1}{2}}\left(\frac{1}{(2i+1)(2i+3)}P_{i+2} - \frac{2}{(2i-1)(2i+3)}P_i\right.$$
$$\left.+\frac{1}{(2i-1)(2i+1)}P_{i-2}\right).$$

After using the disctere-variation method a corresponding scheme has a form:

$$a_{k-4,l}u^{k-4,l} + b_{k-2,l-2}u^{k-2,l-2} + c_{k-2,l}u^{k-2,l} + d_{k-2,l+2}u^{k-2,l+2}$$
$$+e_{k,l-4}u^{k,l-4} + h_{k,l-2}u^{k,l-2} + p_{k,l}u^{k,l} + q_{k,l+2}u^{k,l+2}$$
$$+r_{k,l+4}u^{k+2,l} + s_{k+2,l-2}u^{k+2,l-2} + v_{k+2,l}u^{k+2,j}$$
$$+w_{k+2,l+2}u^{k+2,l+2} + z_{k+4,l}u^{k+4,l} = g_{kl}, \tag{12.15}$$

where

$$a_{k-4,l} = J^{31}_{k-4,l}, \quad b_{k-2,l-2} = J^{21}_{k-2,l-2}, \quad c_{k-2,l} = (J^{22}_{k-2,l} + J^{32}_{k-2,l}),$$

$$d_{k-2,l+2} = J^{23}_{k-2,l+2}, \quad e_{k,l-4} = J^{11}_{k,l-4}, \quad h_{k,l-2} = (J^{12}_{k,l-2} + J^{24}_{k,l-2}),$$

$$p_{kl} = (J^{13}_{kl} + J^{25}_{kl} + J^{33}_{kl}), \quad q_{k,l+2} = (J^{14}_{k,l+2} + J^{26}_{k,l+2}), \quad r_{k,l+4} = J^{15}_{k,l+4},$$

$$s_{k+2,l-2} = J^{27}_{k+2,l-2}, \quad v_{k+2,l} = (J^{28}_{k+2,l} + J^{34}_{k+2,l}),$$

$$w_{k+2,l+2} = J^{29}_{k+2,l+2}, \quad z_{k+4,l} = J^{35}_{k+4,l},$$

$$J^{11}_{k,l-4} = \frac{1}{\sqrt{(2l+1)(2l-7)}(2l-1)(2l-5)(2l-3)},$$

$$J^{12}_{k,l-2} = -\frac{2}{\sqrt{(2l-3)(2l+1)}(2l-1)^2(2l+3)} - \frac{2}{\sqrt{(2l-3)(2l+1)}(2l-1)^2(2l-5)},$$

$$J^{13}_{k,l} = \frac{1}{(2l+5)(2l+1)(2l+3)^2} + \frac{4}{(2l-1)^2(2l+3)^2} + \frac{1}{(2l-3)(2l+1)(2l-1)^2},$$

$$J^{14}_{k,l+2} = -\frac{2}{\sqrt{(2l+1)(2l+5)}(2l+3)^2(2l+7)} - \frac{2}{\sqrt{(2l+1)(2l+5)}(2l-1)(2l+3)^2},$$

$$J^{15}_{k,l+4} = \frac{1}{\sqrt{(2l+1)(2l+9)}(2l+3)(2l+7)(2l+5)},$$

$$J^{21}_{k-2,l-2} = \frac{2}{\sqrt{(2k+1)(2k-3)}(2k-1)\sqrt{(2l+1)(2l-3)}(2l-1)},$$

$$J^{22}_{k-2,l} = -\frac{4}{\sqrt{(2k+1)(2k-3)}(2k-1)(2l+3)(2l-1)},$$

$$J^{23}_{k-2,l+2} = \frac{2}{\sqrt{(2k+1)(2k-3)}(2k-1)\sqrt{(2l+1)(2l+5)}(2l+3)},$$

$$J^{24}_{k,l-2} = -\frac{4}{(2k+3)(2k-1)\sqrt{(2l+1)(2l-3)}(2l-1)},$$

$$J^{25}_{k,l} = \frac{8}{(2k+3)(2k-1)(2l+3)(2l-1)},$$

$$J^{26}_{k,l+2} = -\frac{4}{(2k+3)(2k-1)\sqrt{(2l+1)(2l+5)}(2l+3)},$$

$$J^{27}_{k+2,l-2} = \frac{2}{\sqrt{(2k+1)(2k+5)}(2k+3)\sqrt{(2l+1)(2l-3)}(2l-1)},$$

$$J^{28}_{k+2,l} = -\frac{4}{\sqrt{(2k+1)(2k+5)}(2k+3)(2l+3)(2l-1)},$$

$$J^{29}_{k+2,l+2} = \frac{2}{\sqrt{(2k+1)(2k+5)}(2k+3)\sqrt{(2l+1)(2l+5)}(2l+3)},$$

$$J^{31}_{k-4,l} = \frac{1}{\sqrt{(2k+1)(2k-7)}(2k-1)(2k-5)(2k-3)},$$

$$J^{32}_{k-2,l} = -\frac{2}{\sqrt{(2k-3)(2k+1)}(2k-1)^2(2k+3)} - \frac{2}{\sqrt{(2k-3)(2k+1)}(2k-1)^2(2k-5)},$$

$$J^{33}_{k,l} = \frac{1}{(2k+5)(2k+1)(2k+3)^2} + \frac{4}{(2k-1)^2(2k+3)^3} + \frac{1}{(2k-3)(2k+1)(2k-1)^2},$$

$$J^{34}_{k+2,l} = -\frac{2}{\sqrt{(2k+1)(2k+5)}(2k+3)^2(2k+7)} - \frac{2}{\sqrt{(2k+1)(2k+5)}(2i+3)^2(2k-1)},$$

$$J^{35}_{k+4,l} = \frac{1}{\sqrt{(2k+1)(2k+9)}(2k+3)(2k+7)(2k+5)}.$$

The system (12.15) is the same as the well known 13-point difference scheme and it is divided into four subsystems, so as these systems are independent: the first is for even k and l; the second is for even k and odd l; the third is for odd k and even l; the fourth is odd k and l.

Solving these systems we obtain u^{ij}, $i, j = \overline{2, n}$.

12.2. On solving two-dimentional problems in an unbounded domain

Solving problems of mathematical physics in unbounded domains by a finite-difference method is connected to obvious difficulties. Therefore in some areas of this kind the above stated method of the solution can be applied, if the coordinate functions are chosen appropriately. In the case of the unbounded area this method together with the evident advantage over a finite-defference method (the corresponding systems of algebraic equations are not of a very high order) has one characteristic property: the infiniteness of the area results only in the presence of infinite integrals. Thus the matrix of a system of equations of a type (12.14) considered above have purely a band structure, but, as was already said, lower order, because for approximation of an unknown function number of basic vectors $n^2 << h^{-2}$ (where h is netstep) is usually required.

Schematically for an appropriate technique we consider an example of the solving of a following problem:

$$Lu(x,y) = f(x,y), (x,y) \in D, u|_{\partial D} = 0,$$
$$\lim_{x\to\infty} u(x,y) = \lim_{y\to\infty} u(x,y) = 0, \tag{12.16}$$

$$D = \{(0,\infty) \times (0,+\infty)\},\ \partial D : \{x = 0 \cup y = 0\}$$

Let L be some elliptical differential operator of second order with constant coefficients

$$L = A\partial_{xx} + B\partial_{xy} + C\partial_{yy} + a\partial_x + b\partial_y + cI.$$

An unknown function we shall present approximately in the following form:

$$\overset{n}{u}(x,y) = \sum_{i=0}^{n}\sum_{j=0}^{n} u^{ij}\varphi_{ij}^{\alpha,\beta}(x,y)$$

$$\varphi_{ij}^{\alpha,\beta}(x,y) = e^{-\alpha x - \beta y}\chi L_i(x)\chi L_j(y), \qquad \alpha > 0, \beta > 0, \tag{12.17}$$

$\chi L_n = L_{n+1} - L_n$ and L_n is a Laguerre polynomial of the order n $(0 < \alpha, \beta < 1)$.

We shall define on a set of two variable functions, determined in D, a scalar product and norm as follows:

$$(u,v) = \int\int_D e^{-x-y}u(x,y)v(x,y)dxdy, \qquad \| u \|^2 = \int\int_D e^{-x-y}u^2 d\omega,$$

and we shall consider the appropriate Hilbert space H_{-L}.

Then for the functions $\varphi_{ij}^{\alpha,\beta}$ following properties are true:

a)

$$(\varphi_{i,j}^{\alpha,\beta}, \varphi_{mn}^{1-\alpha,1-\beta}) = (2\delta_{im} - \delta_{i,m-1} - \delta_{i,m+1})(2\delta_{j,n} - \delta_{j,n-1} - \delta_{j,n+1}) \qquad (12.18)$$

b) The system of functions $\varphi_{ij}^{\alpha,\beta}$ is complete and closed in the H_{-L} space (see, for example, Szegö [1959]).

Proceeding from it, we shall reduce the solving of a problem (12.16) to the solving to the following infinite system:

$$(Lu - f, \varphi_{i,j}^{1-\alpha,1-\beta}) = 0, \quad i = \overline{1,\infty}, \quad j = \overline{1,\infty}. \qquad (12.19)$$

We shall be taken by the analysis of a structure of equations (12.19):

$$\begin{aligned}(Lu, \varphi_{ij}^{1-\alpha,1-\beta}) &= A(u_{xx}, \varphi_{ij}^{1-\alpha,1-\beta}) + B(u_{xy}, \varphi_{ij}^{1-\alpha,1-\beta}) \\ &+ C(u_{yy}, \varphi_{ij}^{1-\alpha,1-\beta}) + a(u_x, \varphi_{ij}^{1-\alpha,1-\beta}) \\ &+ b(u_y, \varphi_{ij}^{1-\alpha,1-\beta}) + c(u, \varphi_{ij}^{1-\alpha,1-\beta}).\end{aligned}$$

We shall disassemble in detail calculation of the first summand:

$$\begin{aligned}S_1 &= (u_{xx}, \varphi_{ij}^{1-\alpha,1-\beta}) \\ &= \sum_{s=o}^{m}\sum_{t=0}^{n} u^{st}\int_0^\infty e^{-x}(\alpha^2 L_{s+1}(x) + \alpha(2-\alpha)L_s(x) \\ &-L_s'(x))\chi L_i(x)dx \int_0^\infty e^{-y}\chi L_t(y)\chi L_j(y)dy.\end{aligned}$$

By virtue of (12.18) and the well known expression (Szegö [1959]):

$$L_{k+1}'(x) - L_k'(x) = -L_k(x),$$

we have:

$$\begin{aligned}S_1 &= -4\alpha(1-\alpha)u_{ij} - 2\alpha^2 u_{i-1,j} - 2(1-\alpha)^2 u_{i+1,j} + 2\alpha(1-\alpha)u_{i,j-1} \\ &+\alpha^2 u_{i-1,j-1} + (1-\alpha)^2 u_{i+1,j-1} + 2\alpha(1-\alpha)u_{i,j+1} + \alpha^2 u_{i-1,j+1} \\ &+(1-\alpha)^2 u_{i+1,j+1}.\end{aligned}$$

Now we shall present results of calculation in a form of the matrix, doing the scheme of construction of the discrete-difference analogue for operator ∂_{xx} evidently:

i,j	$j-1$	j	$j+1$
$i-1$	α^2	$2\alpha(1-\alpha)$	$(1-\alpha)^2$
i	$-2\alpha^2$	$-4\alpha(1-\alpha)$	$-2(1-\alpha)^2$
$i+1$	α^2	$2\alpha(1-\alpha)$	$(1-\alpha)^2$

We shall name this table the template for the operation ∂_{xx}. The similar calculations for the other operations give the following results:

i	$j-1$	j	$j+1$
$i-1$	$\alpha\beta$	$(1-2\alpha)\beta$	$-(1-\alpha)\beta$
i	$\alpha(1-2\beta)$	$(1-2\alpha)(1-2\beta)$	$-(1-\alpha)(1-2\beta)$
$i+1$	$-\alpha(1-\beta)$	$-(1-2\alpha)(1-\beta)$	$(1-\alpha)(1-\beta)$

∂_{xy}

i	$j-1$	j	$j+1$
$i-1$	β^2	$-2\beta^2$	β^2
i	$2\beta(1-\beta)$	$-4\beta(1-\beta)$	$2\beta(1-\beta)$
$i+1$	$(1-\beta)^2$	$-2(1-\beta)^2$	$(1-\beta)^2$

∂_{yy}

i,j	$j-1$	j	$j+1$
$i-1$	α	$1-2\alpha$	$-(1-\alpha)$
i	-2α	$-2(1-2\alpha)$	$2(1-\alpha)$
$i+1$	α	$1-2\alpha$	$-(1-\alpha)$

∂_x

i,j	$j-1$	j	$j+1$
$i-1$	β	-2β	β
i	$1-2\beta$	$-2(1-2\beta)$	$1-2\beta$
$i+1$	$-(1-\beta)$	$2(1-\beta)$	$-(1-\beta)$

∂_y

i,j	$j-1$	j	$j+1$
$i-1$	1	-2	1
i	-2	4	-2
$i+1$	1	-2	1

I

It is obvious that the system of equations (12.19) has the form of the well investigated 9-point difference schemes. The matrices of such systems has the band structure, therefore it can successfully be solved, for example, by Saidel's method especially if taking into account in comparison low order of such systems.

The form of a system of equations essentially depends on parameters α and β. They are chosen in view of smoothness of functions u.

All templates are considerably simplified at $\alpha = \beta = 1/2$. Especially simple and elegance is the template for the Laplase's operator:

	$j-1$	j	$j+1$
$i-1$	0.5	0	0.5
i	0	-2	0
$i+1$	0.5	0	0.5

$$\Delta = \partial_{11} + \partial_{22}$$

Now, we shall make some remarks 12.1.–12.8., generalizing the previously results in subsections 12.1., 12.2.

Remark 12.1. The technique is especially expedient for investigation of a system of differential equations, for example, from Chapter I.

Remark 12.2. The area D can be a infinite band, half-plane, plane and etc.. In the latter cases it should as basic functions choose $\exp(-\alpha x^2 - \beta y^2)H_i(x)H_j(y)$, where $H_i(x)$ are Hermite's polynomials.

Remark 12.3. The method of this subsection can be represented in the case of equations (systems of equations) of higher orders than second.

For example, the template for the biharmonic operator in case $\alpha = \beta = 0.5$ looks as follows:

i,j	$j-2$	$j-1$	j	$j+1$	$j+2$
$i-2$	0.25	−2	0.5	0	0.25
$i-1$	0	−2	0	−2	0
i	0.5	0	5	0	0.5
$i+1$	0	−2	0	−2	0
$i+2$	0.25	0	0.5	0	0.25

$$\Delta^2 = (\partial_{11} + \partial_{22})^2$$

We now apply to problems of numerical realization of the method of this section. The following conclusions hold (which we shall formulate as Remarks):

Remark 12.4. The coefficients of differential equations can be variable, for example, in the case of linear functions, for construction of appropriate templates and equations it is necessary also to use such expressions:

$$xL_n(x) = (2n+1)L_n(x) - (n+1)L_{n+1}(x) - nL_{n-1}(x)$$

Obviously, this structure of corresponding equations is worsened. A template for the operator $x^i y^i L(\partial_1, \partial_2)$, where L is a linear differential operator with constant coefficients not higher then the second order, will have a form of the templates by the size $(3+2i) \times (3+2j)$ $(i,j = 0,1,...)$.

We shall consider, for example, the following BVP:

$$\begin{gathered} Lu = [\partial_x(\alpha(x)\beta(y))\partial_x + \partial_y(\alpha(x)\beta(y))\partial_y)]u = f, \quad D: |x| < 1, \ |y| < 1 \\ u|_{\partial D} = 0. \end{gathered} \tag{12.20}$$

Let us profit from a technique of subsection 12.1. Then the application instead of φ_{ij}, determined by equality (12.2), generalized orthogonal functions weighted by $\alpha(x)$ and $\beta(y)$, results in the variation-discrete scheme of a type (12.4). In the case where

$$\alpha(x) = (1-x^2)^{\lambda-1/2}, \beta(y) = (1-y^2)^{\mu-1/2}, \lambda, \mu > -1/2,$$

for the solution of a problem (12.20) we shall find in a form of the Fourier–Jacobi series:

$$u(x,y) = \sum_{ij} u_{ij} \chi q_i^{(\lambda)}(x) \chi r_j^{(\mu)}(y),$$

where

$$\chi q_i^{(\lambda)}(x) = [i] \int_{-1}^{x} q_i^{(\lambda)}(x) dx, \quad \chi r_j^{\mu)}(y) = [j] \int_{-1}^{y} r_j^{(\mu)}(y) dy,$$

q and r are ultra-spherical polynomials, orthogonal with weights $\alpha(x)$ and $\beta(y)$ accordingly and satisfying to conditions:

$$q_i^{(\lambda)}(\pm 1) = r_i^{(\mu)}(\pm 1) = (\pm 1)^i.$$

$[i]_\alpha^{-2}$, depending on q and r, is chosen so:

$$[i]_1^{-2} = \int_{-1}^{1} \alpha(x)[q_i^{(\lambda)}]^2 dx, \quad [i]_2^{-2} = \int_{-1}^{1} \beta(y)[r_i^{(\mu)}(y)]^2 dy.$$

Using known expressions Szegö [1959], the following equalities hold:

$$\begin{aligned}\chi q_i^{(\lambda)}(x) &= \frac{[i]_1}{2(i+\lambda)} \left[\frac{i+2\lambda}{i+1} q_{i+1}^{(\lambda)}(x) \right. \\ &\quad \left. - \frac{i}{i+2\lambda-1} q_{i-1}^{(\lambda)}(x) + \frac{2(i+\lambda)(2\lambda-1)}{(i+1)(i+2\lambda-1)} \right] \\ &= A_i^{(\lambda)} q_{i+1}^{(\lambda)}(x) - B_i^{(\lambda)} q_{i-1}(x) + C_i^{(\lambda)},\end{aligned} \tag{12.21}$$

and, similarly,

$$\chi r_i^{(\mu)}(y) = A_i^{(\mu)} r_{i+1}^{(\mu)}(y) - B_i^{(\mu)} r_{i-1}^{(\mu)}(y) + C_i^{(\mu)} \tag{12.22}$$

Here, e.g.,

$$\begin{aligned}A_i^{(\lambda)} &= \frac{[i]_2(i+2\lambda)}{2(i+1)(i+\lambda)}, \\ B_i^{(\lambda)} &= \frac{[i]_2 i}{2(i+2\lambda-1)(i+\lambda)}, \\ C_i^{(\lambda)} &= \frac{[i]_2(2\lambda-1)}{2(i+\lambda)(i+2\lambda-1)}.\end{aligned}$$

Then the problem (12.20), according to the variational principle of Dirichlet, is equivalent to a following system (lowering in the coefficients A, B and C the indeces μ, λ):

$$\begin{aligned}&-\left(A_m^2[m+1]_1^{-2} + B_m^2[m-1]_1^{-2} + C_m^2\gamma + A_n^2[n+1]_2^{-2} + B_n^2[n-1]_2^{-2} + C_n^2\delta\right)u_{mn} \\ &\quad + \left(A_{m-2}B_m[m-1]_1^{-2} - \gamma C_m C_{m-2}\right)u_{m-2,n}\end{aligned}$$

$$
\begin{aligned}
&+\left(A_m B_{m+2}[m+1]_1^{-2} - \gamma C_m C_{m+2}\right) u_{m+2,n} \\
&+\left(A_{n-2} B_n [n-1]_2^{-2} - \delta C_n C_{n-2}\right) u_{m,n-2} \\
&+\left(A_n B_{n+2}[n+1]_2^{-2} - \delta C_n C_{n+2}\right) u_{m,n+2} \\
&-\gamma C_m{}' \sum_i C_i u_{in} - \delta C_n{}' \sum_j C_j u_{mj} \\
= g^{mn}& \\
= \int_D& f(x,y) \chi q_m^{(\lambda)}(x) \chi r_n^{(\mu)}(y) d\omega. \qquad (12.23)
\end{aligned}
$$

Stroke means, that in sums the indexes do not accept significances, equal $m \pm 2$, $m, n \pm 2, n$ and

$$\gamma = \int_{-1}^{1} \alpha(x) dx, \qquad \delta = \int_{-1}^{1} \beta(y) dy.$$

We shall recall, that to the index m there correspond coefficients dependent upon μ, and an index n upon λ. Besides for $m = 0, 1$ and $n = 0, 1$ it should take into account the remark of subsection 12.1 (see, e.g., (12.4)).

Remark 12.5. In the case of the differential operators considered in sections 7, 8 (with appropriate boundary conditions) do not present difficulties in the study of questions of an error estimation, convergence of corresponding processes by virtue of a possibility of application of variational methods in the unbounded areas (Mikhlin [1970], Toupin [1965]).

Remark 12.6. In the case where the area $D = (0, \infty) \times (0, \infty)$ the question of strong minimum, appropriate coordinate system of functions, is reduced to the study of eigenvalues of the well known finite-difference systems of equations, appropriate to the to differential operators in the recthangular areas.

Remark 12.7. A similar situation is observed when, instead of first quadrant, the area D is band, half-plane, plane etc..

Remark 12.8. As was remarked, the questions of numerical realization are connected, except questions of stability, to a problem of calculation of basic functions and right parts of scalar products of appropriate systems.

The calculation of basic functions connected with the orthogonal polynomials can be calculated in two ways: by three-point recurrence expressions, or by decompositions, with a preliminary finding of the zeros of these polynomials. As numerical experiments have shown, calculation of values of the polynomials by decompositions more preferable especially for the in comparison large orders, than use recurrence expressions. With this purpose the standard program of calculation of zero of orthogonal polynomials of Jacobi, Laguerre, Hermite, giving reliable results in a range of orders $n \approx 300$ are elaborated (Vashakmadze [1982]).

12.3. Discrete-difference schemes of approximate solving

two-dimensional boundary value problems

A special class of algebraic analogues of differential problems, based on simultaneous approximation of an unknown vector in multiple series by polynomial and spline polynomial systems is constructed below.

Thus, the design of a specified class is based on a projection operation of an initial operator equation in various directions by a different rule. At first the problem is discretised respect to unknown moments of the wanted solution, then the full algebrisation is obtained by means of the variational-difference — in Courant's sence — approximation of the same moments which for rather smooth solutions coincides with the finite-difference approximation schemes.

The advantage of such schemes with respect to purely Ritz — or finite-difference ones in a number of cases is quite obvious. For example, if the area is cylindrical and the coefficients of the differetiated problem depend only on some of the independent variables, such schemes for partially curvilinear bounds do not need approximation of the boundary conditions.

They are made up by template, as is observed in a finite-difference method.

The matrix corresponding to the discrete-different scheme has a block type structure, enabling us to apply to its inversion special methods of linear algebra. Usually the number of equations in such schemes is essentially less, than in difference systems.

The discrete-difference schemes by virtue of their nonhomogeneity can be investigated also in classes of functions, which have different orders of smoothing on different coordinate axes.

We shall consider below some examples of the discrete-difference schemes for two-dimensional BVP:

$$L(\partial_1, \partial_2)u(x, y) = f(x, y), \qquad (x, y) \in D, \tag{12.24}$$

$$l(\partial_1, \partial_2)u(x, y) = \varphi(x, y), \qquad (x, y) \in \partial D. \tag{12.25}$$

The operator L is everywhere below taken or operator of the plane theory of elasticity l_2 with Lamé coefficients, which depends on one variable, or Laplace's operator. The boundary operator l is given in a classical sense.

In the case where $D = (-1, 1; -h(x), h(x))$, the discrete-difference scheme (appropriate to equation (12.24) and boundary conditions at $x = \pm 1$) can be made out according to subsection 12.1. and then, depending on differential properties of coefficients, to take advantage or a variational-difference, or finite-difference approximation.

We shall note that in case of the operator l_2 each scalar equation will contain not more than 13-point unknowns, and in case of the Laplace's operator not more than 9.

Let $D = (-1, 1; -1, 1)$ and

$$u(x, \pm 1) = 0. \tag{12.26}$$

Then the discrete-difference scheme is essentially simplified and accepts the following form:

$$-2(\lambda^i+\mu^i)B_{i,j}\overline{u}_2+(\mu^i A_{i,j}+2(\lambda^i+2\mu^i)(2j+1)I)\overline{u}_1=g_1^{ij},$$
$$-2(\lambda^i+\mu^i)B_{i,j}\overline{u}_1+((\lambda^i+2\mu^i)A_{i,j}+2\mu^i(2j+1)I)\overline{u}_2=g_2^{ij}, \tag{12.27}$$

$$i=\overline{1,M+1},\quad j=\overline{1,N+1},\quad (M+1)\tau=2,$$

in the case where $l_2=-\mu(x)\,\mathrm{rot}\;\mathrm{rot}+(\lambda(x)+2\mu(x))\,\mathrm{grad}\;\mathrm{div}$, and

$$(A_{i,j}+2(2j+1)I)\overline{u}=g^{ij}, \tag{12.28}$$

in the case where $\Delta=\partial_{\alpha\alpha}$.

In systems (12.27) and (12.28) designations are entered: $\overline{u}=u-\xi$, where u is exact solution of the problem (12.24)–(12.25),

$$A_{i,j}u=-\Lambda_{\bar{x}x}(a_{s-1}\overset{j-2}{u^i}-c_s u^i+a_{s+1}\overset{j+2}{u^i}),\quad u^i=u((i-1)\tau)$$

$$B_{i,j}u=\Lambda_{\overset{\circ}{x}}(\overset{j+1}{u^i}-\overset{j-1}{u^i}),\quad Iu\equiv u,$$

$$a_j=2(2j+1)^{-1},\qquad c_j=a_{j-1}+a_{j+1}$$

τ is a step on the axis ox; $\Lambda_{\overset{\circ}{x}}, \Lambda_{\bar{x},x}$ are accordingly a divided difference of the first and second orders.

At $i=1, M+1$ the systems (12.27)–(12.28) give the boundary conditions

$$A_{i,j}u=\overset{j+1}{\varphi^i}-\overset{j-1}{\varphi^i},$$

$$A_{i,j}u=-\bar{l}(a_{j+1}\overset{j+2}{u^i}-c_j\overset{j}{u^i}+a_{j-1}\overset{j-2}{u^i}), \tag{12.29}$$

where $\bar{l}$ is the approximate boundary operator .

Systems (12.27) and (12.28), as are easily seen, are split into two independent subsystems. The finding of the solution of these systems can be conducted, for example, by applying the operation factorization scheme.

As in case of the operator l_2, as in the case of the $\partial_{\alpha\alpha}$ with mixed boundary conditions, after calculation of the right parts of equations (12.27) or (12.28) and (12.29), it is required $\approx MN$ of arithmetic operations.

Let $u(x,y)$ belongs to the set $\bigcup_1(\Omega)$ respect x and sets $\bigcup_2(\Omega)$ at y; Then a class of contiguity we shall define(determine) as a direct sum of sets $\bigcup_1(\Omega_1)_x\bigoplus\bigcup_2(\Omega_2)_y\ni u(x,y)$.

We shall introduce $\bigcup_p^{\alpha,\beta}(\Omega)$ a class of vector functions, possessing in $\bar{\Omega}$ the continuous derivatives of order $\beta-1$, and derivatives of order $\alpha+\beta$ belong to $L_p(\Omega)$.

Validity of the following theorem is obvious.

Theorem 12.4. *Let $u(x,y)\in\bigcup_2^{2,2}(-1,1)_x\bigoplus\bigcup_2^{\alpha,2}(-1,1)_y$ $(\alpha>0)$ is the unique solution of BVP (12.24)–(12.25), L is or* $-\mu\,\mathrm{rot}\;\mathrm{rot}+(\lambda+2\mu)\,\mathrm{grad}\;\mathrm{div}$, *or*

Δ; l is any boundary operator in a classical sense when $x = \pm 1$ and coterminous with the operator, determined from (12.26) at $y = \pm 1$.

Then the discrete-finite operator, appropriate to the schemes (12.27) or (12.28) with boundary conditions (12.29), is self-adjoint and positive definite. Its solution converges to the exact solution with the speed $O(\tau^2 + N^{-(1+2\alpha)/2})$.

12.4. Continuous analogue of alternating direction method

Let us consider the boundary value problem for linear strong-elliptic self-adjoint system of differential equation in a domain D with a boundary ∂D:

$$Lu = (k^{ij}\partial_{ij} - q)u(x_1, x_2) = -f(x_1, x_2), \quad (x_1, x_2) \in D \tag{12.30}$$

$$[u + \sigma(x_1, x_2)N(u)]_{\partial D} = \varphi(x_1, x_2), \tag{12.31}$$

where $k^{ij} = ||k^{ij}_{\alpha\beta}(x_1, x_2)||^n_{\alpha,\beta=1}$ is a positive definite symmetrical matrix with constant coefficients,

$$N(u) = ||N_\alpha(u)\delta_{\alpha\beta}||^n_1, \quad N_\alpha(u) = \sum_{\beta=1}^{n} k^{ij}_{\alpha\beta} u_{\beta,i} \cos(\nu, x_j),$$

ν is an external normal, $q = \{q_1(x_1, x_2), ..., q_n(x_1, x_2)\}$ is diagonal nonnegative matrix, $\sigma(x_1, x_2) > 0$ is a scalar function, $u = (u_1, u_2, ..., u_n)^T$ is an unknown, $f = (f_1, f_2, ..., f_n)^T, \varphi = (\varphi_1, ..., \varphi_n)^T$ are known vector functions.

Let us introduce the following definition:

The continuous analogue of the alternating direction of iteration scheme of boundary value problem (12.30)-(12.31) we will define two following serial of expressions (Vashakmadze [1972]):

$$\begin{aligned} B_1 u^{s+1/2} = B_{12}u^s + f &= F_{s+1/2}, \\ u^{s+1/2} + \sigma_1 N(u^{s+1/2})|_{\partial_1 D} &= \varphi_1, \end{aligned} \tag{12.32a}$$

$$\begin{aligned} B_2 u^{s+1} = rIu^{s+1/2} + (B_2 - rI)u^s &= F_{s+1}, \\ u^{s+1} + \sigma_2 N(u^{s+1})|_{\partial_2 D} &= \varphi_2(x_1), \end{aligned} \tag{12.32b}$$

where r is an iterative parameter, I is an unit operator, $\sigma = \sigma_1 + \sigma_2$,

$$B_i = ||(r + \hat{q}^i)I + \delta_{\alpha\beta}\hat{A}_i||^n_{\alpha,\beta=1}, \quad B_{12} = B_1 + I, q^1 + q^2 = q,$$

$$\hat{q}^i \geq q^i_\alpha(x_1, x_2), \qquad \hat{A}_i = -a_i\partial_{ii}, \ a_i > 0.$$

The following theorem of convergence of the iterative process (12.32) is true:

Theorem 12.5. *Let the following conditions be fulfiled:*

i) Domain D is square, $0 \leq x_1, x_2 \leq 1$, $A_i = -\partial_{ii}$, $LB = BL$,

ii) $bB \leq -L < \frac{1}{r}B_1B_2 (b > 0, 0 < r \leq 1)$, $B = (\hat{A}_1 + \hat{A}_2)||\delta_{\alpha\beta}||^n_1$,

iii) $u^0_\alpha \in C^4(0, 1; 0, 1)(\alpha = \overline{1, n})$,

Then the sequence of vector functions u^s converges to the solution $u(x, y)$.

Proof. From the formula (12.32) follows:

$$B_1B_2u^{s+1} = (B_1B_2 + rI)u^s + rf, \;\; [u^{s+1} + \sigma_\alpha N(u^{s+1})]_{\partial_\alpha D} = \varphi_\alpha,$$

where $\sigma = \sigma_1 + \sigma_2$.

From this formula as the inverse operator $(B_1B_2)^{-1}$ exists in the space of vector functions whose coordinates belongs to $\in C^4(0,1;0,1)$, it follows that

$$u^{s+1} = (B_1B_2)^{-1}[(B_1B_2 + rL)u^s + rf], \;\; [u^{s+1} + \sigma_\alpha N[u]|_{\partial_\alpha D} = \varphi_i$$

Then an error $z = u - u^s$ will satisfy the following homogeneous equality:

$$z^{s+1} = (B_1B_2)^{-1}(B_1B_2 + rL)z^s, \;\; [z^{s+1} + \sigma_\alpha N(z^{s+1})]_{\partial_\alpha D} = 0.$$

or

$$z^{s+1} = T^s z^0, \;\; [z^{s+1} + \sigma_\alpha N(z^{s+1})]_{\partial_\alpha D} = 0, \tag{12.33}$$

where

$$T^s = [(B_1B_2)^{-1}(B_1B_2 + rL)]^s.$$

By using the condition ii) there holds the formula

$$B_1B_2 = rbB + C, C > 0 \tag{12.34}$$

and the following inequality:

$$(Tz, z) > 0, \;\; [z^{s+1} + \sigma_\alpha N(z^{s+1})]_{\partial_\alpha D} = 0.$$

On the other hand, from conditions ii) the formula (12.34) and a corollary of theorem 12.4 (4.Y) Kantorovich, Akilov [1959], in the space $L_2(D)$ gives:

$$\begin{aligned}(Tz, z) &= (Iz, z) - r(-L(B_1B_2)^{-1/2}z, (B_1B_2)^{-1/2}z)\\ &\le (Iz, z) - rb(B(B_1B_2)^{-1/2}z, (B_1B_2)^{-1/2}z)\\ &= (Iz, z) - ((B_1B_2 - C)(B_1B_2)^{-1/2}z, (B_1B_2)^{-1/2}z)\\ &< \|z\|^2.\end{aligned}$$

From the last inequality it follows that $\|T\| < 1$. Because of a representation (12.33) and the condition iii) the sequence $z^{s+1} \to 0$. The theorem is proved.

Now we consider some particular cases of this theorem when $N(u) = 0$.

Corollary 12.1. *In the case of differential equations of the plane theory of elasticity, the condition ii) is true if we suppose that*

$$b = \mu, \;\; B = (A_1 + A_2)\|\delta_{\alpha\beta}\|, \;\; B_i = ((rI + (2\lambda + 3\mu)A_i)\|\delta_{\alpha\beta}\|_1^2, r \in (0,1).$$

Corollary 12.2. *In the case when L is the operator of shell theory Vekua [1968], ii) is true, when (compare with subsection 12.1, see (12.9))*

$$b = \min\left\{1 - \varepsilon, \frac{(1-\eta)h^2}{2}\right\}, \;\; B = (A_1 + A_2)\;\|\delta_{ij}\|_1^1,$$

$$B_\alpha = B_\alpha((r + 2\mu)I + (2\lambda^* + 3\mu)A_\alpha)\|\delta_{jk}\|, \;\; r \in (0,1),$$

where $0 < \eta < 1$, $(1 + 2h^2\eta)^{-1} < \varepsilon < 1, 2h$ *is the thickness of shell.*

Let us consider the following differential equations

$$Lu = \partial_1(k_1\partial_1 u) + \partial_2(k_2\partial_2 u) - qu$$
$$= f,\ k_1, k_2 > 0,\ q \geq 0.$$

Instead of the iterative process (12.32) the following scheme also may be used:

$$\begin{aligned}
&((r + q_1)I + A_1)u^{s+1/2} = (rI - A_2)u_s + f = F_{s+1/2},\\
&\left[u^{s+1/2} + \sigma_1 \frac{\partial u^{s+1/2}}{\partial \nu}\right]_{\partial_1 D} = \varphi_1\\
&((r + q_2)I + A_2)u^{s+1} = (rI - A_1)u_{s+1/2} + f = F_{s+1},\\
&\left[u^{s+1} + \sigma_2 \frac{\partial u^{s+1}}{\partial \nu}\right]_{\partial_2 D} = \varphi_2,
\end{aligned} \tag{12.35}$$

where $A_\alpha = -\partial_\alpha(k_\alpha \partial_\alpha)$.

Let $L = \partial_{\alpha\alpha}$, $A_\alpha = -\partial_{\alpha\alpha}$. In this case if the scheme (12.35) is applied, the following theorem is true.

Theorem 12.6. *Let the following conditions be fulfilled:*

i) $\|A_2 u\| < +\infty$

ii) $0 < r < \pi^2$.

Then for $\forall\varepsilon > 0\ \exists S(\varepsilon)$ *such that when* $s > S(\varepsilon)$

$$\|u - u^s\| < \varepsilon.$$

Proof. The correctness of this theorem has been proved in accordance with the scheme of proof of theorem 12.6 (Guilinger [1965]).

The difference consists in the proof of the statement which is analogous to Lemma 4.4 (Guilinger [1965]).

Thus, if we use the spectral expansion of linear operator

$$(I - T_{A_1}T_{A_2})(T_{A_1}T_{A_2})^s,\ T_{A_i} = (rI + A_i)^{-1}(rI + A_i)$$

by eigenvalies of operators A_i (Iosida [1965]), the following estimate is evident:

$$\|(I - TA_1 T_{A_2})(T_{A_1}T_{A_2})^s\| < \frac{1}{s}$$

The last inequality is an analogue of Guilinger's estimation.

We remark that the scheme of alternating directions (12.35), when $(L = \Delta)$, by the substitution

$$A = \{I + \tau A_1, I + \tau A_2\},\ B = \{A_1 + A_2, A_1 + A_2\},\ u = (u^s, u^{s+1/2})^T,$$

$$u_\tau = (\tau^{-1}(u^{s+1/2} - u^s),\ \tau^{-1}(u^{s+1} - u^{s-1/2}))^T,\ \varphi = (F_{s+1/2}, F_{s+1})^T,\ \tau = \frac{1}{r}$$

have reduced to the canonical form of Samarski [1967] for operator equations

$$Au_\tau = Bu + \varphi$$

for two layer iteration processes, when the operators A and B are energetically equivalent. In this theorem 12.6 this condition does not fulfilled.

Let us consider the problem of approximately solving BVP (12.30)–(12.31) and use the iteration process (12.32). Correspondingly, a one dimensional BVP will be solved by Tichonov, Samarski [1962] (TS), or Volkov [1971], or generalized factorization (see Vashakmadze [1972]) (GF) methods. This process defines the alternating direction method of an arbitrary order exactly.

Let $\omega_h \subset D$ is a net domain

$$\omega_h = \{0 \le x_1 < x_2 < \dots < x_{n_1} = 1,\ 0 \le y_1 < \dots < y_{n_2} = 1\}$$

$$h = \max\{x_{i+1} - x_i,\ y_{i+1} - y_i\}$$

and n_1, n_2 are arbitrary positive integers.

In the net ω_h we form the iteration process of BVP (12.32) by (TS) or (GF) methods in the following way. When $s = 0$ in the first serial of BVP (12.32_1) for $y = y_j$, we find the approximate values of $u^{1/2}(x, y_j)$. Further, assume $x = x_i$, and from BVP (12.32_2) we define approximate values of u_{ij}^1 of the vector $u^1(x_i, y_j)$ and so on.

Values $F_{s+1/2}$ and F_{s+1} for this example are calculated to be:

$$F_{s+1/2} = B_{12}u^s + F,$$

$$F_{s+1} = r\sum_{i=0}^{s}(u^{i+1/2} - u^i) + B_2 u^0.$$

To calculate $B_{12}u^s$ it is necessary to use the formulae of numerical differentiations. However, in the particular case where $B_{12} = rI - B_2$ the values $F_{s+\frac{1}{2}}$, are calculated immediately by values of the vectors $u^s, u^{s+1/2}$ and f in such way:

$$F_{s+1/2} = rIu^s - F_s + f$$

The right parts of the iterations schemes (12.35) (when $q = 0$) may be calculated in the following way:

$$F_{s+1/2} = F_{s-1/2} + 2r(u^s - u^{s-1/2}),$$
$$F_{s+1} = F_s + 2r(u^{s+1/2} - u^s).$$

13. To a Numerical Solution of One-Dimensional Boundary Value Problems

Below we consider a problem of numerically solving a boundary value problem for second order ordinary nonlinear differential equation (see Vashakmadze [1964])

$$y''(x) = f(x, y(x), y'(x)) \quad (0 < x < 1), \tag{13.1}$$

$$y(0) = \alpha, \quad y(1) = \beta. \tag{13.2}$$

Separately we consider the case where (13.1) has the form

$$y''(x) = f(x, y(x)) \quad (0 < x < 1), \tag{13.3}$$

where the function $f(x, y)$ satisfies the Lipshitch condition with respect to y with the constant $L \leq 8$. As is known from Picard [1930] this assumption is sufficient for existence and unique of the solution of the problem (13.3), (13.2).

In respect of the problem (13.1), (13.2) we assume that

$$\frac{1}{4}(L + L') < 1, \tag{13.4}$$

where

$$L = \max_x |f_y|, \quad L' = \max_x |f_{y'}| \quad (-\infty \leq y,\ y' \leq \infty).$$

As we shall see below that the inequality (13.4) is a sufficient condition for the existence of the unique solution of the problem (13.1), (13.2).

The approximate solution of the problems (13.3),(13.2) and (13.1),(13.2) is analyzed in many papers by different authors.

One can find the detailed list of literature on the numerical solution of the problem (13.3), (13.2) in the monogragh Henrici [1962]. From this monograph it follows that the convergence of the computational process has no more than the fourth order of the net step.

The differential equation (13.1) with more general boundary conditions than (13.2) has been elaborated in Schröder [1957] (see also Beresin, Zgidkov [1959], where the simple variant of Schröder's results is referenced) with the numerical analyze. From these works it follows that in the occasion of boundary conditions a convergence of a computational process has the second order of the net step. We also have to point out the monographs of Fox [1957], Mikeladze [1951] which are devoted to two-point boundary value problems . There are given several methods of the approximate solution of the problem (13.1), (13.2) without an estimation of the error and a convergence of the computational process in it.

In this section we construct computational schemes based on the new quadrature formulae called $(P) - (Q)$ formulae. With the help of these computational algorithms we obtain an approximate solution of the problem (13.1), (13.2) and a derivative of the solution. We show that these functions converge to the exact solution of the problem (13.1), (13.2) and to the derivative of this solution and if $y(x) \in C^p[0,1]$, then the convergence has $(p-2)$-order of the net step.

13.1. (P) and (Q) formulae

Let us consider the Lagrange interpolation formula

$$y(x) = \sum_{i=1}^{p} L_i(x)y(x_i) + R_{p-1}(x_i), \tag{13.5}$$

where $0 \leq x_1 < x_2 < ... < x_p \leq 1$, and

$$L_i(x) = \prod_{\substack{k=1 \\ i \neq k}}^{p} \frac{x - x_k}{x_i - x_k}, \quad R_{p-1}(x) = y(x, x_1, \ldots, x_p) \prod_{j=1}^{p} (x - x_k).$$

Differentiating twice (13.5) we obtain

$$\begin{aligned} y''(x) = & \sum_{i=1}^{p} L_i''(x)y(x_i) + y(x, x_1, \ldots, x_p)\frac{d^2}{dx_2}\prod_{j=1}^{p}(x - x_j) \\ & + 2y(x, x, x_1, \ldots, x_p)\frac{d}{dx}\prod_{j=1}^{p}(x - x_j) \\ & + 2y(x, x, x, x_1, \ldots, x_p)\prod_{j=1}^{p}(x - x_j). \end{aligned}$$

From this expression, in particular for $x = x_i$ $(i = 2, 3, \ldots, p-1)$, we shall obtain

$$y''(x_i) = \sum_{j=1}^{p} A_{ij}y(x_j) + R''_{p-1}(x_i) \quad (p \geq 3, \quad i = 2, 3, \ldots, p-1), \tag{13.6}$$

where

$$A_{ij} = L_j''(x_i) \quad (i = 1, 2, \ldots, p-1, p)$$

and for the remainder $R''_{p-1}(x_i)$ we have the estimation

$$|R''_{p-1}(x_i)| \leq \frac{c_1 M_p}{p!}h^{p-2} + \frac{c_2 M_{p+1}}{(p+1)!}h^{p-1} (i = 2, 3, \ldots, p-1). \tag{13.7}$$

$$h = \max_{1 \leq i \leq p-1}(x_{i+1} - x_i).$$

In the special case where $p = 2s+1$ and $x_{s+1} - x_i = x_{2s+2-i} - x_{s+1}$ $(i = 1, 2, \ldots, s)$, then for $R''_{p-1}(x_{s+1})$ inequality (13.7) has the form

$$|R''_{p-1}(x_{s+1})| \leq \frac{c_2 M_{p+1}}{(p+1)!}h^{p-1}. \tag{13.8}$$

M_p and M_{p+1} mean the upper boundaries of the absolute values of the function $y^{(p)}(x)$ and $y^{(p+1)}(x)$ for $0 \leq x \leq 1$ in the expressions (13.7), (13.8), c_1 and c_2 are constants depending on the distribution of the points x_i $(i = 1, 2, \ldots, p)$. We can compute them easily. To define $y(x_i)$, $(i = 2, 3, \ldots, p-1)$ from (13.6) we prove the following lemma.

Lemma 13.1. *A determinant of the system (13.6)*

$$det\{A_{ij}\}_2^{p-1} \neq 0.$$

Proof. Before we begin to prove the lemma let us remember that the Lagrange interpolation and the MacLaurin formulae are grouped differently for polynomials. But they can be easily reduced to each other. We do it below. Besides, as $\det\{A_{ij}\}_2^{p-1}$ is independent from the function $y(x)$ then we assume that $y(x)$ is the polynomials of the $(p-1)$ degree.

If we differentiate twice the MacLaurin formula and substitute $x = x_i$ then we have (assuming $x_1 = 0$)

$$y^{(k)}(0) = \sum_{j=2}^{p-1} \beta_{kj} y''(x_j) \quad (k = 2, 3, \ldots, p-1), \tag{13.9}$$

where

$$det\{\beta_{kj}\}_2^{p-1} = \frac{\prod\limits_{k=2}^{p-1}(k-2)!}{W(x_2, x_3, \ldots, x_{p-1})}$$

and W is Vandermonde's determinant. Substituting the expression (13.9) in the MacLaurin formula we obtain

$$y(x_i) - y(0) - x_i y'(0) = \sum_{k=2}^{p-1} \frac{x_i^k}{k!} \sum_{j=2}^{p-1} \beta_{kj} y''(x_j) \quad (i = 2, 3, \ldots, p-1),$$

$$y(x_p) - y(0) - x_p y'(0) = \sum_{k=2}^{p-1} \frac{x_p^k}{k!} \sum_{j=2}^{p-1} \beta_{kj} y''(x_j).$$

From the last equalities we can easily obtain

$$y(x_i) - \frac{x_p - x_i}{x_p} y(0) - \frac{x_i}{x_p} y(x_p) = \sum_{k=2}^{p-1} \frac{x_p x_i^k - x_i x_p^k}{x_p^k!} \sum_{j=2}^{p-1} \beta_{kj} y''(x_j) \tag{13.10}$$

$$(i = 2, 3, \ldots, p-1).$$

If the determinant of the system (13.10) is not equal to zero then the expression for $y''(x_i)$, obtained from (13.10), is equivalent to expression (13.6) without the remainder term. But we can represent the determinant (13.10) in the form

$$\Delta = \frac{1}{(p-1)!(p-2)!W[x_2, x_3, \ldots, x_{p-1}]}$$
$$\times \left\{ x_p^{p-2} \begin{vmatrix} x_2^2 & \ldots & x_2^{p-1} \\ \ldots & \ldots & \ldots \\ x_{p-1}^2 & \ldots & x_{p-1}^{p-1} \end{vmatrix} - x_p^{p-1} \begin{vmatrix} x_2 & x_2^3 & \ldots & x_2^{p-1} \\ \ldots & \ldots & \ldots & \ldots \\ x_{p-1} & x_{p-1}^3 & \ldots & x_{p-1}^{p-1} \end{vmatrix} \right.$$

$$+\cdots+(-1)^{p-2}x_p^{2p-4}\begin{vmatrix} x_2 & \dots & x_2^{p-2} \\ \dots & \dots & \dots \\ x_{p-1} & \dots & x_{p-1}^{p-2} \end{vmatrix}\Bigg\}$$

$$= \frac{(-1)^{p-2}W[x_2, x_3, \dots, x_p]\prod_{i=2}^{p-1} x_i}{(p-1)!(p-2)!x_p^{p-2}W[x_2, x_3, \dots, x_p]}$$

$$= \frac{(-1)^{p-2}\prod_{i=2}^{p-1} x_i \prod_{i=2}^{p-1} x_p - x_i}{(p-1)!(p-2)!} \neq 0,$$

from where

$$det\{A_{ij}\}_2^{p-1} = \frac{1}{\Delta},$$

and thus, the lemma is proved.

From this lemma and the equality (13.10) the equation (13.6) will give

$$y(x_i) = \frac{x_p - x_i}{x_p}y(0) + \frac{x_i}{x_p}y(x_p) + \sum_{j=2}^{p-1} b_{ij}(y''(x_j) - R''_{p-1}(x_j)), \tag{13.11}$$

$$(i = 2, 3, \dots, p-1).$$

As the numbers b_{ij} are independent of the form $y(x)$, and if we assume $y(x) = x^2$, then from the formula (13.11) we obtain

$$\sum_{j=2}^{p-1} b_{ij} = \frac{x_i(x_i - x_p)}{2} < 0, \quad (i = 2, 3, \dots, p-1). \tag{13.12}$$

If $x_i = x_p - x_{p+1-i}$, $i = 1, 2, \dots, [(p+1)/2]$ from the symmetrical formulae about the points x_i and x_p we conclude that

$$b_{ij} = b_{p+1-i,p+1-j} \quad \left(i, j = 2, 3, \dots, \left[\frac{(p+1)}{2}\right]\right). \tag{13.13}$$

We can prove an estimation

$$\left|\sum_{j=2}^{2s} b_{s+1,j}R''_{p-1}(x_j)\right| \leq \frac{c_3 M_{p+1}}{(p+1)!}h^{p+1}, \tag{13.14}$$

where c_3 is a constant, which we compute easily. It is depends on p and the distribution of points x_i $i = 1, 2, \dots, 2s+1$. By using the Taylor formula, equalities (13.12), (13.13), when $p = 2s + 1$ and the symmetries of the points x_{s+1}, for the remainder term of the formula (13.11) when $i = s + 1$, the inequality (13.14) follows.

The formulae (13.11) we call (P) formulae. They are the new quadrature formulae, though for $i = s + 1$, $s = (p-1)/2$, $x_j = (j-1)h$ the formula (13.11) coincides with the formula (10) from Mikeladze [1940] for $n = 2$, $k = 0$.

Note, that when we compute the coefficients b_{ij} we have not to solve the foregoing algebraic system.

If we substitute $y''(x)$ by the interpolate formula

$$y''(x) = \sum_{j=2}^{p-1} l_j(x) y''(x_i) + r_{p-3}. \tag{13.15}$$

in the identity

$$y(x) = \frac{x_p - x}{x_p} y(0) + \frac{x}{x_p} y(x_p) + \frac{1}{x_p}\left(x_p \int_0^x dx \int_0^x y''(x)dx - x \int_0^{x_p} dx \int_0^x y''(x)dx \right),$$

where

$$l_j(x) = \prod_{\substack{i=2 \\ j\neq i}}^{p-1} \frac{x - x_i}{x_j - x_i},$$

$$r_{p-3}(x) = y''(x, x_2, \dots, x_{p-1}) \prod_{i=2}^{p-1} (x - x_i),$$

from the unique of (P) - formulae it follows that

$$b_{ij} = \frac{1}{x_p}\left(x_p \int_0^{x_i} dx \int_0^x l_j(t)dt - x_i \int_0^{x_p} dx \int_0^x l_j(t)dt \right),$$

$$-\sum_{j=2}^{2s} b_{ij} R''(x_j) = \frac{1}{x_p}\left(x_p \int_0^{x_i} dx \int_0^x r_{p-3}(t)dt \right. \tag{13.16}$$

$$\left. - x_i \int_0^{x_p} dx \int_0^x r_{p-3}(t)dt \right), \quad (i = 2, 3, \dots, p-1).$$

By analogy to (P) formulae we can obtain representations for $y'(x_i)$ $(i = 1, 2, \dots, p)$. From the identity

$$y'(x) = \frac{1}{x_p}[y(x_p) - y(0)] - \frac{1}{x_p} \int_0^{x_p} dt \int_x^t y''(t)dt,$$

using (13.15), we have

$$y'(x_i) = \frac{1}{x_p}[y(x_p) - y(0)] - \frac{1}{x_p} \sum_{j=2}^{p-1} c_{ij} y''(x_j) - \frac{1}{x_p} \int_0^{x_p} dx \int_{x_i}^x r_{p-3}(t)dt, \tag{13.17}$$

$$(i = 1, 2, \dots, p),$$

where

$$c_{ij} = \int\limits_{0}^{x_p} dx \int\limits_{x_i}^{x} l_j(t)dt, \quad (j = 2, 3, \ldots, p-1). \tag{13.18}$$

We call the expressions (13.17) as (Q) formulae. If we assume $y(x) = x^2$ in (13.17) then we obtain

$$\sum_{j=1}^{p} c_{ij} = -\frac{x_i^2 - (x_p - x_i)^2}{2}. \tag{13.19}$$

For $x_i = x_p - x_{p+1-i}$, $i = 1, 2, \ldots, [(p+1)/2]$, the coefficients c_{ij} satisfy the conditions

$$c_{ij} = -c_{p+1-i,p+1-j} \quad \left(j = 2, 3, \ldots, \left[\frac{p+1}{2}\right]\right). \tag{13.20}$$

The equalities (13.13) and (13.20) follow from (13.16) and (13.18) immediately. Thus, we can reduce by half the time of spending on the computing of the coefficients b_{ij} and c_{ij}.

13.2. A solution of the boundary value problem (13.3), (13.2)

We construct a nonlinear algorithm for numerically solving the problem (13.3), (13.2). The main part of it is an expression with nonpositive coefficients. For this purpose we use (P) formulae.

We can easily see that for $x_i = (i-1)h$, $i = 1, 2, \ldots, 2k+1$, $p = 2s+1$, $h = 1/2ks$ (P) formulae can be rewritten as:

$$y_{(t-1)s+i} = \frac{2s-(i-1)}{2s} y_{(t-1)s+1} + \frac{i-1}{2s} y_{(t+1)s+1} + \sum_{j=2}^{2s} b_{ij}\Phi_{(t-1)s+j} \tag{13.21}$$

(for $t = 1, 2, \ldots, 2k-1$ and $i = 2, 3, \ldots, s+1$; for $t = 2k-1$ and $i = 2, 3, \ldots, s$), where

$$y(x_{ts+i}) = y_{ts+i}, \quad y''(x_{ts+j}) - R''_{p-1}(x_{ts+j}) = \Phi_{ts+j}.$$

For $i = s+1$, we have:

$$y_{ts+1} = \frac{1}{2} y_{(t-1)s+1} + \frac{1}{2} y_{(t+1)s+1} + \sum_{j=2}^{2s} b_{s+1,j}\Phi_{(t-1)s+j}, \tag{13.22}$$
$$(t = 1, 2, \ldots, 2k-1).$$

Let us rewrite the subsystem (13.22) to the other form. To do it we multiply every t-th and $(2k-t)$th equations from (13.22) by t/k. Addition of the equalities which we have obtained this way, gives:

$$y_{ks+1} = \frac{1}{2} y(0) + \frac{1}{2} y(1) + \sigma_{ks+1}, \tag{13.23}$$

where

$$\sigma_{ks+1} = \sum_{t=1}^{k-1} t \sum_{j=2}^{2s} b_{s+1,j}\Phi_{(t-1)s+j} + k\sum_{j=2}^{2s} b_{s+1,j}\Phi_{(t-1)s+j}$$
$$+\sum_{t=1}^{k-1} t \sum_{j=2}^{2s} b_{s+1,j}\Phi_{(2k-1-t)s+j}. \tag{13.24}$$

Let us assume that $1 \leq s \leq 3$, then $3 \leq p \leq 7$. This condition provides the nonpositive coefficients $b_{s+1,j}$, $j = 2, 3, \ldots, 2s$ (for $p = 9$, $b_{5,5} > 0$). From (13.12) we have

$$\sum_{j=2}^{p-1} b_{ij} = -\frac{h^2(i-1)(p-i)}{2} \geq -\frac{(p-1)^2h^2}{8} \quad (i = 2, 3, \ldots, 2s), \tag{13.25}$$

as $(i-1)(p-1)$ have the maximum value for $i = s+1$, $p = 2s+1$.

From (13.24) and (13.25) we have[1]

$$\sigma_{ks+1} = -\frac{1}{8}\Phi(\xi), \tag{13.26}$$

where $\Phi(\xi)$ is between maximum and minimum numbers Φ_i $(i = 2, 3, \ldots, 2ks)$.

As follows, the formula (13.23) gives the possibility of expressing y_{ks+1} through its boundary values and a linear combination of values $\Phi(x) = y''(x) - R''_{p-1}(x)$.

Analogously to the expression (13.23) formulae we can write for the arbitrary y_{ts+1} $(t = 1, 2 \ldots, k-1, k+1, \ldots, 2k-1)$ using (13.22) and (13.23).

We assume that $t \in [1, k-1]$. Take the first t equations from (13.22), multiply the first of them to $1/t$, the second to $2/t$ etc., multiply the last to $t/t = 1$, add them and we obtain immediately:

$$y_{ts+1} = \frac{t}{t+1} y_{(t+1)s+1} + \frac{1}{(t+1)} y(0) + \Sigma^{[t]}, \tag{13.27}$$

where we shall have

$$\Sigma^{[t]} = \frac{2}{t+1}\sum_{r=1}^{t} r \sum_{j=2}^{2s} b_{s+1,j}\Phi_{(2s-1)s+j} = -\frac{t}{8k^2}\Phi(\xi_t) \quad (t = 1, 2 \ldots, k-1). \tag{13.28}$$

Then (13.23) and (13.27) gives the sought dependence between y_{ts+1} and $y(0)$, $y(1)$ and Φ_i, $i = 2, 3, \ldots, 2ks$:

$$y_{ts+1} = \frac{2k-t}{2k} y(0) + \frac{t}{2k} y(1) + \sigma_{ts+1} \quad (t = 1, 2, \ldots, k-1), \tag{13.29}$$

[1] The equality (13.26) is obtained from the lemma, which is well known. From this lemma we have $\sum_i a_i y(x_i) = y(\xi)\sum_i a_i$ when a has the same signs, and $y(\xi) \in [\min_i y(x_i), \max_i y(x_i)]$.

where

$$\sigma_{ts+1} = \frac{t}{k}\sigma_{ks+1} + \frac{t}{k-1}\Sigma^{[k-1]} + \ldots + \frac{t}{|t+1|}\Sigma^{[t+1]} + \Sigma^{[t]}. \tag{13.30}$$

From the last expression using (13.28) we can easily obtain:

$$\sum_{i=1}^{k-t} \frac{t}{k-i}\Sigma^{[k-1]} = -\frac{(k-1)t}{8k^2}\Phi(\xi_i) == -\frac{(k-1)t}{k^2}\sigma^{I}_{ks+1}.$$

From the expressions using (13.26) and (13.27) it follows that

$$\sigma_{ts+1} = \left[1 - \left(\frac{k-t}{k}\right)^2\right]\sigma^{II}_{ks+1} \quad (t = 1, 2, \ldots, k-1).$$

For the sums σ^{I}_{ks+1} and σ^{II}_{ks+1} we have the equality analogueous with (13.26). For an integer $t \in [k+1, 2k-1]$ by analoguey with (13.29) we also have

$$y_{(2k-t)s+1} = \frac{2k-t}{2k}y(1) + \frac{t}{2k}y(0) + \sigma_{(2k-t)s+1} \quad (t = 1, 2, \ldots, k-1), \tag{13.31}$$

where

$$\sigma_{(2k-t)s+1} = \frac{t}{k}\sigma_{ks+1} + \frac{t}{k-1}\Sigma^{[k+1]} + \ldots + \frac{t}{k+1}\sigma^{[2k-(t+1)]} + \Sigma^{[2k-t]},$$

$$\Sigma^{[2k-t]} = \frac{2}{t+1}\sum_{r=1}^{t} r \sum_{j=2}^{2s} b_{s+1,j}\Phi_{(2k-t-1)s+j} = -\frac{t}{8k^2}\Phi(\xi_{2k-t}),$$

$$\sum_{i=1}^{k-t} \frac{t}{k-i}\Sigma^{[2k-i]} = -\frac{(k-t)t}{k^2}\sigma^{III}_{ks+1},$$

$$\sigma_{(2k-i)s+1} = \left[1 - \left(\frac{k-t}{k}\right)^2\right]\sigma^{IV}_{ks+1} \quad (t = 1, 2, \ldots, k-1).$$

The expressions (13.23), (13.29) and (13.31) give the formula for every y_{ts+1}, $t = 1, 2 \ldots, 2k-1$:

$$y_{ts+1} = \frac{2k-t}{2k}y(0) + \frac{t}{2k}y(1) + \sigma_{ts+1}, \quad (t = 1, 2, \ldots, 2k-1), \tag{13.32}$$

where

$$\sigma_{ts+1} = \left[1 - \left(\frac{k-t}{k}\right)^2\right]\sigma^{t}_{ks+1} \quad (t = 1, 2, \ldots, 2k-1). \tag{13.33}$$

(For values σ'''_{ks+1}, σ^{IV}_{ks+1}, σ^{t}_{ks+1} there are true the equalities same to σ^{I}_{ks+1} and σ''_{ks+1}, i.e., formulae similar to (13.26)).

The system (13.32) is equivalent to the subsystem (13.22). Let us construct the formulae of the form (13.32) for the other $y_{(t-1)s+i}$, when $i \neq s+1$. From (13.21) by (13.32) we obtain immediately

$$y_{(t-1)s+i} = \frac{2ks - st + s - i + 1}{2ks} y(0) + \frac{st - s + i - 1}{2ks} y(1) + \sigma_{(t-1)s+i}, \quad (13.34)$$

where

$$\sigma_{(t-1)s+i} = \frac{i-1}{2s}\sigma_{(t+1)s+1} + \frac{2s-(i+1)}{2s}\sigma_{(t-1)s+1} + \sum_{j=2}^{2s} b_{ij}\Phi_{(t-1)s+j}. \quad (13.35)$$

The formulae (13.34) and (13.35) show that $y(x)$ is expressed through the boundary values (13.2) and sums in every point of the mesh. The main part of the sums has negative coefficients.

The equalities (13.33) and (13.35) if $\Phi = 1$ give

$$|\sigma_{ts+1}| < |\sigma_{ks+1}| \leq \frac{1}{8};$$

$$|\sigma_{(t-1)s+i}| < \frac{1}{8}\max_j\{1, 1 - h^2(4s^2 - \gamma_i)\} = \frac{1}{8} \quad (13.36)$$

($j = 2, 3, \ldots, 2ks$; for $t = 1, 2, \ldots, 2k-2$ and $i = 2, 3, \ldots, s$; for $t = 2k-1$ and $i = 1, 2, \ldots, s, s+2, \ldots, 2s$),
where

$$\gamma_i = \gamma_{p+1-i} = \frac{1}{h^2}\sum_{j=2}^{2s}|b_{ij}|, \quad (i = 2, 3, \ldots, s).$$

The system (13.34) is equivalent to the boundary value problem (13.3), (13.2), if we instead $y''(x)$ substitute $f(x, y(x))$ from (13.3).

Omitting the remainder $R''_{p-1}(x)$ in (13.34) we obtain the nonlinear algebraic system of equations, where the unknowns are $\bar{y}_i = y_i - \eta_i (i = 2, 3, \ldots, 2ks)$. Here η_i denotes an error which we obtain neglecting the remainder terms.

Thus, we have

$$\bar{y}_{(t-1)s+i} = \frac{2ks - st + s - i + 1}{2ks} y(0) + \frac{st - s + i - 1}{2ks} y(1) + \bar{\sigma}_{(t-1)s+i} \quad (13.37)$$

(for $t = 1, 2, \ldots, 2k-2$ and $i = 2, 3, \ldots, s+1$; for $t = 2k-1$ and $i = 2, 3, \ldots, 2s$), where $\bar{\sigma}_{ts+1}$ is obtained from σ_{ts+1} if we omit the remaining term in it.

From the aforesaid it follows that the inequalities (13.36), which correspond to formulae (13.34), are fulfilled for $p = 3, 5, 7$ and $x_i = (i-1)h, \quad i = 1, 2, \ldots, 2ks+1$.

We show below that we can obtain the identical formulae, as (13.34), with the same estimations as (13.36) for every $p \geq 3$ for the special choice of the mesh.

To show that first of all we must construct a formula of the form (13.22) with nonpositive coefficients. We also have to find such a partition of an interval $(0, 1)$, which gives the possibility of writing a formula in the aforesaid form for every central point of the subinterval. Then it is clear that we can rewrite a set of such formulae to the form (13.32).

For this purpose we use an equality

$$y(x_{s+2}) = \frac{1}{2}y(0) + \frac{1}{2}y(x_{p+2}) - \frac{1}{2}\int\limits_0^{x_{s+2}} ty''(t)dt - \frac{1}{2}\int\limits_{x_{s+2}}^{x_{p+2}} (x_{p+2}-t)y''(t)dt$$

$$= \frac{1}{2}[y(0) + y(x_{p-2})] - \frac{1}{2}\int\limits_0^{x_{s+2}} t[y''(t) + y''(x_{p+2}-t)]dt,$$

where

$$p = 2s+1, \quad 2x_{s+2} = x_{p+2} \leq 1.$$

If we substitute the integral in the last expression for Gauss' quadrature formula with s an abscissa, then we have

$$y(x_{s+2}) = \frac{1}{2}[y(0) + y(x_{p+2})] + \sum_{j=2}^{2s+2} b^*_{s+2,j}y''(x_j) + r(x_{s+2}), \tag{13.38}$$

where the numbers

$$b^*_{s+2,j} = b^*_{s+2,2s+4-j} < 0, \quad b^*_{s+2,s+2} = 0 \quad (j = 2,3,\ldots,s+1)$$

are obtained if the Gauss' coefficients which have been computed for the interval $[-1,1]$ are multiplied by $-0.25x_{s+2}x_j$, $j = 2,3,\ldots,s+1$; for the remainder terms we have

$$|r(x_{s+2})| \leq \frac{c_3^* M_{p+1}}{(p+1)!}h_1^{p+1}. \tag{13.39}$$

In (13.39) c_3^* is a constant depending on p and x_i $(i = 1,2,\ldots,p+2)$ and

$$h_1 = \max_{1\leq i\leq s+1} |x_{i+1} - x_i|.$$

In (13.38) the numbers x_j $(j = 2,3,\ldots,s-1)$ are zeros of the Legendre polynomial of the s-th order, which lies in the interval $(0, x_{s+2})$, and x_j, $j = s+3,\ldots,2s+2$, are zeros, which belong to interval (x_{s+2}, x_{2s+2}).

If we divide $(0,1)$ on the k equal partial intervals and use the formula (13.38), then we have

$$y_{t(s+1)+1} = \frac{1}{2}y_{(t-1)(s+1)+1} + \frac{1}{2}y_{(t+1)(s+1)+1}$$
$$+ \sum_{j=2}^{2s+2} b^*_{s+2,j}y''_{(t-1)(s+1)+j} + r_t(x_{s+2}) \quad (t = 1,2,\ldots,2k-1). \tag{13.40}$$

The formulae (13.22) and (13.40) have an identical structure, only the mesh is uniform in (13.22), and it is nonuniform (special) in (13.40). So to the formulae (13.42) we can give the form analogueous to the formulae (13.32).

If we use (P) formulae and assume that $x_{t(s+1)+i}$ $(t = 0, 1, \ldots, 2k-1) i = 2, 3, \ldots, s+2)$ are Gauss' abscissas, then we have

$$y_{(t-1)(s+1)+i} = (1 - kx_i) y_{(t-1)(s+1)+1} + kx_i y_{(t+1)(s+1)+1} +' \sum_{j=2}^{2s+2} b^*_{i,j} \Phi_{(t-1)(s+1)+j}$$

$$= (1 - kx_i) y_{(2t-1)(s+1)+1} + kx_i y_{(t+1)(s+1)+1} +' \sum_{j=2}^{2s+2} b^*_{ij} y_{(t-1)(s+1)+j}$$

$$-' \sum_{j=2}^{2s+2} b^*_{ij} R''_{p-1}(x_{(t-1)(s+1)+j},$$

(for $t = 1, 2, \ldots, 2k-2$ and $i = 2, 3, \ldots, s+1$; for $t = 2k-1$ and $i = 2, 3, \ldots, s+1, s+3, \ldots, 2s+2)$.
Here and below the stroke in the left side means that when summing any two values (better symmetrical) of j are missed.

With the help of the last system and formulae analogueous to formulae (13.32), constructed by modification of the system (13.40) we can obtain

$$y_{(t-1)(s+1)+i} = \alpha_{(t-1)(s+1)+i} y(0) + \beta_{(t-1)(s+1)+i} y(1) + \sigma^*_{(t-1)(s+1)+i}, \qquad (13.41)$$

(for $t = 1, 2 \ldots, 2k-2$ and $i = 2, 3 \ldots, s+2$; for $t = 2k-1$ and $i = 2, 3, \ldots, 2s+2)$, where

$$\alpha_{(t-1)(s+1)+i} = \frac{2k - 2kx_i - t + 1}{2k}, \quad \beta_{(t-1)(s+1)+i} = \frac{2kx_i + t - 1}{2k},$$

and for all $\sigma^*_{t(s+1)+i}$, which have the same structure as σ_{ts+i}, we have

$$|\sigma^*_{t(s+1)+i}| = \frac{1}{8} \max\{1, 1 - \frac{1}{4k^2}(1 - \gamma^*_i)\}, (\Phi \equiv 1). \qquad (13.42)$$

$$(j = 2, 3, \ldots, 2k(s+1)),$$

where

$$\gamma^*_i = 4k^2 \sum_{j=2}^{2s+2} |b^*_{ij}|.$$

Omitting the remainder terms in (13.41) we have

$$\bar{y}_{(t-1)(s+1)+i} = \alpha_{(t-1)(s+1)+i} y(0) + \beta_{(t-1)(s+1)+i} y(1) + \bar{\sigma}^*_{(t-1)(s+1)+i}. \qquad (13.43)$$

Thus we could construct the expression in which the main part has not positive coefficients. We recall, that for a uniform mesh the correctness of such presentation is proved when $p = 3, 5, 7$; for Gauss' net the coefficients of the main part have an identical sign for every odd $p \geq 3$.

With the help of the nonlinear algorithm (13.37) and (13.43) and also the estimation, obtained for R''_{p-1} and r, we can prove the following theorem.

Theorem 13.1. *Let conditions $L < 8$ be fulfilled and $y(x) \in C^{(p+1)}[0,1]$. Then:*

1) The systems (13.37) and (13.43) have an unique solution relatively y_i, and these solutions can be obtained by the method of sequence aproximations;

2) Both in the case of a uniform mesh (p=3,5,7) and in the case of Gauss' net $(p = 2s+1 \geq 3)$ the convergence of the approximate solution of the problem (13.3), (13.2) to the exact solution has (p-1)-th order relatively h and h_1.

Proof. For a proof of part one it is sufficient to show that the operators σ and σ^*, corresponding to the nonlinear algorithms (13.37) and (13.43), are operators of compression. For σ it is true by (13.36), for σ^* this fact follows from (13.42), if we note that using the condition $L < 8$ and the corresponding choice an integer k to ensure the positiveness of the expression

$$8 - L\ \max\{1, 1 - \frac{1}{4k^2}(1-\gamma_i^*\} = 8 - L^*,$$

are always possible.

Let us begin to estimate an error. At first let us consider a case with an uniform mesh. Subtract (13.37) from (13.34) paying attention to (13.8), (13.14) and expression for σ $(i = 2, 3, \ldots, 2ks)$, we obtain

$$\max|\eta_i| \leq \frac{c_4 M_{p+1}}{(8-L)(p+1)!} h^{p-1} \quad (p = 3,5,7). \tag{13.44}$$

For the Gauss mesh we also do the same: subtract (13.43) from (13.41); paying attention to (13.8), (13.12), (13.39) and the expressions for σ_i^* $(i = 2, 3, \ldots, 2k(s+1))$, we obtain

$$\max|\eta_i| \leq \frac{c_5 M_{p+1}}{(8-L^*)(p+1)!} h^{p-1} \quad (p \geq 3). \tag{13.44'}$$

If now a number of partial intervals k tends to infinity then the inequalities (13.44) and (13.44') prove the theorem 13.1 as c_4 and c_5 are independent of k.

If $y_i^{[n]}$ is the n-th approximation of the solution (13.37), (13.43)2 by iterative method and n satisfies these inequalities

$$n \leq \frac{(p-1)\ln h^{-1} + |\ln(1-\lambda)|}{|\ln\lambda|}, \quad \text{where } \lambda = \frac{L}{8} \text{ (for a uniform mesh)}$$

$$n \leq (p-1)\ln h^{-1} + \frac{|\ln(1-\lambda^*)|}{|\ln\lambda^*|}, \quad \text{where } \lambda^* = \frac{L^*}{8} \text{ (for a Gaussian mesh)},$$

then finally we obtain

$$|y_i - y_i^{[n]}| \leq \left[\frac{c_4 M_{p+1}}{(8-L)(p+1)!} + 1\right] h^{p-1} \quad (p = 3,5,7; i = 2,3,\ldots,2ks),$$

$$|y_i - y_i^{[n]}| \leq \left[\frac{c_5 M_{p+1}}{(8-L^*)(p+1)!} + 1\right] h_1^{p-1} \quad (p = 2s+1; i = 2,3,\ldots,2k(s+1)).$$

13.3. A solution of the boundary value problem (13.1), (13.2)

Let us be given an uniform (for p=3,5,7) or Gauss' (for $p = 2s + 1 \geq 3$) partition of the interval [0,1]; if we use the formulae (13.34) and (13.40) then we obtain (substituting $y''(x)$ by $f(x, y(x), y'(x))$:

$$y_{(t-1)z+i} = \alpha_{(t-1)z+i}y(0) + \beta_{(t-1)z+i}y(1) + \sigma_{(t-1)z+i} \tag{13.45}$$

(for $t = 1, 2, \ldots, 2k-2$ and $i = 2, 3, \ldots, z+1$; for $t = 2k-1$ and $i = 2, 3, \ldots, 2z$), where

$$z = \begin{cases} s, & \text{if the uniform mesh,} \\ s+1, & \text{if the Gaussian mesh.} \end{cases}$$

Then the (Q) formulae from subsection 13.1 for the interval $(x_{(k-1)z+1}, x_{(k+1)z+1})$ will have the form

$$y'_{(k-1)z+i} = k[y_{(k+1)z+1} - y_{(k-1)z+1}] - k' \sum_{j=2}^{2z} c_{ij} y''_{(k-1)z+j} - \rho_{p-3,(k-1)z+i}, \tag{13.46}$$

where

$$\rho_{p-3,(k-1)z+i} = k \int\limits_{x_{(k-1)z+1}}^{x_{(k+1)z+1}} dx \int\limits_{x_{(k-1)z+i}}^{x} y''(x, x_{(k-1)z+2}, \ldots$$

$$\ldots, x_{(k+1)z-1}, x_{(k+1)z}) \prod_{j=2}^{2z} (x - x_{(k-1)z+j}) dx.$$

The formula (13.46) includes a difference $y_{(k+1)z+1} - y_{(k-1)z+1}$. A computation of this difference from (13.45) (paying attention to the structure of expressions $\sigma_{(k+1)z+1}$ and $\sigma_{(k-1)z+1}$) gives

$$\begin{aligned} y_{(k+1)z+1} &- y_{(k-1)z+1} \\ &= \frac{1}{k}[y(1) - y(0)] + (\Sigma^{[k+1]} - \Sigma^{[k-1]}) \\ &= \frac{1}{k}[y(1) - y(0)] + \frac{2}{k} \sum_{r=1}^{k-1} r \sum_{j=2}^{2z} b_{z+1,j} [\Phi_{(2k-r-1)z+j} - \Phi_{(r-1)z+j}]. \end{aligned} \tag{13.47}$$

For $\Sigma^{[k+1]} - \Sigma^{[k-1]}$ we have

$$\frac{2}{k} \sum_{r=1}^{k-1} r \sum_{j=2}^{2s} [\Phi_{(2k-r-1)s+j} - \Phi_{(r+1)s+j}] = -\frac{k-1}{8k^2} [\Phi_{2k-t} - \Phi_t] \tag{13.48}$$

where Φ_{2k-t} and Φ_t accordingly are the middles between the numbers Φ_{2ks-i} and Φ_i, $i = 2, 3, \ldots, kz-1$.

Substituting the expression for the difference (13.47) in (13.45) we obtain for $y'(x_i) = y'_i$:

$$y'_{(k-1)z+i}$$
$$= y(1) - y(0) + 2\sum_{r=1}^{k-1} r \sum_{j=2}^{2z} b_{s+1,j}[\Phi_{(2k-r-1)+j} - \Phi_{(r-1)z+j}]$$
$$-k'\sum_{j=2}^{2z} c_{ij} y''_{(k-1)z+j} - \rho_{p-3,(k-1)z+i}, \quad (13.49)$$
$$(i = 1, 2, \ldots, 2z+1).$$

Below the formulae (13.45) and (13.49) are used for solving the problem (13.1), (13.2). At the same time with these expressions let us consider two following Cauchy problems:

$$y_1'(x) = f(x, \lambda(x), y_1(x)) \quad (l_1 \le x \le 1) \quad (13.50)$$
$$y_1(l_1) = \gamma;$$
$$y_1'(x) = f(x, \mu(x), y_1(x)) \quad (l_2 \ge x \ge 0) \quad (13.51)$$
$$y_1(l_2) = \delta,$$

where the definitions are introduced

$$x_{(k-1)z+1} = l_1, \quad x_{(k+1)z+1} = l_2, \quad y'(x) = y_1(x).$$

The functions $\lambda(x)$ and $\mu(x)$ and the numbers γ and δ will be determined lower.

If in the systems (13.45) and (13.49) the remainder terms are omitted and the equation (13.1) is used, then we have

$$\bar{y}_{(t-1)z+i} = \alpha_{(t-1)z+i} y(0) + \beta_{(t-1)z+i} y(1) + \bar{\sigma}_{(t-1)z+i} \quad (13.52)$$

$$\bar{y}'_{(k-1)z+i} = y(1) - y(0) + 2\sum_{r=1}^{k-1} r \sum_{j=2}^{2s} b_{z+1,j}[\bar{f}_{(2k-r-1)z+j} - \bar{f}_{(r-1)z+j}]$$
$$-k'\sum_{j=2}^{2z} c_{ij} \bar{f}_{(k-1)z+j} \quad (i = 1, 2, \ldots, 2z+1), \quad (13.53)$$

where

$$\bar{f}_i = f(x_i, \bar{y}(x_i), \bar{y}'(x_i)).$$

Now we consider the totality of the expressions (13.52), (13.53), (13.50) and (13.51) which are the nonlinear analogue of the problem (13.1), (13.2). An approximate solution $\bar{y}_i = y_i - \eta_i$ and its derivative $\bar{y}'_i = y'_i - \eta'_i$, where η_i and η'_i are errors, can be obtained in the following way. The equations (13.52) and (13.53) give the first approximation $y_i^{[0]}$ $(i = 2, 3, \dots, 2kz)$ and $y_i'^{[0]}$ $(i = (k-1)z+1, \dots, (k+1)z+1)$; assuming in (13.50) and (13.51)

$$\lambda(x_i) = \lambda_i = y_i^{[0]} \quad (l_1 \le x_i \le 1), \quad \gamma = y'^{[0]}(l_1),$$

$$\mu(x_i) = \mu_i = y_i^{[0]} \quad (l_2 \ge x_i \ge 0), \quad \delta = y'^{[0]}(l_2),$$

we solve two Cauchy problems by any known stable method (on this occasion see lower some notes) with the help of which $y_i^{[0]}$ $(i = 2, 3, \dots, (k-1)z, (k+1)z + 2, \dots, 2kz)$ are given. Having zero approximation $y_i^{[0]}$ and $y_i'^{[0]}$, $i = 2, \dots, 2kz$ by means of the systems (13.52) and (13.53) we can obtain $y_i^{[1]}$, $i = 2, 3, \dots, 2kz$ and $y_i'^{[1]}$, $i = (k-1)z + 1, \dots, (k+1)z + 1$. Assuming in (13.50) and (13.51)

$$\lambda(x_i) = \lambda_i = y_i^{[1]} \quad (l_1 \le x_i \le), \quad \gamma = y^{[1]}(l_1),$$

$$\mu(x_i) = \mu_i = y_i^{[1]} \quad (l_2 \ge x_i \ge 0), \quad \delta = y_1^{[1]}(l_2),$$

we can find $y_i'^{[1]}$, $i = 2, 3, \dots, (k-1)z, (k+1)z + 2, \dots, 2kz$.

Continuing this process we consistently find $y_i^{[m]}$ and $y_i'^{[m]}$, $m = 2, 3, \dots$

It is easy to see that $y_i^{[m]}$ and $y_i'^{[m]}$ will tend to $\bar{y}$ and $\bar{y}_i'$ if the operator which corresponding nonlinear schemes, are the operator of compression. For the equations (13.52) and (13.53) this condition can be easily obtained with the help of the estimations, obtained for $\sigma_{(t-1)z+i}$ from subsection 13.2, formulae (13.19), (13.48) and (13.4); for Cauchy problem (13.50) and (13.51) the Cacciopoli–Banach principle is fair.

We must note that we can solve the problems (13.50) and (13.51) sufficiently exactly (i.e., with error $O(h_{z-s}^{p-1})$, where $h_0 = h$), as the formulae (13.53) give the initial segment of the values of the unknown integral in $2z + 1$ of the knot points.

As the nonlinear analogue has an unique solution from where justice of inequalities follows

$$\max |\eta_i| \le c_{z-s+6} M_{p+1} h_{z-s}^{p-1}, \tag{13.54}$$

$$\max |\eta_i'| \le (c_{z-s+6}' M_p + c_{z-s+6}'' M_{p+1}) h_{z-s}^{p-1} \quad (i = 1, 2, \dots, 2kz + 1), \tag{13.55}$$

where c_{z-s+6}, c_{z-s+6}' and c_{z-s+6}'' are some nonnegative constants independed on k.

If the number of partial subsegments k tends to infinity then it follows from inequality (13.54) and (13.55) the aforesaid process converges.

We now make several notes on a numerical solution of Cauchy problems (13.50) and (13.51) by multistep method (Henrici [1962]).

At first, if for the problem (13.50) we have to use a stable formulae, for the problem (13.51) we need essentially, a "nonstable formulae", i.e., whose characteristic polynomials have zeroes with modul ≥ 1 and multiple of zeroes with moduli not equal to one.

Secondly, if the mesh is uniform we shall need only two working formulae, at the same time for the Gauss' partition we must operate with $s+1$ stable and $s+1$ "nonstable" formulae.

Summing up all aforesaid we are convinced of justice of the following theorem.

Theorem 13.2. *Let conditions (13.4) and* $y(x) \in C^{(p+1)}[0,1]$ *are fulfilled for problem (13.1), (13.2).*

Then :

1)The problem (13.1), (13.2) has an unique solution;

2)The nonlinear analogue has an unique solution, and for it the method of sequence approximations are defined the convergence process;

3) As in the case of a uniform mesh (p=3,5,7) and in the case of Gauss' net ($p = 2s + 1 \geq 3$) *the convergence of the solution of nonlinear analogue to the solutions of the problem (13.1), (13.2) and its derivative has (p-1)th order with respect* h *or* h_1 *accordingly.*

From the aforesaid algorithm also follows the theorem.

Theorem 13.3. *Let the conditions:*

a) $|f(x,y,y') - f(x,z,z')| \leq L|y-z| + L'|y'-z'|$,

b) $\frac{1}{4}(L + L') < 1$

be fulfilled.

Then the problem (13.1), (13.2) has on unique solution, which we can construct by an iterative method.

For the proof of this theorem it is sufficient to construct expressions analogue to (P) and (Q) formulae when $y \in C^{(2)}[0,1]$

Actually, if $2kz + 1 = n + 1$, $h = 1/n$, then from the Taylor formula we have

$$y_i = \frac{1}{2}y_{i-1} + \frac{1}{2}y_{i+1} - \frac{h^2}{4}[y''_{i+1} + y''_{i-1}] + \frac{h^2}{4}r_i \qquad (13.56)$$

$$y'_i = \frac{1}{2h}[y_{i+1} - y_{i-1}] - \frac{h^2}{2}[y''_{i+1} - y''_{i-1}] + \frac{h^2}{2}r'_i \quad (i = 2,3,\ldots,n), \qquad (13.57)$$

where

$$|r_i| = |(y''_{i+1} - y''(\xi_{i+1})) + (y''_{i-1} - y''(\xi_{i-1}))| < \varepsilon_i,$$

$$|r'_i| = |(y''_{i+1} - y''(\xi_{i+1})) - (y''_{i-1} - y''(\xi_{i-1}))| < \varepsilon'_i$$

$$x \leq \xi_{i+1} \leq x_{i+1}, \; x_{i-1} \leq \xi_i \leq x_i$$

The expression (13.56) corresponds to (P) formulae and (13.57) to (Q) formulae.

In conclusion we note that when in (13.3) $f(x,y(x)) = q(x)y(x) + f(x)$, then by using (P) and (Q) formulae an ordinary finite-difference scheme is generalized. The following theorem is true (details see in section 14).

Theorem 13.4. *Let the domain net be uniform and the function* $q(x) \geq 0$, $f(x) \in C^{(p-1)}[0,1]$ $(p \geq 2)$.

Then an application of (P) *formulae gives the algorithm by which the convergence of the approximate solution to the exact solution of a linear boundary value problem has (p-1) order with respect to a step.*

As follows from the monograph Bellman, Kalaba [1965], the positivity of the operators (or the fact of monotony) gives new method proofing of convergence and besides also better estimates of a domain of convergence for quasilinear BVPs

$$y''(x) = f(x,y(x)), \quad y(0) = y(1) = 0. \qquad (13.58)$$

In this connection the high accuracy method of numerical solving (13.58) by the simple iterative scheme of Bellman–Kalaba [1965]:

$$y''_{n+1} = f(x, y_n) + (y_{n+1} - y_n) f_y(x, y_n), \quad y_{n+1}(0) = y_{n+1}(1) = 0,$$

when

$$-f_y(x, y_n) \leq \pi^2 - \varepsilon \quad (f_y \equiv \frac{\partial f}{\partial y}),$$

represents important practical and theoretical meaning (see also Henrici [1962]).

14. Generalized Factorization Method

According to the article of Vashakmadze [1972] below for the numerical solution of BVP:

$$-(Au + qu) = \frac{d}{dt}(k(t)u'(t)) - q(t)u(t) = f(t) \quad (k > 0, q \geq 0), \tag{14.1}$$

$$u(0) - k_1 u'(0) = \alpha, \quad u(1) + k_2 u'(1) = \beta \quad (k_1 \geq 0, k_2 \geq 0), \tag{14.2}$$

the method of any order of accuracy, depending on the order of the smooth of the unknown solution $u(t)$ will be given.

Preliminarily we shall put the auxiliary formulae. They are generalized faulte (P) and (Q) formulae of subsection 13.1. Thus we suppose that $u(t) \in C^{p+1}(0, 1)$, $p = 2s + 1$.

(P) formulae have a form (the notation here and below are borrowed from section 13):

$$u(t_i) = \alpha_i^{p,1}(k)u(t_1) + \beta_i^{p,1}(k)u(t_{i+s}) - \sum_{j=2}^{p-1} b_{ij}^{p,1} k(t_j)[Au(t_j) - R_{p-1}(t_i)], \quad i = 2, 3, ..., p-1, \tag{14.3}$$

where

$$\alpha_i^{p,1}(k) = \left(\int_{t_1}^{t_p} k^{-1}(t)dt\right)^{-1} \int_{t_i}^{t_p} k^{-1}(t)dt, \quad \beta_i^{p,1}(k) = 1 - \alpha_i^{p,1}(k).$$

(Q) formulae are presented as follows:

$$u'(t_i) = \gamma_i^{p,1}(k)[u(t_p) - u(t_1)] + \sum_{j=2}^{p-1} c_{i,j}^{p,1} k(t_j)Au(t_j) + \int_{\tau_1}^{\tau_p} dt \int_{\tau_i}^{t} k(t)Ar_{p-3}(t)dt, \quad i = 1, 2, ..., p, \tag{14.4}$$

where

$$\tau_i = \int_{t_i}^{t_p} k^{-1}(t)dt, \qquad \gamma_i^{p,1}(k) = (k(t_i)\int_{t_1}^{t_p} k^{-1}(t)dt)^{-1}.$$

Coefficients $b_{i,j}^{p,1}(k)$ and $c_{i,j}^{p,1}(k)$ are calculated thus:

$$b_{i,j}^{p,1}(\tau) = \frac{1}{\tau_p - \tau_1}[(\tau_p - \tau_1)\int_{\tau_1}^{\tau_i} dt \int_{\tau_1}^{t} l_j(t)dt - (\tau_i - \tau_1)\int_{\tau_1}^{\tau_p} dt \int_{\tau_1}^{t} l_j(t)dt],$$

$$(i, j = 2, 3, ...p-1),$$

$$c_{i,j}^{p,1} = \int_{\tau_1}^{\tau_p} dt \int_{\tau_i}^{t} l_j(t)dt, \quad (i = 1, 2, ..., p,\ j = 2, 3, ..., p-1), \tag{14.5}$$

$$l_j(t) = \prod_{\substack{i=2\\ j\neq i}}^{p-1} \frac{t - t_i}{t_j - t_i}.$$

Now let ω_h designate the net area, determined as follows:

$$\omega_h = \{0 = t_1, t_2, ..., t_n, t_{n+1} = 1; h_i = t_i - t_{i-1}\}.$$

As bounding points of the net ω_n we shall name those t_i knots, for which or $i \le s+1$, or $i \ge n-s+1$.

From relations (14.3) for bounding points it follows that

$$\begin{aligned} u(t_i) &= \alpha_i^{i+s,1}(k)u(0) + \beta_i^{i+s,1}(k)u(t_{i+s}) \\ &- \sum_{j=2}^{2s} b_{i,j}^{i+s,1} k(t_j)Au(t_j) + O(h^{2s+1}), \quad i \le s+1; \\ u(t_i) &= \alpha_i^{n+1,i-s}(k)u(t_{i-s}) + \beta_i^{n+1,i-s}(k)u(1) \\ &- \sum_{j=n-2s+1}^{n} b_{i,j}^{n+1,i-s} k(t_j)Au(t_j) + O(h^{2s+1}), \quad i \ge n-s+1. \end{aligned} \tag{14.6}$$

The relations (14.4) and (14.6) permit to receive expressions of a following form:

$$\begin{aligned} u(t_i) &= \frac{\alpha_i}{1 + k_1\gamma_1}[u(0) - k_1 u'(0)] + \frac{\beta_i + k_1\gamma_1}{1 + k_1\gamma_1} u(t_{i+s}) \\ &- \sum_{j=2}^{2s} (b_{i,j} - \frac{k_1\alpha_i}{1 + k_1\gamma_1} c_{i,j}) k(t_j)Au(t_j) + O(h^{2s+1}), \quad i \le s+1; \\ u(t_i) &= \frac{\beta_i}{1 + k_2\gamma_{n+1}}[u(1) + k_2 u'(1)] + \frac{\alpha_i + k_2\gamma_{n+1}}{1 + k_2\gamma_{n+1}} u(t_{i-s}) \end{aligned} \tag{14.7}$$

$$- \sum_{j=n-2s+1}^{n} (b_{i,j} - \frac{k_2\beta_i\gamma_{n+1}}{1+k_2\gamma_{n+1}} c_{n+1,j}) k(t_j) Au(t_j) + O(h^{2s+1}),$$
$$i \geq n-s+1.$$

Here and below, in the coefficients the top indexes and the dependence of factors on the function $k(t)$ are omitted. In addition the designation $h = \max_i(t_{i+1} - t_i)$ is entered. A feature of the formulae (14.7) is that the right parts contain same expressions, (from conditions (14.3)), as data of a initial problem. Obviously, theapproach of construction of the formulae of a type (14.7) allows generalization for other conditions. Let $t_i - t_{i-j} = t_{i+j} - t_i (s+2 \leq i \leq n-s)$. Then the residual member of the formula (14.3) allows the valuation (see s.s.13.1):

$$| \sum_{j=i-s+1}^{i+s-1} b_{i,j} k(t_j) AR_{p-1}(t_j)| < c_1 M_{p+1} h^{p+1}, \quad M_{p+1} = \max_{(0,1)} |u^{(p+1)}(t)|.$$

For interior knots $t_i \in \omega_h$ from expressions (14.3) there follows:

$$u(t_i) = \alpha_i u(t_{i-s}) + \beta_i u(t_{i+s}) - \sum_{j=i-s+1}^{i+s-1} b_{i,j} k(t_j) Au(t_j) + O(h^{2s+2}). \tag{14.8}$$

If now in the formulae (14.8) we replace the expression Au by $qu + f$ and then omit the remainder term, we shall obtain algebraic system of linear equations, the solution of which shall designate through u_i $(i = 2, 3, ..., n)$.

The matrix appropriate to this system is a multi-diagonal matrix depending on s. For the solution of such a system without labour a factorization method, advanced for systems with three diagonal matrix, is generalized.

A system of equations concerning the values u_i, received from (14.7), for convenience we shall rewrite as:

$$u_i = \frac{\beta_i + k_1\gamma_1}{1+k_1\gamma_1} u_{i+s} + \sum_{j=2}^{2s} d_{ij} u_j + F_i, \qquad i = 2, 3, ..., s+1;$$

$$u_i = \alpha_i u_{i-s} + \beta_i u_{i+s} + \sum_{j=i-s+1}^{i+s-1} d_{ij} u_j + F_j, \qquad i = s+2, ..., n-s; \tag{14.9}$$

$$u_i = \frac{d_i + k_2\gamma_{n+1}}{1+k_2\gamma_{n+1}} u_{i-s} + \sum_{j=n-2s+1}^{h} d_{ij} u_j + F_i, \qquad i = n-s+1, ..., n,$$

where, for example,

$$F_i = \frac{\alpha_i}{1+k_1\gamma_{1,i}} \alpha + \sum_{j=2}^{2s} d_{ij} f(t_j), \qquad i \leq s+1.$$

The first s of the formulae give the following recurrence expression:

$$u_i = A_i u_{i+s} + \sum_{\substack{j=i+1 \\ j \neq i+s}}^{2s} A_{ij} u_j + B_i, \qquad i = 2, 3, ..., s+1, \tag{14.10}$$

where

$$A_{ij} = \frac{e_{ij}}{1 - e_{ij}}, \qquad j = i+1, ..., 2s, \qquad j \neq i+s,$$

$$A_i = A_{i,i+s} = \frac{\beta_i + k_1\gamma_1}{(1 - e_{ii})(1 + k_1\gamma_1)}, \tag{14.11}$$

$$e_{ii} = d_{ij} + \sum_{k=2}^{i-1} d_{ik} \sum_{l=k}^{i-1} A_{ij} \prod_{m=k}^{l-1} A_{m,m+1}, \qquad \prod_{m=k}^{l-1} \cdot = 1, \qquad k > l-1,$$

$$B_i = \frac{F_i + \sum\limits_{k=2}^{i-1} d_{ik} \sum\limits_{l=k}^{i-1} B_l \prod\limits_{m=k}^{l-1} A_{m,m+1}}{1 - e_{ii}}, \qquad i = 2, 3, ..., s+1.$$

Let i be the number of any internal point of the net area ω_h. Then from expressions (14.9), (14.10) and (14.11) follows:

$$u_i = A_i u_{i+s} + \sum_{j=i+1}^{i+s-1} A_{ij} u_j + B_j, \qquad i = s+2, ..., n-s, \tag{14.12}$$

where

$$A_{ij} = \frac{e_{ij}}{1 - e_{ii}}, \qquad A_i = A_{i,i+s} = \frac{\beta_i}{1 - e_{ii}},$$

$$e_{ij} = d_{ij} + \sum_{k=i-s+1}^{i-1} d_{ik} \sum_{l=k}^{i-1} Al_{l_j} \prod_{m=k}^{l-1} A_{m,m+1} + d_i A_{i-s,j}, \tag{14.13}$$

$$B_i = (1 - e_{ii})^{-1}(F_i + \sum_{k=i-s+1}^{i-1} d_{ik} \sum_{l=k}^{i-1} B_l \prod_{m=k}^{l-1} A_{m,m+1} + d_i B_{i-s}).$$

The values $u_i, i = n-s+1, ..., n$ satisfy the following equalities:

$$u_i = \sum_{j=i+1}^{n} A_{ij} u_j + B_i, \qquad i = n-s+1, ..., n-1, \tag{14.14}$$

where

$$A_{ij} = e_{ij}(1 - e_{ii})^{-1},$$

$$e_{ij} = d_{ij} + \sum_{k=i-s+1}^{i-1} d_{ik} \sum_{l=k}^{i-1} A_{ij} \prod m = k^{l-1} A_{m,m+1} + \frac{\alpha_i + k_2\gamma_{n+1}}{1 + k_2\gamma_{n+1}} A_{i-s,j}, \tag{14.15}$$

$$B_i = (1 - e_{ii})^{-1}(F_i + \sum_{k=i-s+1}^{i-1} d_{ik} \sum_{l=k}^{i-1} B_l \prod_{m=k}^{l-1} A_{m,m+1} + \frac{\alpha_i + k_2\gamma_{n+1}}{1 + k_2\gamma_{n+1}} B_{i-s}).$$

At last, the value u_n defines explicitly:

$$u_n = B_n, \tag{14.16}$$

where

$$B_n = (1 - e_{nn})^{-1}(F_n + \frac{\alpha_n + k_2\gamma_{n+1}}{1 + k_2\gamma_{n+1}} B_{n-s} + \sum_{k=n-s+1}^{n-1} d_{n,k} \sum_{l=k}^{n-1} B_l \prod_{m=k}^{l-1} A_{m,m+1}),$$

$$e_{nn} = d_{nn} + \sum_{k=n-s+1}^{n-1} d_{nk} \sum_{l=k}^{n-1} A_{ln} \prod_{m=k}^{l-1} A_{m,m+1} + \frac{\alpha_n + k_2\gamma_{n+1}}{1 + k_2\gamma_{n+1}} A_{n-s,n}. \quad (14.17)$$

Let α_i and β_i satisfy to following bilateral inequalities:

$$\frac{1}{s} < \beta_i, \quad \alpha_{n+1-i} < \frac{1}{2}, \quad i = 2, 3, ..., s+1,$$

$$1 - c_3 h^2 < \alpha_i \beta_i^{-1} < 1 + c_4 h^2, \quad c_3, c_4 > 0, \quad i = s+2, ..., n-s,$$

where c_3, c_4 are constants. Obviously, it is possible by the appropriate choice h (see expressions for α and β from the formula (14.3).

Then from (14.11), (14.13) and (14.15) follow:

$$\begin{aligned} |A_{ij}| &< (1 - c_5 h) \max_{i \le s+1} \{A_i, A_{s+2}\}, \\ |B_i| &< c_6\alpha + c_7\beta + c_8 \max_j |f(t_j)|, \end{aligned} \quad (14.18)$$

where nonnegative constants c_5, c_6, c_7, c_8 do not depend on h.

The conditions (14.18) are the definition of stability of computing process by the formulae (14.11), (14.13) and (14.15) concerning initial data and right part accordingly (Godunov, Riabenki [1962], ch. II, §3).

The stability of process (14.10), (14.12) and (14.14) for calculation of values u_i is also obvious, as the operator appropriate these expressions is an operator of compression.

From the abovestated formulae it follows that the method of generalized factorization is optimum, as the number of arithmetic actions necessary for calculation of approximate solution u_i is directly proportional to the number of points of the net areas ω_h.

15. Nonlinear Case with Newton's Boundary Conditions

We shall consider a nonlinear boundary value problem:

$$u''(x) = f(x, u(x), u'(x)), \quad o < x < 1, \quad -M < u, \quad u' < M, \quad (15.1)$$

$$k_1 u(0) - u'(0) = \alpha, \quad k_2 u(1) + u'(1) = \beta, \quad k_1^2 + k_2^2 > 0. \quad (15.2)$$

With the basis of generalized (P) and (Q) formulae (see subsection 13.1 and also (14.7) for $k(t) = 1$) in this part we shall begin the construction of one-parametrical computing schemes, to an equivalent nonlinear problem (15.1)–(15.2).

Let is given uniform or Gaussian (in a sense of subsection 13.2) lying in the interval [0,1]. We shall make out the formulae for central knots x_{tz+1} :

$$u_{tz+1} = \frac{1}{2}u_{(t-1)z+1} + \frac{1}{2}u_{(t+1z+1} + A_t, \qquad t = \overline{2, 2k-2}, \tag{15.3}$$

where

$$A_t = \sum_{j=2}^{2z} b_{z+1,j} y''_{(t-1)z+j} + O(h_{z-s}^{p+1}).$$

To these formulae we shall attach expressions similar to the formulae (14.7):

$$\begin{aligned} u_{z+1} &= \frac{1}{2}\frac{1}{k+k_1}(k_1 u(0) - u'(0)) + \frac{1}{2}\frac{2k+k_1}{k+k_1}u_{2z+1} + A_1, \\ u_{(2k-1)z+1} &= \frac{1}{2}\frac{1}{k+k_2}(k_2 u(1) - u'(1)) + \frac{2k+k_2}{2(k+k_2)}u_{(2k-2)z+1} + A_{2k-1}, \end{aligned} \tag{15.4}$$

where

$$A_i = \sum_{j=2}^{2z}(b_{z+1,j} - k^2\frac{x_{z+1}}{k+k_1}c_{1,j})u''_j + O(h_{z-s}^{p+1}),$$

$$A_{2k-1} = \sum_{j=2(k-1)z+2}^{2kz}(b_{z+1,j} + k^2\frac{x_{z+1}}{k+k_2}c_{2z+1,j})u''_j + O(h_{z-s}^{p+1}).$$

The formula (15.3) multiplies accordingly on the uncertain multipliers $\alpha_i(i = \overline{1, 2k-1})$ and selects these numbers so that ratios were executed:

$$\begin{aligned} u_{kz+1} &= \frac{2+k_2}{2(k_1+k_2+k_1k_2)}\alpha + \frac{2+k}{2(k_1+k_2+k_1k_2)\beta}\beta + \sigma_{kz+1}, \\ \sigma_{kz+1} &= (k_1+k_2+k_1k_2)^{-1}[(2+k_1)(k+k_1)A_1 \\ &+(2+k_2)\sum_{i=2}^{k-1}(2k+ik_1)A_i + k(2+k_1)(2+k_2)A_k \\ &+(2+k_1)\sum_{i=2}^{k-1}(2k+ik_2)A_{2k-i} + (2+k_1)(k+k_2)A_{2k-1}], \end{aligned} \tag{15.5}$$

$$u_{tz+1} = \frac{\alpha}{2k+(t+1)k_1} + \frac{2k+tk_1}{2k+(t+1)k_1}y_{(t+1z+1} + \Sigma^{[t]}, \qquad t = \overline{1, k-1},$$

$$\Sigma^{[t]} = \frac{2}{2k+(t+1)k_1}[(k+k_1)A_1 + \sum_{i=2}^{t}(2k+ik_1)A_i],$$

$$u_{(2k-t)z+1} = \frac{\beta}{2k+(t+1)k_2} + \frac{2k+tk_1}{2k+(t+1)k_2}u_{(2k-t+1)z+1} + \Sigma^{[2k-t]}, \qquad t = \overline{1, k-1},$$

where

$$\Sigma^{[2k-t]} = \frac{2}{2k+(t+1k_2}[(k+k_2)A_{2k-1} + \sum_{i=2}^{t}(2k+ik_2)A_{2k-i}].$$

From expressions (15.5), after some calculations, follows

$$u_{tz+1} = \frac{2k+(2k-t)k_2}{2k(k_1+k_2+k_1k_2)}\alpha + \frac{2k+tk_1}{2k(k_1+k_2+k_1k_2)}\beta + \sigma_{tz+1}, \qquad t=\overline{1,2k-1}, \tag{15.6}$$

where

$$\sigma_{tz+1} = \frac{2k+tk_1}{2k+kk_1}\sigma_{kz+1} + \sum_{j=t}^{k-1}\frac{2k+tk_1}{2k+jk_1}\Sigma^{[j]},$$

$$\sigma_{(2k-t)z+1} = \frac{2k+tk_2}{2k+kk_2}\sigma_{kz+1} + \sum_{j=t}^{k-1}\frac{2k+tk_2}{2k+jk_2}\Sigma^{[2k-j]}.$$

From the formulae (14.3) and (15.6) we can easily receive expressions, similar to (15.6), appropriate to the other netpoints of the area $\omega_h : x_{(t-1)z+i}$ $(i \neq z)$. We have:

$$u_{(t-1)z+i} = \frac{2k+(2k-2kx_i-t+1)k_2}{2k(k_1+k_2+k_1k_2)}\alpha + \frac{2k+(2kx_i+t-1)k_1}{2k(k_1+k_2+k_1k_2)} + +\sigma_{(t-1)z+1}, \tag{15.5a}$$

where

$$\sigma_{(t-1)z+i} = (1-kx_i)\sigma_{(t-1)z+1} + kx_i\sigma_{(t+1)z+1} + \sum_{j=2}^{2s} b_{ij}\Phi_{(t-1)z+h}$$

$$(t=\overline{2,2k-1},\ i=\overline{2,z+1}).$$

If we use the formulae of type:

$$u_i = k\frac{x_{2z+1}-x_i}{k+k_1}\alpha + k\frac{1+x_ik_1}{k+k_1}y_{2z+1} + \sum_{j=2}^{2s}(b_{ij} - k^2\frac{x_{2z+1}-x_i}{k+k_1}c_{1j})u_j'' + O(h_{z-s}^p)$$

and

$$u_{2kz+1-i} = k\frac{x_{2z+1}-x_i}{k+k_2}\beta + k\frac{1+x_ik_2}{k+k_2}y_{(2k-1)z+1} + \sum_{j=2(k-1)z+2}^{2kz}(b_{2z+2-i,j} + k^2\frac{x_{2z+1}-x_i}{k+k_2}c_{2z+1,j})u_j'' + O(h_{z-s}^p)$$

for bounding points x_i and $1-x_i$, $(i=\overline{2,z})$, analogously to the last formulae we will have:

$$u_i = \frac{1+(1-x_i)k_2}{k_1+k_2+k_1k_2}\alpha + \frac{1+x_ik_1}{k_1+k_2+k_1k_2}\beta + \sigma_i, \tag{15.5b}$$

$$u_{2kz+1-i} = \frac{1+x_ik_2}{k_1+k_2+k_1k_2}\alpha + \frac{1+(1-x_i)k_1}{k_1+k_2+k_1k_2}\beta + \sigma_{2kz+1-i},$$

where

$$\sigma_i = \frac{k + x_i k k_1}{k + k_1}\sigma_{2z+1} + \sum_{j=2}^{2z}(b_{ij} - k^2\frac{x_{2z+1} - x_i}{k + k_1}c_{ij})y_j'' + O(h_{z-s}^p),$$

$$\sigma_{2kz+1-i} = \frac{k + x_i k k_2}{k + k_2}\sigma_{2(k-1)z+1}$$

$$+ \sum_{j=2(k-1)z+2}^{2kz}(b_{2z+2-i,j} + k^2\frac{x_{2z+1} - x_i}{k + k_2}c_{2z+1,j})y_j'' + O(h_{z-s}^p).$$

We therefore will attach the last formulae (15.5a) and (15.5b) to the expressions (15.5)and shall name such set as the formulae of a type (15.5).

The formulae of type (15.5) are difference analogue of Green's function any arbitrary (fixed) degree of exactly concerning ordinates of unknown solution (compare with Berezin, Zgidkov [1958] or Schröder [1957]). To (15.5) should add the difference formulae respect derivatives of first order if the right hand function f depends from $u'(x)$.

It is evident that for these purposes use of the formulae of numerical differentiation there is inconvenient. However, if to take advantage generalized (Q) formulae (14.4) for the points $x_{(i-1)z+i}(i = \overline{1, 2z+1})$ and (15.6) (at $t = k-1, k+1$), for the derivative we receive the following expressions:

$$u'_{(k-1)z+1} = \frac{k_1\beta - k_2\alpha}{k_1 + k_2 + k_1k_2} + \sigma'_{(k-1)z+i}[f], \tag{15.7}$$

where

$$\sigma'_{(k-1)z+i} = \frac{2}{k_1 + k_2 + k_1k_2}$$

$$\times\left\{k_1\left[\frac{1}{2}k(2 + k_2)A_k + \sum_{i=2}^{k-1}(2k + ik_2)A_{2k-i} + (k + k_2)A_{2k-1}\right]\right.$$

$$\left.-k_2\left[\frac{1}{2}k(2 + k_1)A_k + \sum_{i=2}^{k-1}(2k + ik_1)A_i + (k + k_1)A_1\right]\right\} -$$

$$-k\sum_{j=2}^{2z}c_{i,j}y''_{(k-1)z+j} + O(h^{p-1}).$$

The construction of the one-parameter schemes will be completed, if to expressions of a type (15.6) and (15.7) we attach two Cauchy (initial) problems:

$$u_1'(x) = f(x, \lambda(x), u_1(x)), \qquad l_1 \le x \le 1,$$

$$u_1(l_1) = \gamma, \qquad l_1 = x_{(k-1)z+1},$$

$$u_1'(x) = f(x, \mu(x), u_1(x)), \qquad l_2 \ge x \ge 0,$$

$$u_1(l_2) = \delta, \qquad l_2 = x_{(k+1)z+1}.$$

Now we return to study the problem (15.1)–(15.2) and introduce the following values:

$$\begin{aligned}
\omega_1 &= \frac{1}{8} + \frac{1}{4(k_1 + k_2 + k_1 k - 2)} \left(4 + k_1 + k_2 + \frac{(k_2 - k_1)^2}{2(k_1 + k_2 + k_1 k_2)}\right), \\
\omega_2 &= \frac{1}{2(k_1 + k_2 + k_1 k_2)}(k_1 k_2 + 2max\{k_1, k_2\}), \\
\omega_2' &= \frac{1}{2} - \frac{k_1 k_2}{4(k_1 + k_2 + k_1 k_2)}, \\
\omega &= \max\{\omega_2, \omega_2'\}.
\end{aligned} \tag{15.8}$$

The following theorem is true.

Theorem 15.1. *Let the function $f(x, u(x), u'(x))$ be continuous with respect to x, satisfy a Lipschitz's condition relative to u and u' with constant L and L' respectively; in addition, let one of two conditions be executed:*

$$\begin{aligned}
\omega(L + L') &< 1, \\
\omega_1 L + \omega_2 L' &< 1.
\end{aligned} \tag{15.9}$$

At these restrictions the initial problem has the unique solution which can be constructed by an iterative method.

The proof of this theorem coincides with the scheme of the proofs of the theorems 13.2 and 13.3.

Now in the formulae of the type (15.6) and (15.7), we omit the remainder terms. We get the expressions for construction of the initial table. The Cauchy problem we shall replace by the multistep methods (see Henrici [1962]). We shall name the resulting system the difference scheme.

Following theorem is true.

Theorem 15.2. *For the problem (15.1)-(15.2) let one of conditions (15.9) be true.*

Then:

1) the difference scheme has a unique solution and the iteration method converges;

2) as in the case of the uniform grid ($p = 3, 5, 7$), as in the case of Gaussian grid ($p \geq 3$) convergence of the solution of the algebraic analogue to the solution of a problem (15.1)-(15.2) and its derivative has ($p - 1$)- degree respect h.

Proof of this theorem is similar to the proof of the theorem 13.2.

The following theorem is true.

Theorem 15.3. *The number of arithmetic operations which is necessary for the calculation of approximate solution $\overline{u}(x)$ and its derivative $\overline{u}'(x)$ has the order $k \ln k$.*

A proof of this theorem is based on the specific character of sums σ_{tz+1}. If we calculate σ_{kz+1}, then $\sigma_{tz+1} \forall t \neq k$ will be calculated, as it is contained in σ_{kz+1} as subsums.

Development of methods of solving of problems (15.1)–(15.2) has basic significance (e.g., see Keldish [1985]; Tichonov, Samarski [1961]) in the study multidimentional problems of mathematical physics. They arise by a natural image, if to take advantage any projective method as a Fourier's method, methods of semi-discretizations for evolutionary equations, methods of Kantorovich type in stationary problems, simplification of models using physical and geometrical hypothesises, etc..

Remark 15.1 To a system of differential problems of a form (15.1)-(15.2) are reduced for example: of the axial symmetry problems in the theory of shells and rod systems, application of methods of group B to two-dimentional equations, to the appropriate various theories of plates and shells (see, e.g., Ciarlet [1997]; Cruse [1978]; Friedrichs, Dresller [1961]; Goldenvaiser [1976]; Grigorenko [1973]; Vlasov [1959]; Yeh Kai-yuan, Rimrott [1993]; survey article of Filin [1974],...).

Further, gravity of development of methods of account of such problems are discussed in monograph of Doolan, Miller, Schilders [1983]; Mikhlin [1966], in case, when $k(t)$ in (14.1) is small parameter. As is known, in problems the mechanics of continuum media the problems (15.1)–(15.2) describe of layer effect phenomena.

The practical convenience of a generalized factorization method constructing algebraic analogue is the following: in difference from other high accuracy methods (Tichonov, Samarski [1961]; Volkov [1971]) *is not present necessity to make up the table of multiple integrals or derivatives from of initial data.*

16. To an Analysis of Numerical Methods for Solving Boundary Value Problems for Second Order Linear Differential Equations with a Small Parameter

16.1. Estimations of derivatives of a solution by using asymptotic expansion

Let us consider a singular perturbation two-point boundary value problem of the following form (Vashakmadze , Muradova [1996]):

$$L_\varepsilon y \equiv \varepsilon^2 y''(x) - q(x)y(x) = f(x), \quad 0 < x < 1, \tag{16.1}$$

$$y(0) = \alpha, \; y(1) = \beta, \tag{16.2}$$

$$q(x) > 0, \quad 0 < \varepsilon << 1.$$

In this subsection we studied the problem of asymptotic estimations of derivatives of a solution $y(x)$ of an initial BVP (16.1), (16.2) by means of some results of the article Vishik, Lusternik [1968].

Let the solution of (16.1), (16.2) be of the following form:

$$\begin{aligned} y(x) &= w_0(x) + v_{00}(x) + v_{01}(x) + \varepsilon(w_1(x) + v_{10}(x) + v_{11}(x)) \\ &\quad + \varepsilon^2(w_2(x) + v_{20}(x) + v_{21}(x)) + \ldots . \end{aligned}$$

Here $w_0(x)$ is a solution of a reduced problem, i.e., of a problem

$$L_0 y = -q(x)y(x) = f(x), \; (\varepsilon = 0). \tag{16.3}$$

$v_{i0}(x)(i \geq 0)$ are boundary layer functions of the zero order in the neighbourhood of 0, $v_{i1}(x)(i \geq 0)$ are boundary layer functions of the zero order in the neighbourhood of 1.

Let us find the first approximation $w_0 + v_{00} + v_{01}$. Then, decomposing $q(x)$ by the Taylor formula in neighbourhood of zero, introducing a new variable $t = \dfrac{x}{\varepsilon}$, and substituting this decomposition into (16.1), we have:

$$L_\varepsilon y = y''(t) - q(0)y(t)$$
$$- \left[q'(0)t\varepsilon + q''(0)\frac{(t\varepsilon)^2}{2!} + \ldots + q^{(m)}(0)\frac{(t\varepsilon)^m}{m!} + q^{(m+1)}(\xi)\frac{(t\varepsilon)^{(m+1)}}{(m+1)!}\right] y(t).$$

By introducing the definitions[1]:

$$L_{q(0)}y \equiv y''(t) - q(0)y(t), \quad Q_{i0}y \equiv -q^{(i)}(0)\frac{t^i}{i!}y(t), \ i = \overline{1,m},$$

$$Q_{m+1,0}y \equiv -q^{(m+1)}(\xi)\frac{t^{m+1}}{(m+1)!}y(t),$$

for the last expressions follows:

$$L_\varepsilon y = L_{q(0)}y + \sum_{i=1}^{m+1}(Q_{i0}y)\varepsilon^i. \tag{16.4}$$

Analogously, by decomposing $q(x)$ in the neighbourhood of $x = 1$, introducing a variable $t_1 = \dfrac{1-x}{\varepsilon}$ and substituting the result into (16.1), we obtain:

$$L_\varepsilon y = y''(t_1) - q(1)y(t_1)$$
$$+ \left[q'(1)t_1\varepsilon - q''(1)\frac{(t_1\varepsilon)^2}{2!} + \ldots \right.$$
$$\left. + (-1)^{m+1}q^{(m)}(1)\frac{(t_1\varepsilon)^m}{m!} + (-1)^{m+2}q^{(m+1)}(\eta)\frac{(t_1\varepsilon)^{(m+1)}}{(m+1)!}\right] y(t_1).$$

Assuming

$$L_{q(1)}y \equiv y''(t_1) - q(1)y(t_1), \quad Q_{i1}y \equiv (-1)^{i+1}q^{(i)}(1)\frac{t_1^i}{i!}y(t_1), \ i = \overline{1,m},$$

$$Q_{m+1,1}y \equiv (-1)^{m+2}q^{(m+1)}(\eta)\frac{t_1^{m+1}}{(m+1)!}y(t_1),$$

we have

$$L_\varepsilon y = L_{q(1)}y + \sum_{i=1}^{m+1}(Q_{i1}y)\varepsilon^i. \tag{16.5}$$

Now we define the first approximation as a sum of the solutions the following equations

$$L_0 y = f, \quad L_{q(0)}y = 0, \quad L_{q(1)}y = 0, \tag{16.6}$$

[1] When a number is complex, a comma is put for clarity.

and denote the corresponding solutions of these equations (16.6) as w_0, v_{00}, v_{01}. Below, of course, we demand also that $w_0 + v_{00} + v_{01}$ satisfies the conditions (16.2). In addition, v_{00} and v_{01} must have a character of the boundary layer , i.e., $v_{00}(\infty) = 0,\ v_{01}(\infty) = 0$. Thus, we must solve the problem:

$$\begin{aligned} L_{q(0)}v_{00} = 0, \quad & \left[v_{00}\left(\frac{x}{\varepsilon}\right) + v_{01}\left(\frac{1-x}{\varepsilon}\right)\right]_{x=0} = \alpha - w_0(0), \\ L_{q(1)}v_{01} = 0, \quad & \left[v_{00}\left(\frac{x}{\varepsilon}\right) + v_{01}\left(\frac{1-x}{\varepsilon}\right)\right]_{x=1} = \beta - w_0(1), \\ & v_{00}(\infty) = 0,\ v_{01}(\infty) = 0. \end{aligned} \tag{16.7}$$

Evidently, a solution of (16.7) is:

$$v_{00} = c_{00}\exp(-x\sqrt{q(0)}/\varepsilon),\ v_{01} = c_{01}\exp(-(1-x)\sqrt{q(1)}/\varepsilon). \tag{16.8}$$

Now, let us consider

$$y = w_0 + v_{00} + v_{01} + \varepsilon(w_1 + v_{10} + v_{11}) + \ldots + \varepsilon^m(w_m + v_{m0} + v_{m1}) + z_m, \tag{16.9}$$

where z_m denotes a remainder term.
Substitute (16.9) in (16.1), follows:

$$L_\varepsilon\left(w_0 + v_{00} + v_{01} + \sum_{i=1}^{m}\varepsilon^i(w_i + v_{i0} + v_{i1}) + z_m\right) = f$$

or

$$L_\varepsilon\left(w_0 + \sum_{i=1}^{m}\varepsilon^i w_i\right) + L_\varepsilon\left(v_{00} + \sum_{i=1}^{m}\varepsilon^i v_{i0}\right) + L_\varepsilon\left(v_{01} + \sum_{i=1}^{m}\varepsilon^i v_{i1}\right) + L_\varepsilon z_m = f.$$

If we represent the operator L_ε for the first summand in the form: $L_\varepsilon = L_0 + \varepsilon^2 L_1$, and for the second and the third summands we use (16.4), (16.5) correspondingly, we obtain:

$$\begin{aligned} &(L_0 + \varepsilon^2 L_1)\left(w_0 + \sum_{i=1}^{m}\varepsilon^i w_i\right) + \left(L_{q(0)} + \sum_{i=1}^{m+1}\varepsilon^i Q_{i0}\right)\left(v_{00} + \sum_{i=1}^{m}\varepsilon^i v_{i0}\right) \\ &+ \left(L_{q(1)} + \sum_{i=1}^{m+1}\varepsilon^i Q_{i1}\right)\left(v_{01} + \sum_{i=1}^{m}\varepsilon^i v_{i1}\right) + L_\varepsilon z_m = f, \end{aligned} \tag{16.10}$$

where

$$L_0 w_0 = f,\ L_{q(0)}v_{00} = 0,\ L_{q(1)}v_{01} = 0.$$

Unite terms for the identical degrees of ε in the every summand of the formula (16.10) and equate them to zero. Obviously, we have:

$$L_0 w_1 = 0, \quad L_0 w_i = -L_1 w_{i-2},\ i = \overline{2, m}, \tag{16.11}$$

$$L_{q(0)}v_{i0} = -\sum_{l=1}^{i} Q_{l0}v_{i-l,0}, \quad L_{q(1)}v_{i1} = -\sum_{l=1}^{i} Q_{l1}v_{i-l,1}, \; i = \overline{1,m}. \tag{16.12}$$

Solving the equations (16.11), (16.12) we construct the following approximations $w_i + v_{i0} + v_{i1}$ taking into account that $w_i + v_{i0} + v_{i1} = 0$ in the points 0 and 1. The solutions v_{i0}, v_{i1} must have again a character of the boundary layer, i.e., $v_{i0}(\infty) = 0$, $v_{i1}(\infty) = 0$. Evidently, these solutions v_{i0}, v_{i1} could be found as a sum of a partial solution of the nonhomogeneous equation (16.12) and a general solution corresponding homogeneous one. Thus,

$$v_{i0} = c_{i0}\exp(-x\sqrt{q(0)}/\varepsilon), \quad v_{i1} = c_{i1}\exp(-(1-x)\sqrt{q(1)}/\varepsilon),$$

where c_{i0}, c_{i1} are polynomials of the $2i$-th degree relative to t and t_1 correspondingly. Then from (16.10) we have:

$$f + \varepsilon^{m+1}g_m + L_\varepsilon z_m = f, \tag{16.13}$$

where for brevity a such definition is introduced:

$$g_m \equiv L_1 w_{m-1} + \varepsilon L_1 w_m + \sum_{i=1}^{m+1} \varepsilon^i Q_{i0} \sum_{k=m-i+1}^{m} \varepsilon^{k-m-1} v_{k0} +$$
$$+ \sum_{i=1}^{m+1} \varepsilon^i Q_{i1} \sum_{k=m-i+1}^{m} \varepsilon^{k-m-1} v_{k1}.$$

Now from (16.13) immediately follows:

$$L_\varepsilon z_m = -\varepsilon^{m+1} g_m. \tag{16.14}$$

In addition from (16.9) we have:

$$z_m(0) = 0, \; z_m(1) = 0. \tag{16.15}$$

Let us consider characteristic equations $\lambda^2 - q(0) = 0$ and $\lambda^2 - q(1) = 0$ for $L_{q(0)}y = 0$, $L_{q(1)}y = 0$ correspondingly. As every equation has two roots, each one of every pair has a negative real part, therefore, the conditions of a regular degeneration the problem (16.1), (16.2) to (16.3) in Vishik, Lusternik sense [1957] are fulfilled.

There is true the following statement.

Lemma 16.1 *Let $q, f \in C^{m+p}[0,1]$, then, for $z_m^{(i)}$, $i = \overline{0,p}$ we have:*

$$z_m^{(i)} = \varepsilon^{m+1-i}\overline{z}_{m,i}, \tag{16.16}$$

where $\overline{z}_{m,i}$ is bounded with respect to ε.

Proof. Let us consider the equation (16.14). For $\varepsilon \to 0$ it follows that $\varepsilon^{m+1}g_m \to 0$, and thus, $z_m \to 0$.

Moreover,

$$z_m = \varepsilon^{m+1}\overline{z}_{m,0}, \tag{16.17}$$

$\overline{z}_{m,0}$ is bounded respect ε.

Further, from (16.14) we have:

$$(L_\varepsilon z_m, z_m) = -(\varepsilon^{m+1} g_m, z_m)$$

or

$$\varepsilon^2 \int_0^1 z_m'' z_m dx - \int_0^1 q z_m^2 dx = -\varepsilon^{m+1} \int_0^1 g_m z_m dx.$$

If we integrate by parts the first intergal in the left side and use the boundary conditions (16.15) we obviously obtain:

$$\varepsilon^2 \int_0^1 (z_m')^2 dx + \int_0^1 q z_m^2 dx = \varepsilon^{m+1} \int_0^1 g_m z_m dx.$$

Now, if we apply the theorem to a middle value to the second integral in the left part of the last expression and use (16.17), it then immediatelly follows:

$$\varepsilon^2 ||z_m'||^2 = \varepsilon^{2(m+1)}\big((g_m, \overline{z}_{m,0}) - \mu||\overline{z}_{m,0}||^2\big),$$

where a norm is taken in the space $L_2[0,1]$, μ denotes a middle value of the function $q(x)$. $\overline{z}_{m,0}$, g_m are bounded with respect to ε. Therefore, $||z_m'||^2$ must have an order $2m$ with respect to ε, i.e., $z_m' = \varepsilon^m \overline{z}_{m,1}$, where $\overline{z}_{m,1}$ is bounded with respect to ε.

Further, from (16.14) it follows:

$$\varepsilon^2 z_m'' - q z_m = -\varepsilon^{m+1} g_m. \tag{16.18}$$

From here using (16.17), we have:

$$z_m'' = \varepsilon^{m-1}(q\overline{z}_{m,0} - g_m) = \varepsilon^{m-1}\overline{z}_{m,2}, \qquad \overline{z}_{m,2} \equiv q\overline{z}_{m,0} - g_m.$$

Let us prove (16.16) for $i = 3, 4, \ldots, p$ by an inductive method. Assume that (16.16) is fulfilled for $i = 2, 3, \ldots, n$ $(n < p)$. Then, from (16.18) we have:

$$\begin{aligned} z_m^{(n+1)} &= \frac{1}{\varepsilon^2}(q z_m - \varepsilon^{m+1} g_m)^{(n-1)} \\ &= \frac{1}{\varepsilon^2} \sum_{i=0}^{n-1} C_{n-1}^i q^{(n-1-i)} z_m^{(i)} - \frac{1}{\varepsilon^2} \varepsilon^{m+1-(n-1)} \overline{g}_{m,n-1}, \end{aligned}$$

where $\varepsilon^{-(n-1)}\overline{g}_{m,n-1}$ is a derivative of the $(n-1)$–th order of the function g_m. Then

$$z_m^{(n+1)} = \sum_{i=0}^{n-1} C_{n-1}^i q^{(n-1-i)} \varepsilon^{m-1-i} \overline{z}_{m,i} - \varepsilon^{m-n} \overline{g}_{m,n-1}$$

or

$$z_m^{(n+1)} = \varepsilon^{m-n}\Big(\sum_{i=0}^{n-1} C_{n-1}^i q^{(n-1-i)} \varepsilon^{n-1-i} \overline{z}_{m,i} - \overline{g}_{m,n-1}\Big).$$

As $\overline{z}_{m,i}$ and $\overline{g}_{m,n-1}$ are bounded respect ε, then the expression in the brackets also is bounded with respect to ε. Now, if we introduce a definition:

$$\overline{z}_{m,n+1} \equiv \sum_{i=0}^{n-1} C_{n-1}^{i} q^{(n-1-i)} \varepsilon^{n-1-i} \overline{z}_{m,i} - \overline{g}_{m,n-1},$$

then the expression (16.16) is true also for $n+1$.

Thus, the lemma is proved.

Let us estimate $y^{(i)}(x)$. The following theorem is true:

Theorem 16.1. *Let $f, q \in C^{p-2}[0,1]$.*

Then for the derivatives of the solution $y(x)$ of the problem (16.1), (16.2) are true inequalities:

$$|y'(x)| \le |w_0'(x)| + \frac{C_1}{\varepsilon}\Big(\exp\big(-x\sqrt{q(0)}/\varepsilon\big) + \exp\big(-(1-x)\sqrt{q(1)}/\varepsilon\big) + \varepsilon|\overline{z}_{0,1}(x)|\Big),$$

$$|y^{(i)}(x)| \le \frac{N_{i-2}(x) C_i}{\varepsilon^i}\Big(\exp\big(-x\sqrt{q(0)}/\varepsilon\big) + \exp\big(-(1-x)\sqrt{q(1)}/\varepsilon\big) + \varepsilon\overline{\overline{z}}_{0,i-2}(x)\Big), \tag{16.19}$$

where

$$i = \overline{2,p}, \quad N_{i-2}(x) = \max_{0\le j\le i-2} |q^{(j)}(x)|, \quad \overline{\overline{z}}_{0,i-2}(x) = \max_{0\le j\le i-2} |\overline{z}_{0,j}(x)|,$$

$$C_1 = \max\left\{ |c_{00}|\sqrt{q(0)}, |c_{01}|\sqrt{q(1)}, 1\right\},$$

$$C_i = \frac{(i-1)!}{\big([(i-2)/2]!\big)^2} \max\left\{ |c_{00}|(q(0))^{\frac{j}{2}}, |c_{01}|(q(1))^{\frac{j}{2}}, 1\right\}, \quad j = \begin{cases} 0, & q(0) \le 1, \\ i-2, & q(0) > 1 \end{cases}.$$

Moreover, if $f, q \in C^{m+p}[0,1]$, then for $y^{(i)}(x)$ the asymptotic expansion holds:

$$y^{(i)}(x) = \sum_{j=0}^{m} \varepsilon^j w_j^{(i)}(x) + \sum_{j=0}^{m} \varepsilon^{j-i}\big(\widetilde{v}_{j0,i}(x) + \widetilde{v}_{j1,i}(x)\big) + \varepsilon^{m+1-i}\overline{z}_{m,i}(x), \tag{16.20}$$

where

$$\widetilde{v}_{j0,i} = \widetilde{c}_{j0,i}\exp(-x\sqrt{q(0)}/\varepsilon), \quad \widetilde{v}_{j1,i} = \widetilde{c}_{j1,i}\exp(-(1-x)\sqrt{q(1)}/\varepsilon),$$

$\widetilde{c}_{j0,i}$, $\widetilde{c}_{j1,i}$ are polynomials of the $2i$-th degree relative to t and t_1 correspondingly.

Proof. Let us consider (16.9). For $m = 0$: $y(x) = w_0(x) + v_{00}(x) + v_{01}(x) + z_0(x)$. Futher, from (16.8) and the lemma 16.1 we have:

$$|y'(x)| \le |w_0'(x)| + \frac{1}{\varepsilon}\Big(|c_{00}|\sqrt{q(0)}\exp\big(-x\sqrt{q(0)}/\varepsilon\big) + |c_{01}|\sqrt{q(1)}\exp\big(-(1-x)\sqrt{q(1)}/\varepsilon\big) + \varepsilon|\overline{z}_{0,1}(x)|\Big).$$

If we substitute the last expression for $y(x)$ into the identity:

$$y^{(i)}(x) = \frac{1}{\varepsilon^2}\left(q(x)y(x) + f(x)\right)^{(i-2)},$$

as $q(x)w_0(x) + f(x) = 0$, then by (16.8) and the lemma 16.1 we have:

$$\begin{aligned}
|y^{(i)}(x)| &= \frac{1}{\varepsilon^2}\left|\left(q(x)\left(v_{00}(x) + v_{01}(x) + z_0(x)\right)\right)^{(i-2)}\right| \\
&\le \frac{1}{\varepsilon^2}\left|\sum_{j=0}^{i-2} C_{i-2}^{j} q^{(i-2-j)}(x)\left(v_{00}^{(j)}(x) + v_{01}^{(j)}(x) + z_0^{(j)}(x)\right)\right| \\
&\le \frac{N_{i-2}(x)(i-2)!}{\varepsilon^i\left([(i-2)/2]!\right)^2} \\
&\quad\times \sum_{j=0}^{i-2}\Big(|c_{00}|(q(0))^{\frac{j}{2}}\exp\left(-\,x\sqrt{q(0)}/\varepsilon\right) \\
&\qquad + |c_{01}|(q(1))^{\frac{j}{2}}\exp\left(-\,(1-x)\sqrt{q(1)}/\varepsilon\right) + \varepsilon|\overline{z}_{0,j}(x)|\Big).
\end{aligned}$$

From this inequality immediately follows (16.19).

(16.20) is obtained by differentiating (16.9) i-times and using the lemma 16.1. The theorem is proved.

16.2. On estimating the remaindler term of a multipoint method

Let us use a generalized finite difference method (developing in section 14 for $k(t) \equiv 1$, $k_1 = k_2 = 0$) for solving the problem (16.1), (16.2) with the uniform mesh. If we divide the segment [0,1] into the n equal parts, then, according to formulae (3)–(5) from Vashakmadze [1966], we have for our case the following system of the difference equations:

$$\begin{aligned}
\varepsilon^2 y_i &= \varepsilon^2\frac{2s-i}{2s-1}y(0) + \varepsilon^2\frac{i-1}{2s-1}y_{2s} + \sum_{j=2}^{2s-1} b_{ij}^{(2s)}[q_j y_j + f_j] \\
&\quad -\varepsilon^2\sum_{j=2}^{2s-1} b_{ij}^{(2s)} R''_{2s-1}(x_j), \quad i = \overline{2,s}\,,
\end{aligned} \tag{16.21}$$

$$\begin{aligned}
\varepsilon^2 y_i &= \frac{\varepsilon^2}{2}y_{i-s} + \frac{\varepsilon^2}{2}y_{i+s} + \sum_{j=i-s+1}^{i+s-1} b_{ij}^{(2s+1)}[q_j y_j + f_j] \\
&\quad -\varepsilon^2\sum_{j=i-s+1}^{i+s-1} b_{ij}^{(2s+1)} R''_{2s}(x_j), \quad i = \overline{s+1, n-s+1},
\end{aligned} \tag{16.22}$$

$$\varepsilon^2 y_i = \varepsilon^2 \frac{2s-n+i-2}{2s-1} y(1) + \varepsilon^2 \frac{n-i+1}{2s-1} y_{n-2s+2} + \sum_{j=n-2s+3}^{n} b_{i,j}^{(2s)} [q_j y_j + f_j] - \varepsilon^2 \sum_{j=n-2s+3}^{n} b_{ij}^{(2s)} R''_{2s-1}(x_j),$$
$$i = \overline{n-s+2, n}. \qquad (16.23)$$

Here recall again that the equations (16.21), (16.23) correspond to the bounding knots $x_i \leq (s-1)h$, $1 - x_i \leq (s-1)h$, $h = 1/n$, and formulae (16.22) correspond to the interior points,

Below let us use some results of section 13 and, for this purpose, rewrite the formula (13.16_2) again:

$$-\sum_{j=2}^{p-1} b_{ij}^{(p)} R''_{p-1}(x_j) = \frac{1}{x_p}\left(x_p \int_0^{x_i} dx \int_0^{x} r_{p-3}(t)dt - x_i \int_0^{x_p} dx \int_0^{x} r_{p-3}(t)dt \right), \quad (16.24)$$

$$r_{p-3}(x) = y''(x, x_2, \ldots, x_{p-1}) \prod_{l=2}^{p-1} (x - x_l), \; i = \overline{2, s}, \; p = 2s. \qquad (16.25)$$

For the interior and right bounding knots, obviously, we have expressions analogous to (16.24) and (16.25).

The inequalities (13.7) for convinience we rewrite in our case in the following form:

$$\left| \sum_{j=2}^{p-1} \varepsilon^2 b_{ij}^{(p)} R''_{p-1}(x_j) \right| \leq \varepsilon^2 \frac{c_1^1 M_p}{p!} h^p + \varepsilon^2 \frac{c_2^1 M_{p+1}}{(p+1)!} h^{p+1}, \quad p = 2s, \qquad (16.26)$$

$$\left| \sum_{j=2}^{p-1} \varepsilon^2 b_{s+1,j}^{(p)} R''_{p-1}(x_j) \right| \leq \varepsilon^2 \frac{c_3 M_{p+1}}{(p+1)!} h^{p+1}, \quad p = 2s+1, \; s \geq 1. \qquad (16.27)$$

Here again $M_p = \max\limits_{0 \leq x \leq 1} |y^{(p)}(x)|$.

On the other hand for M_p from (16.19) we have the following estimation:

$$M_p \leq \frac{\widetilde{N}_{p-2} C_p}{\varepsilon^p} (2 + \varepsilon \widetilde{z}_{0,p-2}), \qquad (16.28)$$

where

$$\widetilde{N}_{p-2} = \max_{0 \leq x \leq 1} N_{p-2}(x), \quad \widetilde{z}_{0,p-2} = \max_{0 \leq x \leq 1} \overline{\overline{z}}_{0,p-2}(x).$$

By (16.28) from (16.26), (16.27) it follows:

$$\left| \sum_{j=2}^{p-1} \varepsilon^2 b_{ij}^{(p)} R''_{p-1}(x_j) \right| \leq \frac{c_1^1 \widetilde{N}_{p-2} C_{p1}}{\varepsilon^{p-2}} (2 + \varepsilon \widetilde{z}_{0,p-2}) h^p$$

$$+\frac{c_2^1\widetilde{N}_{p-1}C_{p+1,1}}{\varepsilon^{p-1}}(2+\varepsilon\widetilde{z}_{0,p-1})h^{p+1},\quad p=2s,$$

$$\left|\sum_{j=2}^{p-1}\varepsilon^2 b_{s+1,j}^{(p)}R_{p-1}''(x_j)\right|\le\frac{c_3\widetilde{N}_{p-1}C_{p+1,1}}{\varepsilon^{p-1}}(2+\varepsilon\widetilde{z}_{0,p-1})h^{p+1},\quad p=2s+1,\quad s\ge 1,$$

where

$$C_{p1}=\frac{1}{p([(p-2)/2]!)^2}\max\left\{|c_{00}|\sqrt{q(0)}^j,|c_{01}|\sqrt{q(1)}^j,1\right\}.$$

According to the conclusions of Vashakmadze [1966], for an error $\eta=\max_i|\eta_i|$, where $\eta_i=\widetilde{y}_i-y_i$, $\widetilde{y}_i$ is an approximate solution, y_i is the exact solution, there is an estimation:

$$\eta\le ch^{2s}M_{2s},$$

where c is a constant, which is not depend on h.

Now, by (16.28) we have:

$$\eta\le ch^{2s}\frac{\widetilde{N}_{2s-2}C_{2s}}{\varepsilon^{2s}}(2+\varepsilon\widetilde{z}_{0,2s-2}).$$

Thus, the error depends also on ε and may be sufficiently large in the neighbourhoods zero and $x=1$. Below we can obtain necessary accuracy by increasing the number of knots.

We try to improve the results without decreasing of the step. Let us return to the formula (16.25). The function $y''(x,x_2,\dots,x_{p-1})$ we write in the following form:

$$\begin{aligned}&y''(x,x_2,\dots,x_{p-1})\\&\quad=\frac{y''(x)}{(x-x_2)(x-x_3)\dots(x-x_{p-1})}\\&\qquad+\sum_{i=2}^{p-1}\frac{y''(x_i)}{(x_i-x)(x_i-x_2)\dots(x_i-x_{i-1})(x_i-x_{i+1})\dots(x_i-x_{p-1})}.\end{aligned}\tag{16.29}$$

As $y''(x)=\dfrac{1}{\varepsilon^2}\{q(x)y(x)+f(x)\}$, then using the asymptotic expansion (16.9) for $y''(x)$ we obtain:

$$y''(x)=\frac{1}{\varepsilon^2}\left\{q(x)\left(\sum_{j=0}^{m}\varepsilon^j(w_j(x)+v_{j0}(x)+v_{j1}(x))+z_m(x)\right)+f(x)\right\}.$$

Then, as $w_0q=-f$, immediatly follows:

$$y''(x)=\frac{1}{\varepsilon^2}\left\{q(x)\left(v_{00}(x)+v_{01}(x)+\sum_{j=1}^{m}\varepsilon^j(w_j(x)+v_{j0}(x)+v_{j1}(x))+z_m(x)\right)\right\}.$$

Substitute the last expression for $y''(x)$ into (16.29), we have:

$$
\begin{aligned}
&y''(x, x_2, \dots, x_{p-1}) \\
&= \frac{1}{\varepsilon^2}\frac{u(x)}{(x-x_2)(x-x_3)\dots(x-x_{p-1})} + \frac{1}{\varepsilon^2}\frac{q(x)z_m(x)}{(x-x_2)(x-x_3)\dots(x-x_{p-1})} \\
&+\frac{1}{\varepsilon^2}\sum_{i=2}^{p-1}\frac{u(x_i)}{(x_i-x)(x_i-x_2)\dots(x_i-x_{i-1})(x_i-x_{i+1})\dots(x_i-x_{p-1})} \\
&+\frac{1}{\varepsilon^2}\sum_{i=2}^{p-1}\frac{q(x_i)z_m(x_i)}{(x_i-x)(x_i-x_2)\dots(x_i-x_{i-1})(x_i-x_{i+1})\dots(x_i-x_{p-1})},
\end{aligned}
$$

where

$$
u(x) \equiv q(x)\left(v_{00}(x) + v_{01}(x) + \sum_{j=1}^{m} \varepsilon^j (w_j(x) + v_{j0}(x) + v_{j1}(x))\right).
$$

From (16.25) we have:

$$
\begin{aligned}
r_{p-3}(x) = \frac{1}{\varepsilon^2}\Bigg(&u(x) - \sum_{i=2}^{p-1} l_i(x)u(x_i) \\
&+[(qz_m)(x, x_2, \dots, x_{p-1})]\prod_{i=2}^{p-1}(x-x_i)\Bigg),
\end{aligned}
\tag{16.30}
$$

where $[(qz_m)(x, x_2, \dots, x_{p-1})]$ denotes a divided difference of the $(p-2)$th order for the function $q(x)z_m(x)$.

According to the lemma 16.1 we have:

$$
\begin{aligned}
&[(qz_m)(x, x_2, \dots, x_{p-1})]\prod_{i=2}^{p-1}(x-x_i) \\
&= \frac{(qz_m)^{(p-2)}(\zeta)}{(p-2)!}\prod_{i=2}^{p-1}(x-x_i) \\
&= \frac{1}{(p-2)!}\sum_{i=0}^{p-2} C_{p-2}^i q^{(p-2-i)}(\zeta) z_m^{(i)}(\zeta)\prod_{i=2}^{p-1}(x-x_i) \\
&= \frac{1}{(p-2)!}\sum_{i=0}^{p-2} C_{p-2}^i q^{(p-2-i)}(\zeta)\varepsilon^{m+1-i}\overline{z}_{m,i}(\zeta)\prod_{i=2}^{p-1}(x-x_i).
\end{aligned}
$$

Then from (16.30) we obtain:

$$r_{p-3}(x)=\frac{1}{\varepsilon^2}\left(u(x)-\sum_{i=2}^{p-1} l_i(x)u(x_i)\right.$$

$$\left.+\frac{1}{(p-2)!}\sum_{i=0}^{p-2} C_{p-2}^i q^{(p-2-i)}(\zeta)\varepsilon^{m+1-i}\overline{z}_{m,i}(\zeta)\prod_{i=2}^{p-1}(x-x_i)\right).$$

Substituting the obtained expression for $r_{p-3}(x)$ in (16.24) and multiplying on ε^2 we evidently have:

$$-\varepsilon^2\sum_{j=2}^{p-1} b_{ij}^{(p)} R_{p-1}''(x_j) = -\sum_{j=2}^{p-1} b_{ij}^{(p)} u(x_j) + \int_0^{x_i} dx \int_0^{x} u(t)dt$$

$$-\frac{x_i}{x_p}\int_0^{x_p} dx\int_0^x u(t)dt + R_{z,1},$$

where

$$R_z^- = \frac{1}{(p-2)!}\sum_{j=0}^{p-2} C_{p-2}^j \varepsilon^{m+1-j}\left(\int_0^{x_i} dx\int_0^x q^{(p-2-j)}(\zeta)\overline{z}_{m,j}(\zeta)\prod_{l=2}^{p-1}(t-x_l)dt-\right.$$

$$\left.-\frac{x_i}{x_p}\int_0^{x_p} dx\int_0^x q^{(p-2-j)}(\zeta)\overline{z}_{m,j}(\zeta)\prod_{l=2}^{p-1}(t-x_l)dt\right).$$

Analogously, we obtain the same expressions for the remaider terms: R_z for the interior knots and R_z^+ for right bounding knots.

Thus, (16.21), (16.22) and (16.23) may be rewritten in another form, where remainder terms have estimations correspondingly:

$$|R_z^-| \le \frac{C_{0,1}\widetilde{N}_{2s-2}\widetilde{z}_{m,2s-2}}{([s-1]!)^2}\varepsilon^{m-2s+3}h^{2s},$$

$$|R_z| \le \frac{C_{0,2}\widetilde{N}_{2s-1}\widetilde{z}_{m,2s-1}}{([(2s-1)/2]!)^2}\varepsilon^{m-2s+2}h^{2s+1},$$

$$|R_z^+| \le \frac{C_{0,3}\widetilde{N}_{2s-2}\widetilde{z}_{m,2s-2}}{([s-1]!)^2}\varepsilon^{m-2s+3}h^{2s}.$$

When $s=1$, $m=0$ we obtain the estimation $O(h^3)$ for the interior knots and $O(\varepsilon h^2)$ for bounding knots. Besides, with increase of m the exactness is improved by increasing the order with respect to ε and in this case it is not necessary to decrease the step value.

17. Some Numerical Realizations

17.1. Numerical results for solving boundary value problems with boundary layers for second order ordinary differential equations

In this subsection we consider some examples of numerical solving the problem (16.1), (16.2) with the help of the formulae (16.21)–(16.23). Further, with the purpose of comparing data of these examples are solved (16.1), (16.2) by also exponentially fitted methods, Doolan, Miller, Schilders [1980] (scheme *II*.6.3):

$$-\varepsilon\sigma_i(p)D_+D_-y_i + q(x_i)y_i = f(x_i),\ 2 \le i \le n,$$
$$y_1 = \alpha,\ y_{n+1} = \beta, \tag{17.1}$$

where

$$\sigma_i(\rho) = \frac{\rho^2 q(x_i)}{4sh^2\frac{1}{2}\rho\sqrt{q(x_i)}},\quad \rho = \frac{h}{\sqrt{\varepsilon}},$$

$$D_+y_i = \frac{y_{i+1}-y_i}{h},\quad D_-y_i = \frac{y_i - y_{i-1}}{h},$$

$$|z_i| = |\widetilde{y}_i - y_i| \le ch^2.$$

Here $\sigma_i(\rho)$ is a fitted coefficient, $\widetilde{y}_i$ is an approximate solution, y_i is an exact solution.

The behaviour of the functions $q(x), f(x)$ essentially influences the exactness of the computations. Obviously, when $q(x)$ is rapidly oscillates, the fitted coefficient, which gives a good approximation on some subintervals, may be rough on the others.

Below, there are introduced the following designations:

$$z = \max_m |\widetilde{y}_m - y_m|,\quad m = \overline{2(1), 2^{k+1}n_0},\quad r = \max_{i(l)j(l)} |\widetilde{y}_{i(l)} - \widetilde{y}_{j(l)}|,$$

$$i(1) = 2^k + 1,\qquad i(l) = i(l-1) + 2^k,\qquad i(l) \le 2^k n_0,$$
$$j(1) = 2^{k+1} + 1,\qquad j(l) = j(l-1) + 2^{k+1},\qquad j(l) \le 2^{k+1}n_0,$$
$$l = 2, 3, \dots,\qquad k = 0, 1, 2, \dots.$$

Example 17.1.1.

$$\varepsilon y''(x) - (2 + \sin w\pi x)y(x) = \varepsilon(2 + w^2\pi^2 \sin w\pi x) - (2 + \sin w\pi x)(x^2 - \sin w\pi x),$$

$$y(0) = 0,\ y(1) = 1.$$

The exact solution: $y(x) = x^2 - \sin w\pi x$.

a) $w = 5$

ε	n	z		r	
		1	2	1	2
10^{-3}	50	$2.5 \cdot 10^{-5}$	0.38956	$9.7 \cdot 10^{-4}$	0.02196
10^{-5}	50	$6.1 \cdot 10^{-7}$	0.00227	$1.3 \cdot 10^{-5}$	0.00225
10^{-7}	50	$6.2 \cdot 10^{-9}$	$2.4 \cdot 10^{-5}$	$1.3 \cdot 10^{-7}$	0

Tab.17.1.1a. Number 1 corresponds to the multi-point method (16.21)–(16.23) ($s = 3$); — 2 corresponds to the exponentially fitted method (17.1).

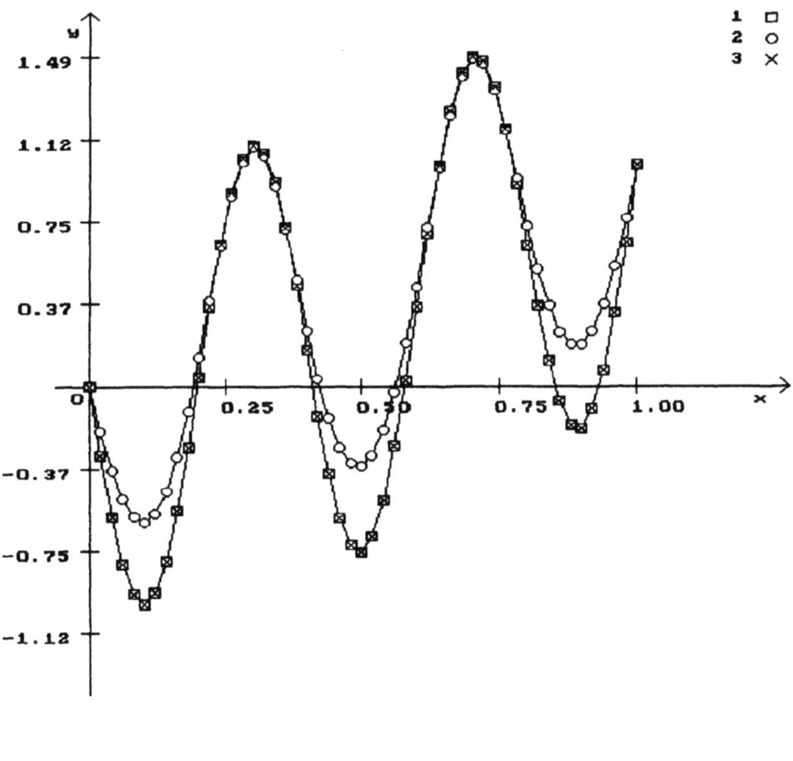

Fig.17.1.1a. $\varepsilon = 0.001$, $n = 50$. Number 1 corresponds to the multi-point method ($s = 3$); — 2 corresponds to the exponentially fitted method; — 3 corresponds to the exact solution.

b) $w = 10$

ε	n	z		r	
		1	2	1	2
10^{-3}	100	$4.3 \cdot 10^{-5}$	0.90713	0.00279	0.02264
10^{-5}	100	$3.1 \cdot 10^{-6}$	0.01855	$9.3 \cdot 10^{-5}$	0.01086
10^{-7}	100	$4.3 \cdot 10^{-5}$	$9.8 \cdot 10^{-5}$	$9.6 \cdot 10^{-7}$	$9.4 \cdot 10^{-10}$

Tab.17.1.1b. Number 1 corresponds to the multi-point method ($s = 3$); — 2 corresponds to the exponentially fitted method.

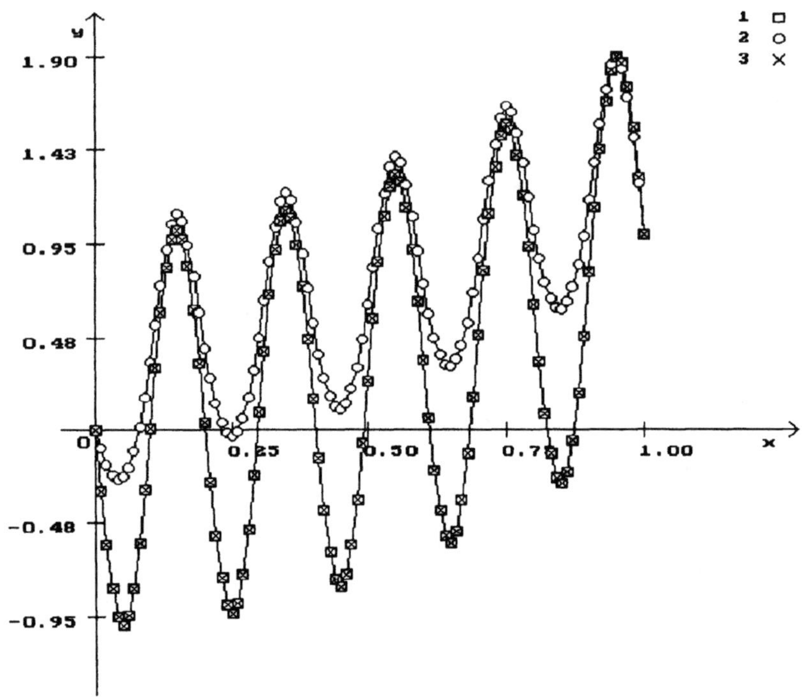

Fig.17.1.1b. $\varepsilon = 0.001$, $n = 100$. Number 1 corresponds to the multi-point method ($s = 3$); — 2 corresponds to the exponentially fitted method; — 3 corresponds to the exact solution.

Example 17.1.2.

$$\varepsilon y''(x) - (2 + \sin \pi x) y(x) = 2\varepsilon + \pi^2 \sin \frac{\pi}{\sqrt{\varepsilon}} x - (2 + \sin \pi x)\left(x^2 - \sin \frac{\pi}{\sqrt{\varepsilon}} x\right),$$

$$y(0) = 0, \; y(1) = 1 - \sin\left(\frac{\pi}{\sqrt{\varepsilon}}\right).$$

The exact solution: $y(x) = x^2 - \sin\left(\frac{\pi}{\sqrt{\varepsilon}}x\right)$.

ε	n	s	z 1	z 2	r 1	r 2
10^{-1}	50	3	$2.8 \cdot 10^{-6}$	1.13134	$2.7 \cdot 10^{-4}$	0.00270
		4	$1.4 \cdot 10^{-7}$		$5.1 \cdot 10^{-5}$	
		5	$5.6 \cdot 10^{-9}$		$6.0 \cdot 10^{-6}$	
10^{-2}	100	3	$5.9 \cdot 10^{-5}$	0.90354	0.00435	0.00614
		4	$6.9 \cdot 10^{-6}$		0.00222	
		5	$6.1 \cdot 10^{-7}$		$2.6 \cdot 10^{-4}$	
10^{-3}	300	3	$1.1 \cdot 10^{-4}$	1.04745	0.00590	0.01080
		4	$4.1 \cdot 10^{-5}$		0.01438	
		5	$1.0 \cdot 10^{-6}$		$6.4 \cdot 10^{-4}$	

Tab.17.1.2. Number 1 corresponds to the multi-point method; — 2 corresponds to the exponentially fitted method.

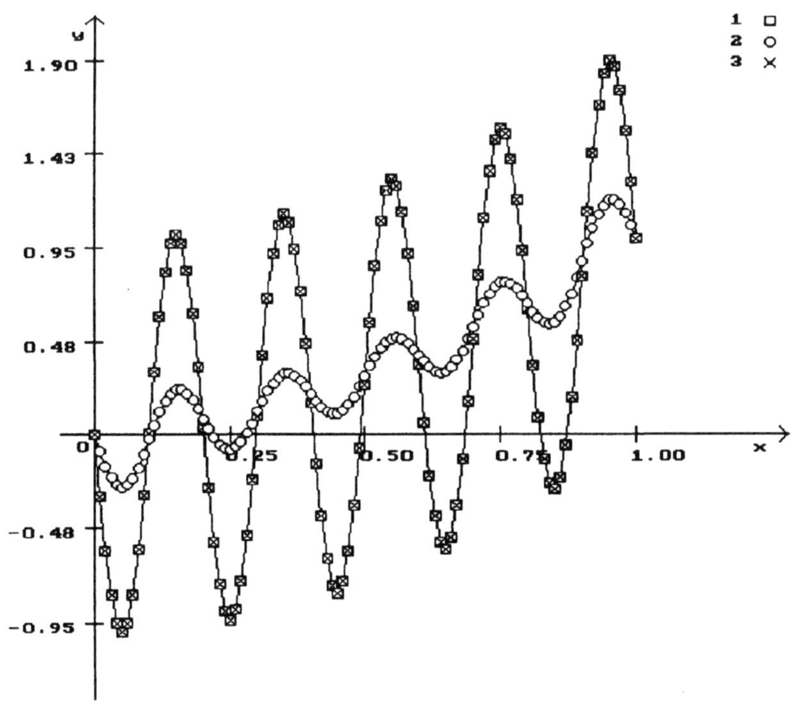

Fig.17.1.2. $\varepsilon = 0.01$, $n = 100$. Number 1 corresponds to the multi-point method ($s = 3$); — 2 corresponds to the exponentially fitted method; — 3 corresponds to the exact solution.

Example 17.1.3.

$$\varepsilon y''(x) - \left|x - \frac{1}{2}\right| y(x) = \left(1 - \left|x - \frac{1}{2}\right|\right) g(x),$$

$$g(x) = \frac{\exp(-(1-x)/\sqrt{\varepsilon}) + \exp(-x/\sqrt{\varepsilon})}{1 + \exp(-1/\sqrt{\varepsilon})}, \quad y(0) = y(1) = 1.$$

The exact solution: $y(x) = g(x)$.

ε	n	s	$z, 1$	$r, 1$
10^{-3}	100	3	$4.1 \cdot 10^{-5}$	0.00108
		4	$6.1 \cdot 10^{-5}$	$5.5 \cdot 10^{-4}$
		5	$1.8 \cdot 10^{-7}$	$4.6 \cdot 10^{-5}$
10^{-5}	300	3	0.00796	0.03798
		4	0.01963	0.11053
		5	0.00130	0.02204
10^{-7}	100	3	0.03723	0.01004
		4	0.07972	0.75232
		5	0.02958	0.00921

Tab.17.1.3. Number 1 corresponds to the multi-point method.

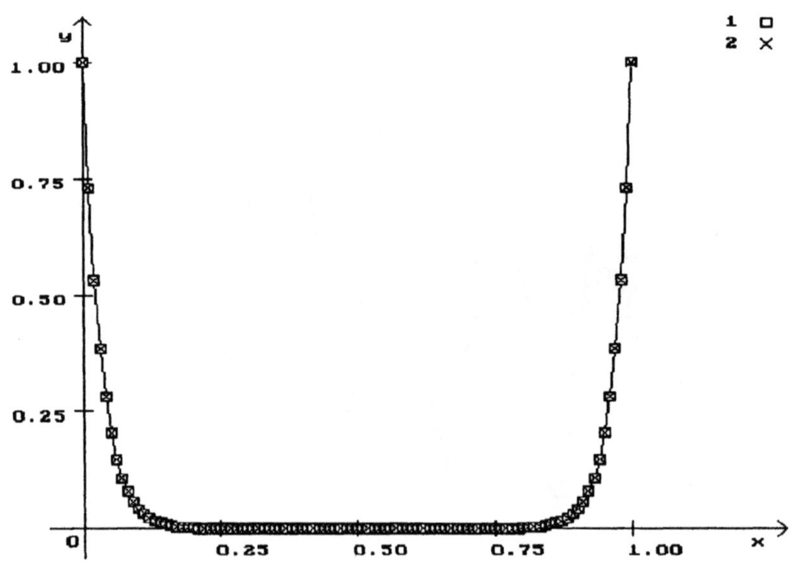

Fig.17.1.3. $\varepsilon = 0.001$, $n = 100$. Number 1 corresponds to the multi-point method ($s = 3$); — 2 corresponds to the exact solution.

17.2. Numerical design of shearing forces for elastic plates

In the case where the boundary conditions in the plate bending problem are split then with respect to a deflection we have to solve the biharmonic problem; The shearing forces satisfy Helmholtz's equations (see section 2):

$$\varepsilon^2 \Delta Q_\alpha(x,y) - Q_\alpha(x,y) = F_\alpha(x,y), \quad x,y \in D : (-1,1)^2, \tag{17.2}$$

where $\varepsilon^2 = \dfrac{h^2}{3}(1+2\gamma)$,

$$F_\alpha(x,y) = \varepsilon^2 D \Delta u^*_{3,\alpha}(x,y) - \frac{\varepsilon^2 h^2}{3(1-\nu)}(1+2\gamma)\partial_\alpha[g_3^+(x,y) - g_3^-(x,y)]$$
$$-\frac{\varepsilon^2(1+\nu)}{2(1-\nu)}\int_{-h}^{h}(h^2 - x^2) f_{3,\alpha}(x,y,t)dt$$
$$+\int_{-h}^{h} t f_\alpha(x,y,t)dt - h(g_\alpha^+(x,y) + g_\alpha^-(x,y))^1.$$

For defining a deflection u_3^* we consider the first BVP (12.4) and use discrete-variation scheme (12.15) if the coordinate functions are given by (12.2), at $\alpha = \beta = 2$.

Further, of course, for the shearing forces there hold the following BVP:

$$\varepsilon_1 \Delta Q_1(\xi,\eta) - Q_1(\xi,\eta) = F_1(\xi,\eta), \tag{17.3}$$

$$\begin{gathered} \partial_1 Q_1(\xi,\eta)\big|_{\xi=0} = \mu_0(0,\eta), \quad \partial_1 Q_1(\xi,\eta)\big|_{\xi=1} = \mu_1(1,\eta), \\ Q_1(\xi,0) = 0, \quad Q_1(\xi,1) = 0, \end{gathered} \tag{17.4}$$

$$\mu_0(0,\eta) = D_0(g_3^+(0,\eta) - g_3^-(0,\eta)),$$
$$\mu_1(1,\eta) = -D_0(g_3^+(1,\eta) - g_3^-(1,\eta)),$$

$$\varepsilon_1 \Delta Q_2(\xi,\eta) - Q_2(\xi,\eta) = F_2(\xi,\eta), \tag{17.5}$$

$$\begin{gathered} Q_2(0,\eta) = 0, \quad Q_2(1,\eta) = 0, \\ \partial_2 Q_2(\xi,\eta)\big|_{\eta=0} = \nu_0(\xi,0), \quad \partial_2 Q_2(\xi,\eta)\big|_{\eta=1} = \nu_1(\xi,1), \end{gathered} \tag{17.6}$$

$$\nu_0(\xi,0) = D_0(g_3^+(\xi,0) - g_3^-(\xi,0)),$$
$$\nu_1(\xi,1) = -D_0(g_3^+(\xi,1) - g_3^-(\xi,1)).$$

Here we change variables x, y by: $\xi = 0.5(x+1)$, $\eta = 0.5(y+1)$ and introduce $\varepsilon_1 = 0.25\varepsilon^2$.

[1] Here nonlinear terms are neglected.

The problems (17.3), (17.4) and (17.5), (17.6) are two-dimensional problems with boundary layer. For numerical solving of these BVP we have used the following iterative schemes defining in subsection 12.4:

$$
\begin{aligned}
&(r+1)\overset{(k+\frac{1}{2})}{Q_1}(\xi,\eta_j)+\varepsilon_1 A_1\overset{(k+\frac{1}{2})}{Q_1}(\xi,\eta)\big|_{\eta=\eta_j}\\
&\qquad = r\overset{(k)}{Q_1}(\xi,\eta_j)-\varepsilon_1 A_2\overset{(k)}{Q_1}(\xi,\eta)\big|_{\eta=\eta_j}-F_1^*(\xi,\eta_j),\\
&\partial_1\overset{(k+\frac{1}{2})}{Q_1}(\xi,\eta)\big|_{\xi=0,\eta=\eta_j}=\mu_0(0,\eta_j),\quad \partial_1\overset{(k+\frac{1}{2})}{Q_1}(\xi,\eta)\big|_{\xi=1,\eta=\eta_j}=\mu_1(1,\eta_j),\\
&\qquad j=\overline{2,n},
\end{aligned}
\tag{17.7}
$$

$$
\begin{aligned}
&(r+1)\overset{(k+1)}{Q_1}(\xi_i,\eta)+\varepsilon_1 A_2\overset{(k+1)}{Q_1}(\xi,\eta)\big|_{\xi=\xi_i}\\
&\qquad = r\overset{(k+\frac{1}{2})}{Q_1}(\xi_i,\eta)-\varepsilon_1 A_1\overset{(k+\frac{1}{2})}{Q_1}(\xi,\eta)\big|_{\xi=\xi_i}-F_1^*(\xi_i,\eta),\\
&\overset{(k+1)}{Q_1}(\xi_i,0)=0,\quad \overset{(k+1)}{Q_1}(\xi_i,1)=0,\quad i=\overline{2,n},
\end{aligned}
\tag{17.8}
$$

$$
\begin{aligned}
&(r+1)\overset{(k+\frac{1}{2})}{Q_2}(\xi,\eta_j)+\varepsilon_1 A_1\overset{(k+\frac{1}{2})}{Q_2}(\xi,\eta)\big|_{\eta=\eta_j}\\
&\qquad \cdot = r\overset{(k)}{Q_2}(\xi,\eta_j)-\varepsilon_1 A_2\overset{(k)}{Q_2}(\xi,\eta)\big|_{\eta=\eta_j}-F_2^*(\xi,\eta_j),\\
&\overset{(k+\frac{1}{2})}{Q_2}(0,\eta_j)=0,\quad \overset{(k+\frac{1}{2})}{Q_2}(1,\eta_j)=0,\quad j=\overline{2,n},
\end{aligned}
\tag{17.9}
$$

$$
\begin{aligned}
&(r+1)\overset{(k+1)}{Q_2}(\xi_i,\eta)+\varepsilon_1 A_2\overset{(k+1)}{Q_2}(\xi,\eta)\big|_{\xi=\xi_i}\\
&\qquad = r\overset{(k+\frac{1}{2})}{Q_2}(\xi_i,\eta)-\varepsilon_1 A_1\overset{(k+\frac{1}{2})}{Q_1}(\xi,\eta)\big|_{\xi=\xi_i}-F_2^*(\xi_i,\eta),\\
&\partial_1\overset{(k+1)}{Q_2}(\xi,\eta)\big|_{\xi=\xi_i,\eta=0}=\nu_0(\xi_i,0),\quad \partial_1\overset{(k+1)}{Q_2}(\xi,\eta)\big|_{\xi=\xi_i,\eta=1}=\nu_1(\xi_i,1),\\
&\qquad i=\overline{2,n},
\end{aligned}
\tag{17.10}
$$

where $A_1=-\partial^2/\partial x^2$, $A_2=-\partial^2/\partial y^2$, r is an iterative parameter.

Obviously, the formulae (17.7), (17.8) and (17.9), (17.10) represent BVPs with boundary layers for second order ordinary differential equations. For solving these problems the generalized factorization method and exponentially fitted methods (Doolan, Miller, Schilders [1980]) may be used.

Below, some examples of the numerical solution of bending problem (12.4), (17.3), (17.4) and (17.5), (17.6) are given. For solving (17.7), (17.8) and (17.9), (17.10) the generalized finite-difference method (see details in section 14) is applied.

Below n denotes a number of knots, K denotes a number of the iterations,

$$\underline{u_3^*}=\min_{i,j}u_{3,ij}^*,\ \overline{u_3^*}=\max_{i,j}u_{3,ij}^*,$$

$$\underline{Q}_\alpha = \min_{i,j} Q_{\alpha,ij},\ \overline{Q}_\alpha = \max_{i,j} Q_{\alpha,ij},\ \overline{z} = \max_{i,j} | \overset{(K)}{Q}_{\alpha,i,j} - \overset{(K-1)}{Q}_{\alpha,i,j} |.$$

In these examples we have taken: the length of the plate is $100cm$, $\nu = 0.17$, $E = 2 \cdot 10^5$, $f_i = 0$, $i = 1,2,3$, $g_i^{\pm} = 0$, $i = 1,2$, a number of $\chi^2 P(x_1)$ and $\chi^2 P(x_2)$ by subsection 12.1 is 10. In tables $D_0 = 1$, $n = 150$ and in corresponding figures for Q_α $h = 3$, $\gamma = 3.5$, $D_0 = 10000$, $n = 50$.

Example 17.2.1.

$g_3^+ = 0.05$, $g_3^- = -0.05$.

h	γ	K	$\overline{z}$	$\underline{u}_3^*$	$\overline{u}_3^*$	$\underline{Q}_\alpha$	$\overline{Q}_\alpha$
3	0.1	6	$7.885 \cdot 10^{-3}$	0	0.00344	−403.57	403.57
	0	6	$5.784 \cdot 10^{-3}$	0	0.00344	−418.18	418.18
	−0.1	6	$4.937 \cdot 10^{-3}$	0	0.00344	−435.26	435.26
3	0.5	8	$8.802 \cdot 10^{-3}$	0	0.00344	−360.40	360.40
1.5		6	$5.485 \cdot 10^{-3}$	0	0.02751	−467.91	467.91

Tab. 17.2.1.

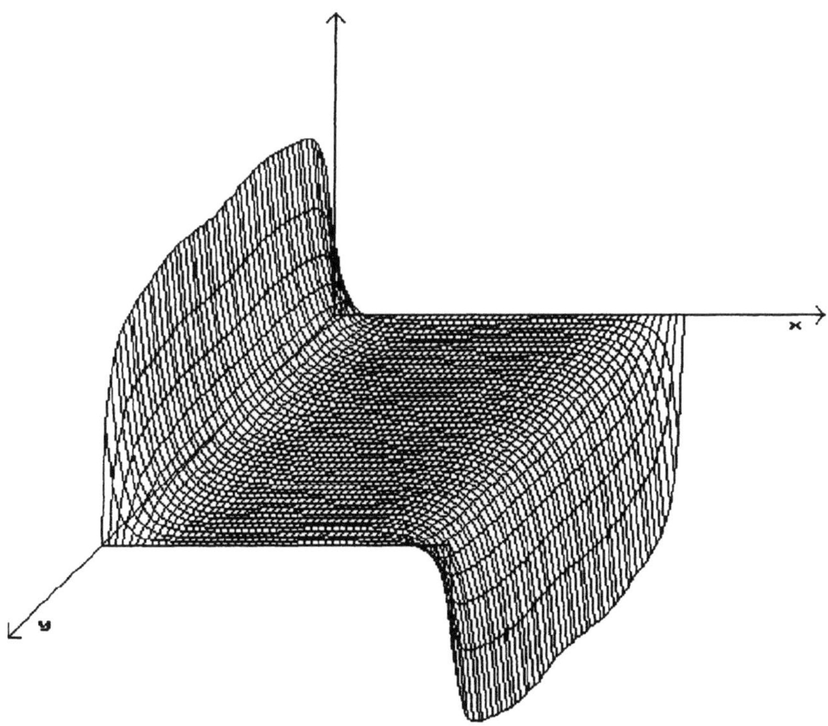

Fig. 17.2.1a. $\alpha = 1$, $K = 17$, $\overline{z} = 9.83 \cdot 10^{-2}$, $\underline{u}_3^* = 0$, $\overline{u}_3^* = 0.00344$, $\underline{Q}_1 = -5144.8$, $\overline{Q}_1 = 5144.8$.

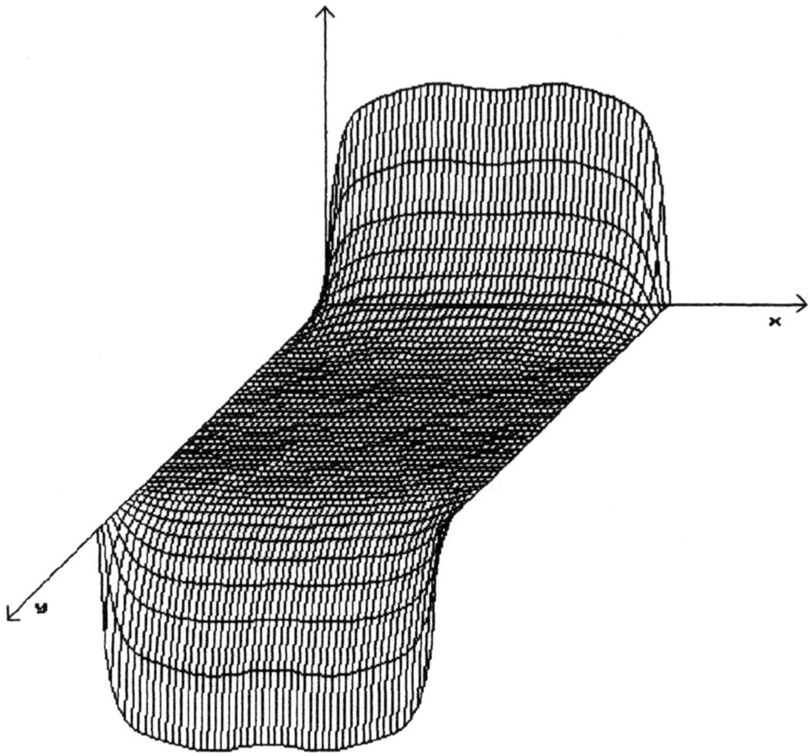

Fig. 17.2.1b. $\alpha = 2$, $K = 19$, $\overline{z} = 7.91 \cdot 10^{-2}$, $\underline{u}_3^* = 0$, $\overline{u}_3^* = 0.00344$, $\underline{Q}_2 = -5144.8$, $\overline{Q}_2 = 5144.8$.

Example 17.2.2.

$g_3^+ = 3.5P_3(x_1)P_4(x_2), \quad g_3^- = -3.5P_3(x_1)P_4(x_2).$

h	γ	K	$\overline{z}$	$\underline{u}_3^*$	$\overline{u}_3^*$	$\underline{Q}_1$	$\overline{Q}_1$
3	0.1	11	$8.189 \cdot 10^{-3}$	−0.00194	0.00194	−303.14	375.59
	0	10	$7.019 \cdot 10^{-3}$	−0.00190	0.00190	−301.15	372.49
	−0.1	8	$7.818 \cdot 10^{-3}$	−0.00187	0.00187	−299.17	369.39
3	0.5	15	$9.571 \cdot 10^{-3}$	−0.00207	0.00207	−311.07	388.11
1.5		7	$7.850 \cdot 10^{-3}$	−0.01455	0.01455	−296.20	364.89

Tab. 17.2.2a.

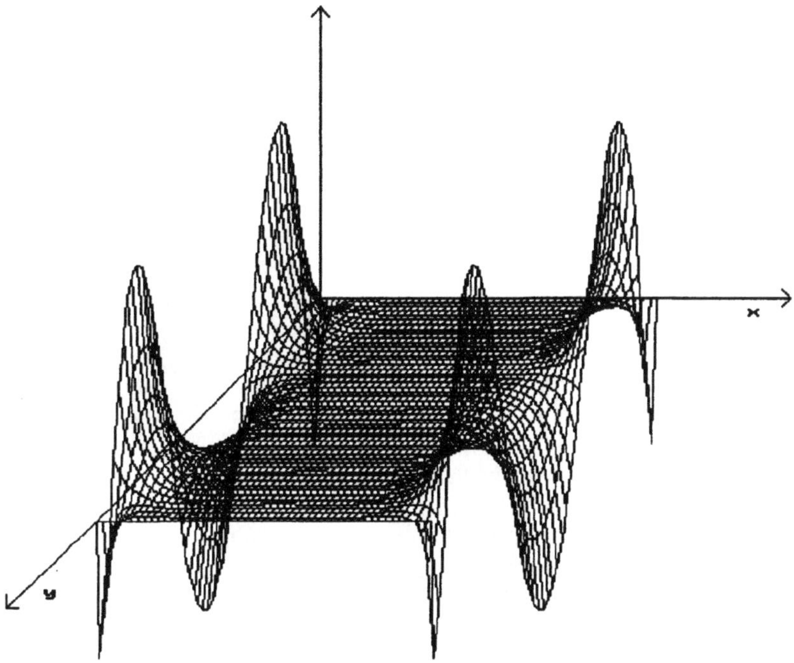

Fig. 17.2.2a. $\alpha = 1$, $K = 17$, $\overline{z} = 8.58 \cdot 10^{-1}$, $\underline{u}_3^* = -0.00307$, $\overline{u}_3^* = 0.00307$, $\underline{Q}_1 = -117402$, $\overline{Q}_1 = 126654$.

h	γ	K	$\overline{z}$	$\underline{u}_3^*$	$\overline{u}_3^*$	$\underline{Q}_2$	$\overline{Q}_2$
3	0.1	10	$9.704 \cdot 10^{-3}$	−0.00194	0.00194	−525.55	525.55
	0	9	$9.593 \cdot 10^{-3}$	−0.00190	0.00190	−511.54	511.54
	−0.1	8	$9.074 \cdot 10^{-3}$	−0.00187	0.00187	−497.84	498.84
3	0.5	14	$8.702 \cdot 10^{-3}$	−0.00207	0.00207	−587.64	587.64
1.5		6	$7.128 \cdot 10^{-3}$	−0.01040	0.01034	−342.98	342.98

Tab. 17.2.2b.

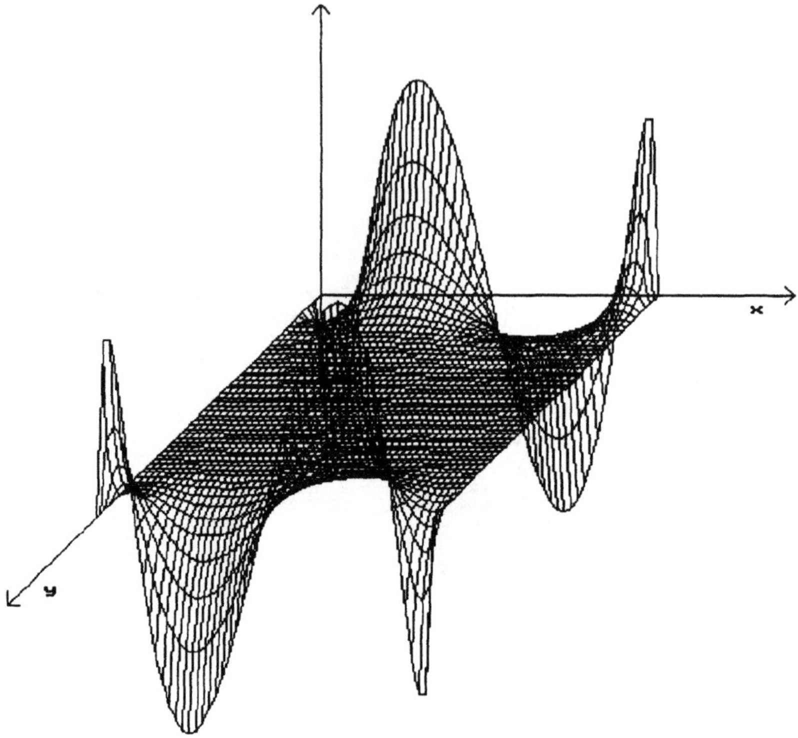

Fig. 17.2.2b. $\alpha = 2$, $K = 17$, $\overline{z} = 9.84 \cdot 10^{-1}$, $\underline{u}_3^* = -0.00307$, $\overline{u}_3^* = 0.00307$, $\underline{Q}_2 = -142014$, $\overline{Q}_2 = 142014$.

Example 17.2.3.

$g_3^+ = 0.05(1 - x_1^2)(1 - x_2^2)$, $g_3^- = -0.05(1 - x_1^2)(1 - x_2^2)$.

$Q_1(0, x_2) = 0$, $Q_1(1, x_2) = 0$, $Q_1(x_1, 0) = 0$, $Q_1(x_2, 1) = 0$,

h	γ	K	$\overline{z}$	$\underline{u}_3^*$	$\overline{u}_3^*$	$\underline{Q}_\alpha$	$\overline{Q}_\alpha$
3	0.1	9	$8.023 \cdot 10^{-3}$	−0.00001	0.00266	−222.61	222.61
	0	8	$9.836 \cdot 10^{-3}$	−0.00001	0.00266	−242.93	242.93
	−0.1	8	$7.497 \cdot 10^{-3}$	−0.00001	0.00265	−268.22	268.22
3	0.5	10	$8.688 \cdot 10^{-3}$	−0.00001	0.00269	−187.06	187.06
1.5		7	$6.036 \cdot 10^{-3}$	−0.00007	0.02113	−320.77	320.77

Tab. 17.2.3.

17.3. Standard programs for design boundary layer effects, one-dimentional case

In this subsection a program complex for the numerically realization of the algorithms, constructing in section 16 is represented.

The package of the programs contains the following parts: **PMETII, LJ, BIJ, FUNCT, AIJ, SOLSYST, GRAPHIC**.

PMETII is a control program and **LJ, BIJ, FUNCT, AIJ, SOLSYST, GRAPHIC** represent subprograms.

PMETII is meant for inputing data: ε, n, s, $y0$, $y1$, con and outputing data: y_i, $i=\overline{2,n}$, $con1$. ε is a small parameter, n is a number of points, s is a number of boundary knots, con is a convergence. Computing y_i, $i=\overline{2,n}$ as $y[i]$, $i=\overline{2,n}$ are fulfilled with double counting. If $con1 = \max_{i(l),j(l)} |y_{i(l)} - y_{j(l)}| < con$ then the process is interrupted, if this condition is not fulfilled then we continue computing for $2n$ points and etc. Obtained data $y[i]$, $i=\overline{2,n}$, $con1$ are outputed on the screen of computer or printer. It is foreseen constructing graphics of the obtained solution y_i, $i=\overline{1,n+1}$ on the screen of computer.

PMETII carries out a call aforesaid subprograms.

LIJ computes the polynomials of an interpolation formula.

BIJ computes coefficients b_{ij} as $b[i,j]$.

FUNCT computes $q(x)$ and $f(x)$.

AIJ forms the matrice of coefficients a_{ij} of the generalized finite difference method (16.21)-(16.23) which has a band structure. The matrice a_{ij}, $i,j=\overline{2,n}$ is written to memory of computer in the rectangle form as $a[j]\char`^[i]$, $j=\overline{2,2s+3}$, $i=\overline{2,n}$. Beginning with row $s+3$ the elements of the matrice are in memory of computer beginning with the first column. For all rows from $s+3$ to $n-2s+3$ the initial number of column of every element is decreased on 1 from every new row.

SOLSYST solves the algebraic system (16.21)-(16.23) by the Gauss exception method. In consequence, values $f0[i]$, $i=\overline{2,n}$ are obtained.

GRAPHIC constructs graphics of the obtained solution $y[i]$, $\overline{2,n}$ with the boundary conditions $y[1]=y0$, $y[n+1]=y1$.

```
program PMETII(input,output);
uses crt,graph;
const nn=1001;
ns=21;
type matrix=array[1..ns,1..ns] of double;
type avec=array[1..nn] of double;
adress=^avec;
amat=array[1..ns] of adress;
type vector=array[1..ns] of double;
label 1,2;

var
b,l : matrix;
```

```
b1 : vector;
a : amat;
y,f0 : avec;
eps,y0,y1 : double;
con,con1 : double;
i,j,n0,n,s,p,i1,ii,j1,m : integer;
u,v : string[2];

$I LJ
$I BIJ
$I FUNCT
$I AIJ
$I SOLSYST
$I GRAPHIC

BEGIN
for i:=1 to ns do
NEW (a[i]);
repeat
window(1,1,80,24);
clrscr;
writeln ('Input epsilon');readln(eps);
writeln ('Input y0');readln(y0);
writeln ('Input y1');readln(y1);
repeat
writeln ('Input the first n (n<=500)');readln(n0);
until n0<=500;
repeat
writeln ('Input s (s<=9)');readln(s);
until s<=9;
writeln ('Input convergence');readln(con);
n:=n0;
p:=2*s;
LJ;
for i:=2 to p-1 do
begin
for j:=2 to p-1 do
b[i,j]:=bij;
end;
p:=2*s+1;
LJ;
i:=s+1;
for j:=2 to p-1 do
b1[j]:=bij;
AIJ;
SOLSYST;
```

```
ii:=1;
1: if n>(nn-1)/2 then goto 2;
j:=ii+1;
for i:=2 to n0 do
begin
y[i]:=f0[j];
j:=j+ii
end;
n:=n*2;ii:=ii*2;
AIJ;
SOLSYST;
j:=ii+1;
con1:=abs(y[2]-f0[j]);
for i:=3 to n0 do begin
j:=j+ii;
if abs(y[i]-f0[j])>con1 then con1:=abs(y[i]-f0[j]);
end;
if con1>con then goto 1;
2: y[1]:=y0;y[n+1]:=y1;
for i:=2 to n do y[i]:=f0[i];
clrscr;
j:=ii+1;
writeln(' Numerical solution of the problem');
writeln;
writeln(' eps=',eps:2:10,', n=',n,', s=',s);
writeln;
writeln(' convergence=',con1:2:10);
writeln;
writeln(' The points of comparison ',n0-1);
writeln;
i1:=0;
window(1,10,80,24);
for i:=2 to n0 do
begin
i1:=i1+1;
if i1>12 then
begin
gotoxy(10,i1+1);
writeln(' Press any key to continue the view');
readkey;
clrscr;
i1:=1;
end;
gotoxy(1,i1);
writeln (' y[',j,']=',y[j]:2:15);
j:=j+ii;
```

```
end;
gotoxy(5,i1+2);
writeln(' Do you want to watch the grafics ?(y/n)');
readln(u);
if u='y' then
GRAPHIC;
writeln (' Do you want to continue the input epsilon,s,n,... ?(y/n)');
readln(u);
until u='n'
END.

procedure LJ;
var i,j,i1,m : integer;
fak : double;
begin
for j:=2 to p-1 do
begin
i1:=0;l[j,1]:=1;fak:=1;
for i:=2 to p-2 do l[j,i]:=0;
for m:=2 to p-1 do
begin
if j<>m then begin
fak:=fak*(j-m);
for i:=m-i1 downto 2 do
l[j,i]:=l[j,i-1]-(m-1)*l[j,i];
l[j,1]:=l[j,1]*(1-m)
end
else
i1:=1
end;
for m:=1 to p-2 do
l[j,m]:=l[j,m]/(m*(m+1)*fak)
end;
end;

function bij:double;
var r1,r2 : double;
function integral(t:double):double;
var ii,v : integer;
per :double;
begin
per:=l[j,p-2];
for ii:=p-3 downto 1 do
per:=l[j,ii]+per*t;
integral:=per;
end;
```

```
begin
r1:=integral(i-1);
r2:=integral(p-1);
bij:=((i-1)*((i-1)*r1-(p-1)*r2));
end;

{FUNCT}1
function q(x:double):double;
var w:double;
begin
w:=10;
q:=2+sin(w*pi*x);
end;
function f(x:double):double;
var w,ff : double;
begin
w:=10;
ff:=eps*(2+sqr(w*pi)*sin(w*pi*x));
f:=ff-(2+sin(w*pi*x))*(sqr(x)-sin(w*pi*x));
end;

procedure AIJ;
var
i,j,i1,j1,v1 : integer;
h,f1 : double;
label 3;
begin
for i:=1 to n do begin
f0[i]:=0;
for j:=1 to 2*s+3 do
a[j]^[i]:=0
end;
h:=1/n;
for i:=2 to s do begin
f1:=0;
for j:=2 to 2*s-1 do begin
a[j]^[i]:=-h*h*b[i,j]*q((j-1)/n);
f1:=f1+h*h*b[i,j]*f((j-1)/n)
end;
a[2*s]^[i]:=-eps*(i-1)/(2*s-1);
a[i]^[i]:=eps+a[i]^[i];
if n+1=2*s then f0[i]:=eps*(i-1)/(2*s-1)*y1+eps*(2*s-i)/(2*s-1)*y0+f1
else
f0[i]:=f1+eps*(2*s-i)/(2*s-1)*y0;
end;
```

[1]Here $q(x)$ and $f(x)$ are defined as in subsection 16.1.

```
i1:=0;
for i:=s+1 to n-s+1 do
begin
f1:=0;j1:=2;
if i>s+2 then i1:=i1+1;
for j:=i-s+1 to i+s-1 do begin
a[j-i1]^[i]:=-h*h*b1[j1]*q((j-1)/n);
f1:=f1+h*h*b1[j1]*f((j-1)/n);
j1:=j1+1;
end;
a[i-i1]^[i]:=eps+a[i-i1]^[i];
if i=s+1 then begin
a[i+s-i1]^[i]:=-eps/2;goto 3;
end;
if i=n-s+1 then begin
a[i-s-i1]^[i]:=-eps/2;goto 3;
end;
a[i-s-i1]^[i]:=-eps/2;a[i+s-i1]^[i]:=-eps/2;
3: f0[i]:=f1;
end;
if n=2*s then f0[s+1]:=f1+eps/2*(y0+y1)
else
begin
f0[s+1]:=f0[s+1]+eps/2*y0;
f0[n-s+1]:=f0[n-s+1]+eps/2*y1;
end;
for i:=n-s+2 to n do begin
f1:=0;j1:=3;
for j:=n-2*s+3 to n do begin
a[j1]^[i]:=-h*h*b[n-i+2,n-j+2]*q((j-1)/n);
f1:=f1+h*h*b[n-i+2,n-j+2]*f((j-1)/n);
if i=j then begin a[j1]^[i]:=eps+a[j1]^[i];end;
j1:=j1+1;
end;
a[2]^[i]:=-eps*(n-i+1)/(2*s-1);
if n+1=2*s then f0[i]:=eps*(n-i+1)/(2*s+1)*y0+eps*(2*s-n+i-2)/(2*s-1)*y1+f1
else f0[i]:=f1+eps*(2*s-n+i-2)/(2*s-1)*y1;
end;
end;

procedure SOLSYST;
var
i,j,m,i1,j1,j2,v1 : integer;
f1 : double;
label 1;
begin
```

```
i1:=s+2;j1:=0;
for m:=1 to n-2 do begin
f0[m+1]:=f0[m+1]/a[2]^[m+1];
for i:=m+2 to i1 do
begin
f0[i]:=f0[i]-f0[m+1]*a[2]^[i];
end;
for j:=3 to 2*s+2-j1 do begin
a[j]^[m]:=a[j]^[m+1]/a[2]^[m+1];
for i:=m+2 to i1 do begin
a[j]^[i]:=a[j]^[i]-a[2]^[i]*a[j]^[m];
end;
end;
for i:=m+2 to i1 do begin
for j:=3 to 2*s+3-j1 do
a[j-1]^[i]:=a[j]^[i];
end;
if i1>=n-s+1 then i1:=n
else i1:=i1+1;
if i1>n-s then j1:=j1+1;
end;
f0[n]:=f0[n]/a[2]^[n];
j1:=n-3;i1:=n-1;
for i:=n-1 downto 2 do begin
j1:=j1-1;f1:=0;
if (i<n-s) and (i>s) then i1:=i1-1;
for j:=i1 downto i do begin
j2:=j-j1;
f1:=f1+a[j2]^[i-1]*f0[j+1];
end;
f0[i]:=f0[i]-f1;
end;
end;

procedure GRAPHIC;
var
gr_treiber,gr_modus,errorcode : integer;
xmax,xmin,ymax,ymin,i : integer;
h,h1,hh,x1,x2,y1,y2,y11,y22,y111,y222,max,x : double;
t,t1 : string[5];
begin
gr_treiber:=detect;
initgraph(gr_treiber,gr_modus,'');
errorcode:=graphresult;
if errorcode <> grok then
begin
```

```
write(' Mistake in graphics');
writeln(grapherrormsg(errorcode));
closegraph;
halt;
end;
setbkcolor(0);
xmax:=530;xmin:=120;
ymax:=230;ymin:=20;
setcolor(3);
line (xmin,ymin,xmin,2*ymax-20);
line (xmin-20,ymax,xmax,ymax);
h1:=(ymax-50)/4;h:=h1;
max:=abs(y[1]);
for i:=2 to n+1 do
begin
if max<abs(y[i]) then
max:=abs(y[i]);
end;
x:=0;hh:=max/4;
for i:=1 to 4 do begin
line(xmin-5,round(ymax-h),xmin+5,round(ymax-h));
line(xmin-5,round(ymax+h),xmin+5,round(ymax+h));
x:=x+hh;
str(x:2:2,t);
outtextxy(xmin-41,round(ymax-h),t);
t1:='-'+t;
outtextxy(xmin-50,round(ymax+h),t1);
h:=h1+h;
end;
h1:=(xmax-210)/4;h:=h1;hh:=0.25;
x:=0;
for i:=1 to 4 do begin
line(round(xmin+h),round(ymax-5),round(xmin+h),round(ymax+5));
x:=x+hh;
str(x:2:2,t);
outtextxy(round(xmin+h-2),round(ymax+11),t);
h:=h+h1;
end;
outtextxy(xmax-20,ymax+5,'x');
outtextxy(xmin-10,ymax+5,'0');
outtextxy(xmin-13,ymin+9,'y');
line (xmin-5,ymin+5,xmin,ymin);
line (xmin,ymin,xmin+5,ymin+5);
line (xmax-5,ymax-5,xmax,ymax);
line (xmax-5,ymax+5,xmax,ymax);
moveto(xmin,ymax);
```

```
h1:=1/n*(xmax-210);x1:=xmin;
for i:=2 to n+1 do
begin
y1:=y[i-1]*(ymax-50)/max;
x2:=x1+h1;
y2:=y[i]*(ymax-50)/max;
setcolor(5);
line(round(x1),ymax-round(y1),round(x2),ymax-round(y2));
setfillstyle(0,0);
bar(round(x1)-3,ymax-round(y1)+3,round(x1)+3,ymax-round(y1)-3);
rectangle(round(x1)-3,ymax-round(y1)+3,round(x1)+3,ymax-round(y1)-3);
x1:=x2;
y1:=y2;
end;
setcolor (5);
bar(round(x1)-3,ymax-round(y2)-3,round(x1)+3,ymax-round(y2)+3);
rectangle(round(x1)-3,ymax-round(y2)-3,round(x1)+3,ymax-round(y2)+3);
readkey;
closegraph;
end;
```

Bibliography

Agmon, S.; Douglis, A.; Nirenberg, L. 1959. Estimates near the boundary for solution of elliptic partial differential equations, satisfying general boundary conditions. *Comm. Pure Appl. Math.* **12**, *N4*: 623–727.

Akhiezer, N. 1961. *The classic problem of moments.* Moscow: Physmathgiz.

Aksentian, O. 1967. The peculiarities of stress–strain state of the plate near the rib. *Appl. Math. and Mech.* **31**, *N1*: 178–186.

Ambartsumian, S. 1967. *The theory of anisotropic plates.* Moscow: Nauka.

Ambartsumian, S. 1974. *The general theory of anisotropic shells.* Moscow: Nauka.

Ambartsumian, S.; Bagdasarian, G.; Belubekian, M. 1977. *Magnetoelasticity of thin plates and shells.* Moscow: Nauka.

Antman, S.S. 1995. *Nonlinear Problems of Elasticity.* Springer–Verlag, Berlin.

Babuška, I.; Li, L. 1992. The problem of plate modeleing: Theoretical and computational results. *Comp. Meth. Appl. Mech. Engrg.* **100**: 249–273.

Bellman, R.E.; Kalaba, R.E. 1965. *Quasilinearization and non-linear boundary value problems.* Amer. Elsener Publ. Co. Inc. New York.

Berdichevski, L. 1973. The approach to the dynamic theory of thin elastic plates. *Mechanika Tviordogo Tela N6*: 99–109.

Berdichevski, L. 1983. *Variational principles of mechanics for continuous media.* Moscow: Nauka.

Berezin, I.; Zgidkov, N. 1959. *Numerical methods.* v.II, Moscow: Physmathgiz.

Besov, O.V.; Iliin, V.P.; Nikolski, S.M. 1975. *Integral representation for functions and embedding theorems.* Moscow: Nauka.

Bolotin, V. 1963. To the theory of layer plates. *Izvestia AS USSR. Tech. Depart.* **3**: 67–72.

Bourquin, F.; Ciarlet, P. 1989. Modeling and justification of eigenvalue problems for junctions between elastic structures. *J. Funct. Anal.* **87**: 392–427.

Brebbia, C.A. 1978. *The boundary element method for engineers.* Pentech Press. London.

Carleman, F. 1926. *Les functions quasi analitiques.* Paris.

Casti, J.; Kalaba, R. 1973. *Imbedding methods in Applied Mathematics.* Addison-Wesley Publ. Comp., Amsterdam.

Cheng Zhen–giong; Wang Xin–xi; Huang Mao–guang. 1993. A new technique applied to Reissner's plate theory with large deflection. *Proceeding 2-nd Inter. Conference in Nonlinear Mechanics, Beijing.* Peking University Press: 224–229.

Chien Wei zang. 1988. Variational principles and generalized variational principles for nonlinear elasticity with finite displacement. *Applied Math. Mech.*, **9**(1).

Ciarlet, P. 1978. *A method of finite elements for elliptical problems.* North–Holland Publ. Co.

Ciarlet, P. 1988. *Mathematical Elasticity, v.1: Three-Dimensional Elasticity,* North–Holland, Amsterdam.

Ciarlet, P. 1990. *Plates and junctions in elastic multi-structures.* RMA **14**, Springer Verlag, Heidelberg.

Ciarlet, P. 1994. *Mathematical Shell Theory. A linearly Elastic Introduction.* Université Pierre et Marie Curie, Paris.

Ciarlet, Ph. 1997. *Mathematical Elasticity:* v.II, Theory of Plates. Elsevier. Amsterdam–London–New York.

Ciarlet, P.; Rabier, P. 1980. *Les équation de von Kármán.* Springer Verlag, Heidelberg.

Constanda, C. 1990. *A mathematical analysis of bending of plates with transverce shear deformation.* Longman: Sci.& Tech. New York.

Courant, R. 1943. Variational methods for the solutions of problems of equilibrium and vibration.*Bull. Amer. Math. Soc.***49**, N1: 1–23.

Courant, R.; Hilbert, D. 1961. *Methods of mathematical physics.* v. **2**, New York.

Cruse, T.A. 1978. *The method of boundary integral equations. Numerical aspects and the application in mechanics.* Moscow: Mir.

Dafermos, C.M. 1968. Some remarks in Korn's inequality. *Z. Angew. Math. Phys.* **19**: 913–920.

Destyunder, P. 1981. Comparison entre les modeles tridimensionnels et bidimensionnels de plaques en élasticité. *RAIRO Analyse Numérique.* **15**: 331–369.

Destyunder, P.; Salaun, M. 1996. Mathematical Analisys of Thin Plate Models. *Mathematics and Applications,* v.**24.**

Dudchenko, A.; Lurie, S.; Obraztsov, I. 1984. *Anisotropic many-layered plates and shells. Reviews of sciences and technics.* Series of Mechanics of solids. **15**. Moscow: VINITI.

Dudnikov, V.; Nazarov, S. 1982. Asymptotically exact equations of thin shells, based on Cosserat's theory. *Proceed. of Tbilisi St. University, Math., Mech., Astr. 13-14.* Tbilisi University Press: 115–150.

Donnel, L. 1976. *Beams, plates and shells,* Mc. Grew–Hill Book Cor.

Doolan, E.P.; Miller, J.J.H.; Schilders, W.H.A. 1980. *Uniform numerical methods for problems with initial and boundary layers.* Bool Press, Dublin.

Dorodnitsin, A. 1960. To one way of the solution of laminar boundary layer. *Appl. Mech. and Tech. Phys. N3*: 111–118.

Duvaut, G.; Lions, J. 1972. *Les inéqualites en méchanique et en physique.* Paris,

Dunod.

Eidus, D. 1951. To the mixed problem of the theory of elasticity. *Soviet. Math. Dokl.. v.* **76**, N2: 181–184.

Faddeev, D.; Faddeeva, V. 1960. *Numerical methods of linear algebra.* Moscow: Physmathgiz.

Fellers, J.; Soler, A. 1970. Approximate solution of the finite cylinder problem using Legendre polynomials. *AJAA. v.* **8**, *N11*: 2037–2042.

Fichera, G. I973.*Existence theorems in elasticity.* Handb. d. Physik. Bd. YI 2. N3. Heidelberg: Springer–Verlag.

Filin, A. 1974. *Modern problems using of computers in solid mechanics.* Leningrad: Stroiizdat.

Filon, N. 1903. On an approximate solution for the bending at a beam of rectangular cross-section etc. *Phylosoph. Trans. of the Roy. Soc. of London. ser. A., v.* **201**: 63–155.

Fox, L. 1957. *The numerical solution of two-point boundary problems in ordinary differential equations.* Oxford, University Press.

Friedrichs, K. 1928. Randwert und Eigenwertproblems und theorie der Elastischen Platten. *Math. Ann.* **98**: 206–248.

Friedrichs, K. 1947. *On the boundary-value problems and Korn's inequality.* Ann. Math. v.**47** N2.

Friedrichs, K. 1950. Kirhhoff's boundary conditions and the edge effect for elastic plates. *Proc. Simpos. App. Mech. 3. Elasticity American Math. Soc.* New York: Mc-Graw-Hill.

Friedrichs, K., Dressler R. 1961. A boundary-layer theory for elastic plates. *Communs Pure Appl. Math.* **14**, N1.

Galinsh A., 1967. The calculation of plates and shells by refined theories. *Investigation by the theory of plates and shells.* Sbornik **5**. Kazan University Press: 66–92.

Gegelia, T. 1962. On some basic space boundary problems of the theory of elasticity. *Proceed. of Tbilisi Math. Inst.* **28:** 53–72.

Gilbert, R.; Hsiao, G.; Shneider, M. 1983. The two-dimensional linear optotropic plate. *Aplicable analysis,***15**: 147–169.

Godunov, S.; Prokopov, G. 1968. *The variational method to solution large systems of linear equations, arising in strong-elliptic problems.* Moscow: Rotaprint of Keldish Institute of Applied Mathematics.

Godunov, S.; Riabenki, V. 1962. *Introduction into the theory of differential schemes.* Moscow: Phys.–Math. Press.

Goldenveizer, A. 1976. *The theory of elastic plates and shells.* Moscow: Nauka.

Goldenveizer, A.; Koloss, A. 1965. To the construction of two-dimensional equations of the theory of elastic thin plates. *Prikladnaia Matematika i Mekhanika.* **29**: 141–155.

Gordeziani, D. 1974. To the exactness of one variant of the theory of thin shells. *Soviet. Math. Dokl..* **215, 4**: 751–754.

Gordeziani, D. 1974. On solvability of some boundary value problems for the single variant of the theory of thin shells. *Soviet. Math. Dokl..* **215, 6**: 1289–1292.

Gordeziani, D. 1975. Some inequalities for a variant of the theory of thin shells. *Sem. Inst. of Appl. Math. Annotation to Papers.* **10**: 7–11.

Gordeziani, D., Evseev, E. 1977. *On the numerical solution of one model of the arched dam.* Tbilisi University Press.

Gordesiani, D.; Samarski, A. 1978.To certain problems of thermoelasticity of plates and shells and the method of summarized approximation. *Complex analysis and its applications.* Moskow: Nauka: 62–71.

Gotsuliac, E.; Guliaev, V.; Tchibiriakov, V. 1973. Differential equations of thermoelastic state of shells with heat shock on the surface. *Appl. Mech. v.* **9**, 2: 32–41.

Grenander, U.; Szegö, G. 1958. *Teoplitz forms and their applications.* Berkley, Los–Angeles.

Grigoliuk, E.; Selezov, I. 1973. *Nonclassic theories of oscillations of beams, plates and shells.* V.**5**. Moscow: Nauka.

Grigorenko, I. 1973. *Isotropic and anisotropic layered shells of rotation with variable regidity.* Kiev: Naukova Dumka.

Guilinger, W.H. 1965. The Peaceman-Rachford method for small mesh increments. *Journal of Math. Analysis and Applications,* **11**: 261–277.

Guliaev, V.; Bagenov, V.; Lizunov, P. 1978. *Nonclassic theory of shells.* Lvov: Vischa shkola.

Guz, A.; Nemish, J. 1984. *Statics of elastic bodies of noncanonical form.* Kiev: Naukova Dumka.

Hellinger, E. 1914. Die allgemeine Ansatze der Mechanik der Kontnua. Leipzig: *Encyclopädie der Mathematishen Wissenschaften, v.IV/4*: 602–694.

Henrici, P. 1962. *Discrete variable methods in ordinary differential equation.* New York–London: J. Wiley and Sons Inc.

Hutchins, G.; Soler, A. 1973. Approximate elasticity solution for moderately thick shells of revolution. *J. App. Mech. APMW.* **9**, N73: 1–6.

Ianenko, N. 1966. *Method of fractional steps to the solution of poly- dimensional problems of mathematical physics.* Novosibirsk University Press.

Jackson, D. 1941. *Fourier series and orthogonal polynomials.* The Carus Math. Monographs. N6.

Jaiani, G. 1982. *Solution of some Problems for a Degenerate Elliptic Equation of Higher Order and Their Aplications to Prismatic Shells.* Tbilisi University Press.

Jaiani, G. 1996. Elastic Bodies with Non-smooth Boundaries-Cusped Plates and Shells. *ZAMM Z. angew. Math; Berlin 76 Suppl.* **2**, 117–120.

Kalandia, A. 1969. Remarks on peculiarities of elastic solution near angles. *Prikladnaia Matematika i Mekhanika. 33, iss.* **1**: 132–135.

Kalandia, A. 1973. *Mathematical methods in two-dimensional theory of elasticity.* Moscow: Nauka.

Kaldani, N. 1983. *Expansions of kernels of Karleman-Vekua's equations into series of generalized degree functions and their application.* Boundary value problems of the theory of generalized analytical functions and their application. Tbilisi University Press.

Kantorovich, L. 1933. One direct method of approximate solution of problem and on minimum of bi-integral. *Izvestia AS USSR. OMEN.* **5**: 647–652.

Kantorovich, L.; Akilov, G. 1959. *The functional analysis in normal spaces.* Moscow: Physmathgiz.

Kantorovich, L.; Krilov, V. 1962. *Approximate methods of high analysis.* Moscow: Physmathgiz.

Keldish, M. 1942. On Galiorkin's method for the solution of boundary problems. *Izvestia AS USSR. Math. Ser. v.* **66**:309–330.

Keldish, M.; Babenko, K.; Gelfond, I.; Dmitriev, N.; Lokitsievski, O.; Tchentsov, N. 1985. The solution of the problem with axially-symmetric flow of gas with shock wave. *M.Keldish: Select. Works. Math. Moscow: Nauka*: 377–420.

Kennard, B. 1953. The new approach to shell theory: circular cylinders.*J. App. Math., v.* **10.** N1.

Khatiashvili, G. 1983. *Almansi-Mitchel's problems for homogeneous and compound bodies,* v.**1**. Tbilisi: Metsniereba.

Khatiashvili, G. 1985. *Almansi-Mitchel's problems for homogeneous and compound bodies,*v.**2**. Tbilisi: Metsniereba.

Khoma, I. 1985. To fundamental matrix of the solution of a single system of differential equations in partial derivatives. *Differential Equations. v.* **21**, *N8*: 1459–1461.

Khoma, I. 1986. *The generalized theory of anizotropic shells.* Kiev: Naukova Dumka.

Khoroshun, L. 1978. To the construction of equations for layer plates and shells.

Appl. Mech. v. XIY, N **10**: 3–21.

Khvoles, A. 1971. The general representation for solutions of equilibrium equations of prismatic shell with variable thickness. *Seminar of the Inst. of Appl. Math. Annot. of reports, v.* **5** : 19–21.

Kiknadze, L. 1960. To the application of conformal mappings to the plane problems of the theory of elasticity. *Proceed. of Inst. Comp.Centre of Acad.Sci.GSSR. v.* **1**: 75–85.

Knops, R. J.; Payne, L.E. 1971. *Uniqueness theories in linear elasticity.* Springer Tracts in natural Philosophy. v.**19**.

Kohn, A.; Vogelius, M. 1984. A new model for thin plates with rapidly varying thickness. *Ins. J. Solids and Structures.* **20**: 333–350.

Koiter, W. 1971. On the mathematical foundation of shell theory. *Actes du Congress International de Mathematique. Paris. v.***3**: 123–130.

Koldorkina, V. 1973. On three-dimensional problems of elasticity theory in piecewise-smooth domains. *Izvestia. Mathematics. N1*: 65–70.

Koltunov, M.; Kravchuk, A.; Maiboroda, V. 1983. *Applied mechanics of strained elastic body.* Moscow: Visshaia shkola.

Kondratiev, V. 1967. Boundary value problems for elliptical equations in domains with conic and angular points. *Works of Moscow Math. Soc.* Moscow University Press: 209–292.

Konovalov, A. 1964. To one iterative scheme of solving statical boundary value problems of the elasticity theory. *Journal vichislitelnoi matematiki i matematicheskoi physiki, v.* **3**, N5: 942–946.

Korn, A.1907. Sur les équations de l'éstisite. *École Norm. Sup., ser. 3.* **24**: 9–75.

Korn, A. 1914. Uber Lösungen der Grundprobleme der Elastizitatstheorie. *Math. Annal.* **75**: 497–549.

Koselev, A. 1958. A priori estimastions in L_p and generalized solutions of elliptic equations and systems. *Uspekhi Matem. Nauk*, **13**, N4: 29–88.

Kovalenko, A. 1975. *Thermoelasticity.* Kiev: Visha Shkola.

Krasnoselski, M.; Vainikko, G.; Zabrieko, P.; Rutitski, J.; Stetsenko, V. 1969. *Approximate solution of operator equations.* Moscow: Nauka.

Kravchuk, A.S. 1974. Note to a proof of a convergence of the finite element method *Prikladnaia Matematika i Mekhanika, v.* **38**, *5*: 958–959.

Krilov, N.; Bogoliubov, N. 1929. La solution approchée du probleme de Dirichlét. *Soviet. Math. Dokl..* **11**: 283–288.

Kupradze, V. 1965. *Potential method in the theory of elasticity.* Jerusalem.

Kupradze, V. 1967. On approximate solution of problems of mathematical physics.

Uspekhi Matem.Nauk. **22**. *N2*: 59–107.

Kupradze, V.; Gegelia, T.; Basheleishvili, K.; Burchuladze, T. 1976. *Space problems in mathematical theory of elasticity.* Moscow: Nauka.

Lagnese, I.E., Lions, J.-L.1988. *Modelling, Analysis and Control of Thin Plates.* Paris: Masson, RMA 6.

Landau, L., Lifchitz, E. 1970. *Theory of Elasticity.* Pergamon Press.

Lavrentiev, M. 1956. On Cauchy's problem for Laplacian equation. *Izvestia AS USSR. Math. Ser.* **20**: 819–842.

Lax, P. 1957. A Phragman-Lingelöf theorem in harmonic analysis and its application to some questions in the theory of elliptic equations.*Communs Pure and Appl.* **10.**

Lax, P.; Milgram, An. 1954. *Parabolic equations, in Controbutions to the theory of partial differential equations.* Prinston.

Lekhnitski, S. 1947.*Anisotropic plates.* Moscow: Otdel. Gosizdat.

Lekhnitski, S. 1977. *The theory of elasticity of anisotropic body.* Moscow: Nauka.

Lembo, M. I.; Podio-Guidugli, P. 1990. Plate theory as an exact conseqnence of three-dimensional linear elasticity. *European J. Mech., A/Solids* **10**: 1–32.

Lions, J.-L.; Magenes, E. 1970. Problémes aux limites non homogénes et applications. *v. 1. Paris, Dunod.*

Lomakin, V. 1976. *The theory of elasticity of nonhomogeneous bodies.* Moscow University Press.·

Love, A. 1959. *A treatise of the mathematical theory of elasticity.* IY edit. Cambridge University Press.

Lukasiewicz, S. 1979.*Local loads in plates and shells.* Warszawa, Leyden: Noordhoff Inter. Publishing.

Lurie, A. 1942. To the theory of thick plates. *Prikladnaia Matematika i Mekhanika.6* **2–3** : 151–168.

Lurie, A. 1955. *Space problems of the theory of elasticity.* Moscow: Gosizdat.

Magnaradze, L. 1937. Base problems of the theory of elasticity for circuit with angular points. *Soviet Math.Dokl.* **16**: 165–186.

Magnaradze, L. 1984. The development of mathematical theory of elasticity of Vekua's shells, based on methods of complex analysis. *Theoretical and numerical methods for calculation of plates and shells. Proceed. All Union Conference.* Tbilisi University Press, v.**1**: 165–186.

Marchuk, G. 1977. *Methods of numerical mathematics.* Moscow: Nauka.

Mazija, V.; Nazarov, S.; Plamenievski, B. 1981. *Asymptotics of the solution of elliptic boundary problems by singular disturbance of the domain.* Tbilisi University

Press.

Meunargia, T. 1987. *Development of method of Vekua for the three-dimensional couple elasticity theory.* Tbilisi University Press.

Meunargia, T. 1991. Non-shallow shells. *Geometric function theory... Pitman Research Notes in Math Series,* **257**, *Longman*: 186–200.

Mikeladze, Sh. 1940. A generalization of a numerical integration method of differential equations with the help of mechanical quadrature formulas, *Proceedings of Tbilisi Math.Ins.* **7**: 47–63.

Mikeladze, Sh. 1951. *New methods for the integrating of differential equations.* Moscow-Leningrad: Gosizdat Fiz.–Matem. Literaturi.

Mikhlin, S. 1951. To Schwarz algorithm. *Soviet Math.Dokl.* **77**, N4: 569–571.

Mikhlin, S. 1952. *The problem of minimum of quadratic functional.* Moscow: Gostekhizdat.

Mikhlin, S. 1966. *Numerical realization of variational methods.* Moscow: Nauka.

Mikhlin, S. 1970. *Variational methods in mathematical physics.* Moscow: Nauka.

Mikhlin, S. 1977. *Linear equations in partial derivatives.* Moscow: Vishaia shkola.

Mindlin, R. 1951. Influence of rotatory inertia and shear on flexsual motions of izotropic elastic plate.*J.Appl.Mech.* **18:** 31–38.

Mindlin, R.; Medick, M. 1959. Extensional vibrations of elastic plates. *J. Appl. Math.,* **26.**

Morgenstern, D. 1959. Herleiting der Plattentheorie und der dreidimensionalen Elastizitatstheorie. *Arch. Ration. Mech. and Analysis,* **3**:

Morozov, N. 1978. *Selected two-dimensional problems of the theory of elasticity.* Leningrad University Press.

Morozov, N. 1984.*Mathematical problems on the theory of cracks.* Moscow: Nauka.

Morrey, C. 1959.'Second order equations in several variables and Holder continuity.*Math. Zeit.,* **72,** *N2*: 146–164.

Mosolov, P.; Miasnikov, V. 1971. The proof of Korn's inequality. *Soviet Math.Dokl.*. **1**: 36–39.

Muskhelishvili, N. 1916. On thermal stresses in the planar BVP of theory elasticity. Petrograd: *Izvestia elektrotekhnicheskogo Instituta,v.* **13**: 23–37.

Muskhelishvili, N. 1966. *Some basic problems of mathematical theory of elasticity.* Moscow: Nauka.

Muskhelishvili, N. 1968. *Singular integral equations.* Moscow: Nauka.

Nečas, J.; Hlaváček, I. 1981. *Mathematical Theory of Elastic and Elasto-Plastic*

Bodies: An Introduction. Elsevier, Amsterdam.

Nikolski, S. 1979. *Quadrature formulas*. Moscow: Nauka.

Novacki, W. 1970.*Theoria sprezistości*. Warszava: PWM.

Novožilov, V. 1951. *The theory of thin shells*. Leningrad: Sudpromgiz.

Oleinik, O.; Iosifian, G. 1978. Saint–Venant's principle in the plane theory of elasticity and boundary problems for biharmonic equation in unlimited domain. *Ciberian Math. Magaz. XIX, N5*: 1154–1165.

Oleinik, O.; Iosifian, G.; Shamaev, A. 1990. *Mathematical problems of theory of strong non-homogeneous elastic medium*. Moscow University Press.

Oliveira, E.R.A.E. 1970. Completeness and convergence in the finite element method. *Tecnica*, **33**, *N 403*.

Ovtchinikov, E.E., Xantis, L.S. 1977. The Korn's type inequality in subspaces and thin elastic structures, *Royal Society Proceedings: Math., Phys., Engin Sci.*, *v.* **453**, *N 1968*: 2003–2016.

Palmov, V. 1983. To one simplest variant of non-contradictory theory of shells. *XIII All Union Conf. "The theory of plates and shells", v. IV. Tallin.*

Panč, V. 1975. *Theories of elastic plates*. Prague: Academia.

Parton, V.; Perlin, P. 1981. *Methods of mathematical theory of elasticity*. Moscow: Nauka.

Parton, V.; Kudriavtsev, B. 1988.*Electromagnetoelasticity of piezoelectrics and electrically conductive solids*. Moscow: Nauka.

Peaceman, D.W. ; Rachford, H. H. 1955. The numerical solution of parabolic and elliptic differential equations. *J. Soc. Ind. Appl. Math.* **3**: 28–41.

Pelekh, B.; Lazko, V. 1982. *Layer anisotropic plates and shells with stress concentrators*. Kiev: Naukova Dumka.

Perlin, P.I. 1982. *Application of a method of the theory of potential to space problems of the theory of elasticity*, Moscow: Manuscript of dissertation. Moscow Inst. Electr. Industr. Engineering. 178p.

Picard, E. 1930. *Lessons sur quelques problèms aux limites de la theorie des equations differentialles*. Gauthier-Villars, Paris.

Pollard, H. 1947. The mean convergence of orthogonal series.*Trans. Amer. Math. Soc.*, **62**: 387–403.

Polski, N. 1955. To convergence of some approximate methods of analysis. *Ukr. Math. Magaz.*, **7**, *N1*: 55–70.

Poniatovski, V. 1962. The theory of plates.*Prikladnaia Matematika i Mekhanika*. **26**: 335–341.

Prager, W.; Singe, Y. 1947. Approximations in elasticity based on the concept of function space. *Quart. Appl. Math., v.* **5**: 241–269.

Rayleigh, J. 1977. *Theory of sound.* London.

Reissner, E. 1944. On theory of bending of elastic plates. *J. Math. Phys.,v.* **23:** 184–191.

Reissner, E. 1950. On a variational theorem of elasticity. *J. Math. Phys.* **v. 29**: 90–95.

Rektorys, K. 1980. *Variational methods in mathematics.* Science and Engineering. Prague: Academie.

Rodrigues, J.F. 1987. *Obstacle Problems in Math. Physics.*North–Holland, Amsterdam.

Saint-Venant, B. de. 1855. Mémoire sur la torsion de prizmes, etc. *Mém. des Savants Etrangers, Paris, t. XIV:* 233–560.

Samarski, A. 1967. On reguliration difference schemes. *Journal Vichislitelnoi Matematiki i matematicheskoi physiki,v.* **7** N1: 62–93.

Sard, A. 1949. Best approximate integration formulas, best approximate formulas.*Amer. J. of Math., LXXI:* 80–91.

Savin, G.; Khoma, I. 1971. To the theory of anisotropic shells free from kinematic hypotheses of normal element. *Appl. Math. v. 7, iss.* **3**: 9–15.

Schröder, I.J. 1957. Über das differenzverfahren bei nichtlinearen Randwertaufgaben. *I.Z. Angew Math.Mech.*,**36**: 319–331.

Schröder, I.J. 1957. Über das differenzverfahren bei nichtlinearen Randwertaufgaben. *II.Z. Angew Math.Mech.*, **37**: 443–455.

Shleniov, M. 1975. Asymptotic method for the solution of boundary problems at the theory of Vekua plates. *I All-Union school on the theory and numercal calculation of plates and shells.* Tbilisi University Press: 269–290.

Sobolev, S. 1936. Schwarz algorithm in the theory of elasticity. *Soviet Math.Dokl..* 4(13), **6**: 235–238.

Sobolev, S. 1950. *Some applications of functional analysis in mathematical physics.* Leningrad University Press.

Sobolev, S. 1975. *Introduction into the theory of volumetric formulas.* Moscow: Nauka.

Soler, A. 1968. Higher order effects in thick rectangular elastic beams Inter. *J. Solid Struct., v.* **4.** Pergamon Press: 723–739.

Strang, G.; Fix, G. 1973. *An analysis of the finite element method.* Prentice-Hall II.

Szegö, G. 1959. Orthogonal polynomials. *New York: Amer. Math. Society.*

Trabucho, L.; Viaño; J.M. 1988. A derivation of generalized torsion theory from three-dimensional elasticity by asymptotic expansion methods. *Applicable Analy-*

sis **31**: 129–148.

Thompson, J. 1969. Some existence theorems for traction boundary value problem of linearized elastostatics. *Arch. Rational Mech. and Anal. v.* **35**, N5: 369–399.

Tikhonov, A.; Arsenin, V. 1974. *Methods of solution of incorrect problems.* Moscow: Nauka.

Timoshenko, S. 1957. *Science history on material resistence. With brief data from the history of elasticity and theory of construction.* New York–Toronto–London, Mc. Graw-Hill Co.

Timoshenko, S.; Voinovski–Krieger, S. 1959. *Theory of plates and shells.* New York–Toronto–London, Mc. Graw-Hill Co.

Todhanter, I.; Pearson, K. 1886. *A history of theory of elasticity and of the strength of materials, v.* **1.** Cambridge.

Todhanter, I.; Pearson, K. 1893. *A history of theory of elasticity and of the strength of materials, v.* **2.** Cambridge.

Toupin, H. 1965. Saint-Venant's principle. *Arch. Ration. Mech. and Anal.* **18**, *N2:* 83–96.

Tskhadaia, F. 1977. *Numerical solution of some boundary problems of the theory of elastic shells.* Tbilisi University Press.

Tvalchrelidze, A. 1984. Theory of elastic shells using some basic surfaces. *Theory and numerical methods of design plates and shells. Proceed. All-Union Conference-Seminar in Tbilisi, v.* **2**: 315–322.

Vaindiner, A. 1973. Some problems of approximation of functions with many variables and effective direct results of the solution of problems on the theory of elasticity. *Sbornic "Elasticity and non-elasticity", iss.* **3.**: 16–46.

Vallée–Poussin de la, Ch. 1933. *Course of Analysis*, v.II, Moscow–Leningrad: Theor.Tekh.Izdat. (translated from English).

Vashakmadze, Tamara. 1989. System of Vekua for anisotropic nonhomogeneous elastic plates. *Proceedings of Vekua Inst.of Appl. Mathematics. v.* **37**: 73–79.

Vashakmadze, Tamaz. 1964. On numerical solution of boundary value problems. *Journal Vichislitelnoi Matematiki i Matematicheskoi Physiki.* v.**4**, N4: 623–637.

Vashakmadze, T. 1966. A generalized finite-difference method. *Differential equations, v. II, N5*: 614–618.

Vashakmadze, T. 1972. Method of generalized factorization and its application to numerical solution some problems mathem. physics. *Proceedings of Sump. Continuum Mechanics and Related Problems of Analysis. Tbilisi: Metsniereba, v.* **1**: 36–45.

Vashakmadze, T. 1973. Three-point operator equations of the theory of shells. *Seminar Inst.of Appl. Math. Reports.* **8**: 23–29.

Vashakmadze, T. 1978. To the investigation of operator of the theory of Vekua's elastic shells. *Complex analysis and its application. Moscow: Nauka:* 102–107.

Vashakmadze, T. 1981a. On the error estimation of transition from some three-dimensional problems of the elasticity theory to two-dimensional models. *Seminar Vekua Inst. of Appl. Math. Reports.* **15**: 57–66.

Vashakmadze, T. 1981b. On the accuracy of approximation of a problem in elasticity theory. *Soviet Math. Dokl. v.* **24**, N3: 568–570.

Vashakmadze, T. 1981c. On some numerical processes solving of linear BVPs theory of elasticity. *Computation with rarely matrices. Material of IV All Union Conf.: Varitional-difference methods in Math. Phys. Novosibirsk:* 5–13.

Vashakmadze, T. 1982. (Editor) *Package of applied program design of spatial construction.* Tbilisi.University Press. (part 1:165 p., part 2:162 p.).

Vashakmadze, T. 1985. On the theory of elastic plates. *Seminar Vekua Inst. of Appl. Math. Reports.* **19**: 55–62.

Vashakmadze, T. 1988. Theory of Elastic Plates. *Advances in Mechanics, v.* **11**, *N3*: 43–76.

Vashakmadze, T. 1991. On the mathematical theory of piezoelectrically and conductive elastic plates. *Trends in Appl. of Math. to Mechanics. Longman Sci & Tech.*: 236–241.

Vashakmadze, T. 1995. Some Mathematical Problems of the Theory of Nonlinear Elasticity. *Trends in Appl. of Math. to Mech. Pitman. Monographs & Surv. Pure and Appl. Math.* **77** *Longman*: 348–357.

Vashakmadze, T. 1996. Some Problems of Computation Mechanics of Rigid Deformed Body.*Facta Universitatis. Series Mechanics,...,v.* **2**, N6: 39–56.

Vashakmadze, T.; Muradova, A. 1996. To an analysis of numerical methods of boundary value problems with a small parameter. *Applied Mathematics and Informatics, v.* **1**, *N1*: 180–192.

Vashakmadze, Tamaz; Vashakmadze, Tamara. 1993. On homogeneity of Vekua theory of plates and shells for anizotropic nonhomogeneous mediums. *Reports of enlarged session of the seminar of VIAM,v.***8**, *N2:* 107–110.

Vekua, I. 1948. . *New methods of the solution of elliptic problems.* Moscow–Leningrad: Gostekhizdat.

Vekua, I. 1955. To one way of calculation of prismatic shells and plates. *Proceed. of Razmadze Inst. of Math.* **21**: 191–259.

Vekua, I. 1959. *Generalized analytical functions.* Moscow: Physmatgiz.

Vekua, I. 1965. The theory of thin shellow shells of variable thickness. *Proc. Razmadze Inst. of Math.* **30**: 5–103.

Vekua, I. 1968. On construction of approximate solution of equations of the shallow spherical shell. *Inter. J. Solid and Structures,* **10**: 991–1003.

Vekua, I. 1970. *Variational principle of construction of theory of shells.* Tbilisi

University Press.

Vekua, I. 1975. On two ways of costruction of non-contradictory theory of elastic shells. *Inf. of I All Union Shool on Theory and Numerical Methods of Calculation of Shells and Plates. Tbilisi: Metsniereba.*

Vekua, I. 1985. *Shell theory: General methods of construction,* Pitman Advance Publ. Prog. Berlin–London–Montreal.

Vekua, I.; Muskhelishvili, N. 1962. Methods of the theory of analytical functions in the theory of elasticity. *Proc. of All Union Cong. on Theor. and Appl. Mech. Review reports. Moscow–Leningrad:* 310-338.

Veriužski, I. 1978. *Numerical methods of potential in some problems of applied mathematics.* Kiev: Vishaia Shkola.

Vishik, M.; Lusternik, L. 1957. Regular degeneration and boundary layer for linear differential equations with a small parameter. *Uspekhi Matematicheskikh nauk, XII, 5(77)* :3–122.

Vladimirov, V. 1961. *Mathematical problems of a single-speed theory for particle carrying. Moscow: AS USSR.*

Vladimirov, V. 1976. *Equations of mathematical physics.* Moscow: Nauka.

Vlasov, V. 1959. *Thinwall elastic rods.* Moscow: Physmathgiz.

Vogelius, M.; Babuška, I. 1981a. On a dimensional reduction method. I. The optimal selection of basic functions. *Math. of computation,v.***37***,N155:* 31–46.

Vogelius, M.; Babuška I. 1981b. On a dimensional reduction method. II. Some approximation–theoretic results. *Math. of computation, v.***37***, N155*: 47–68;

Vogelius, M.; Babuška, I. 1981c. On a dimensional reduction method. III. A posterior–error estimation and adaptive approach. *Math. of computation, v.* **37**, *N156*: 361–384.

Volkov, E. 1965. On differential properties of solution of boundary problems for Laplacian and Poisson equations in a rectangle. *Proceed. Steklov Math. Inst. v. XXVII*: 89–112.

Vorovich, I. 1965. Some mathematical problems of the theory of plates and shells. *Proc. of the II All Union Cong. on Theor. and Appl. Mech. Review reports, iss. 3. Moscow–Leningrad:* 116–136.

Vorovich, I. 1975. Some results and problems of asymptotic theory of plates and shells. *Proceed. of the I All Union School for Theory and Numer. Methods of Calc. of Shells and Plates.*: 51–149.

Vorovich, I. 1975. Bubnov-Galiorkin method, its development and role in applied mathematics. *"Progress in Mechanics of Solids". Moscow: Nauka.*

Washizu, K. 1982. *Variational Methods in Elasticity and Plasticity.* Pergamon

Press. Oxford.

Whittaker, E.T., Watson, G.N. 1935. *A course of modern analysis.* Cambridge.

Williams, M. 1952. Stress singularities resulting from various boundary conditions in singular corners of plates in extension. *J. Appl. Mech. v.* **19**, N4: 526–528.

Yeh Kai–yuan; Rimroft, F.P. 1993. On nonlinear stabilities in thin shallow elastic shells of revolution- a Review. *Proceed. 2-nd Inter. Conference in Nonlinear Mechanics.* Peking University Press.

Zgenti, V. 1988. Investigation of the Stress State Non-homogeneous according to thickness etc.*Prikladnaia Mechanika,* v.**28**, N6: 9–16.

Zgenti, V. 1991. To investigation of stress state of isotropic thick-walled shells of nonhomogeneous structure. *Applied Mechanics, v.* **27**, N5:37–44.

Zgenti, V.; Khvoles, A. 1982. The general solution of one system of equations in partial derivatives. *Differential Equations. v. XXIII, N1*:17–29.

Index

Other *Mathematics and Its Applications* titles of interest:

F. Langouche, D. Roekaerts and E. Tirapegui: *Functional Integration and Semiclassical Expansions*. 1982, 328 pp. ISBN 90-277-1472-X

N.E. Hurt: *Geometric Quantization in Action. Applications of Harmonic Analysis in Quantum Statistical Mechanics and Quantum Field Theory*. 1982, 352 pp. ISBN 90-277-1426-6

C.P. Bruter, A. Aragnal and A. Lichnerowicz (eds.): *Bifurcation Theory, Mechanics and Physics. Mathematical Developments and Applications*. 1983, 400 pp. *out of print*, ISBN 90-277-1631-5

N.H. Ibragimov: *Transformation Groups Applied to Mathematical Physics*. 1984, 414 pp. ISBN 90-277-1847-4

V. Komkov: *Variational Principles of Continuum Mechanics with Engineering Applications. Volume 1: Critical Points Theory*. 1986, 398 pp. ISBN 90-277-2157-2

R.L. Dobrushin (ed.): *Mathematical Problems of Statistical Mechanics and Dynamics. A collection of Surveys*. 1986, 276 pp. ISBN 90-277-2183-1

P. Kree and C. Soize: *Mathematics of Random Phenomena. Random Vibrations of Mechanical Structures*. 1986, 456 pp. ISBN 90-277-2355-9

H. Triebel: *Analysis and Mathematical Physics*. 1987, 484 pp. ISBN 90-277-2077-0

P. Libermann and Ch.M. Marle: *Sympletic Geometry and Analytical Mechanics*. 1987, 544 pp. ISBN 90-277-2438-5 (hb), ISBN 90-277-2439-3 (pb).

E. Tirapegui and D. Villarroel (eds.): *Instabilities and Nonequilibrium Structures*. 1987, 352 pp. ISBN 90-277-2420-2

W.I. Fushchich and A.G. Nikitin: *Symmetries of Maxwell's Equations*. 1987, 228 pp. ISBN 90-277-2320-6

V. Komkov: *Variational Principles of Continuum Mechanics with Engineering Applications. Volume 2: Introduction to Optimal Design Theory*. 1988, 288 pp. ISBN 90-277-2639-6

M.J. Vishik and A.V. Fursikov: *Mathematical Problems in Statistical Hydromechanics*. 1988, 588 pp. ISBN 90-277-2336-2

V.I. Fabrikant: *Applications of Potential Theory in Mechanics. A Selection of New Results*. 1989, 484 pp. ISBN 0-7923-0173-0

E. Tirapegui and D. Villarroel (eds.): *Instabilities and Nonequilibrium Structures (II). Dynamical Systems and Instabilities*. 1989, 328 pp. ISBN 0-7923-0144-7

V.G. Bagrov and D.M. Gitman: *Exact Solutions of Relativistic Wave Equations*. 1990, 344 pp. ISBN 0-7923-0215-X

O.I. Zavialov: *Renormalized Quantum Field Theory*. 1990, 560 pp. ISBN 90-277-2758-9

S.S. Horuzhy: *Introduction to Algebraic Quantum Field Theory*. 1990, 360 pp. ISBN 90-277-2722-8

Other *Mathematics and Its Applications* titles of interest:

V.G. Makhankov: *Soliton Phenomenology*. 1990, 461 pp. ISBN 90-277-2830-5

S. Albeverio, PH. Blanchard and L. Streit: *Stochastic Processes and their Applications in Mathematics and Physics*. 1990, 416 pp. ISBN 0-9023-0894-8

V.A. Malyshev and R.A. Minlos: *Gibbs Random Fields. Cluster Expansions*. 1991, 248 pp. ISBN 0-7923-0232-X

E. Tirapegui and W. Zeller (eds.): *Proceedings of the 3rd Workshop on Instabilities and Nonequilibrium Structures*. 1991, 370 pp. ISBN 0-7923-1153-1

C.A. Marinov and P. Neittaanmaki: *Mathematical Models in Electrical Circuits. Theory and Applications*. 1991, 160 pp. ISBN 0-7923-1155-8

C. Bartocci, U. Bruzzo and D. Hernandez-Ruiperez: *The Geometry of Supermanifolds*. 1991, 242 pp. ISBN 0-7923-1440-9

E. Goles and S. Martinez: *Statistical Physics, Automata Networks and Dynamical Systems*. 1992, 208 pp. ISBN 0-7923-1595-2

A. van der Burgh and J. Simonis (eds.): *Topics in Engineering Mathematics*. 1992, 266 pp. ISBN 0-7923-2005-3

L. Aizenberg: *Carleman's Formulas in Complex Analysis*. 1993, 294 pp. ISBN 0-7923-2121-9

E. Tirapegui and W. Zeller (eds.): *Instabilities and Nonequilibrium Structures IV*. 1993, 371 pp. ISBN 0-7923-2503-6

G.M. Dixon: *Division Algebras: Octonions, Quaternions, Complex Numbers and the Algebraic Design of Physics*. 1994, 246 pp. ISBN 0-7923-2890-6

A. Khrennikov: p-*Adic Valued Distributions in Mathematical Physics*. 1994, 270 pp. ISBN 0-7923-3172-9

Z. Haba, W. Cegła and L. Jakóbczyk (eds.): *Stochasticity and Quantum Chaos*. (Proceedings of the 3rd Max Born Symposium, Sobótka Castle, September 15–17, 1993) 1995, 224 pp. ISBN 0-7923-3230-X

L. Stupelis: *Navier–Stokes Equations in Irregular Domains*. 1995, 582 pp. ISBN 0-7923-3509-0

B. Jefferies: *Evolution Processes and the Feynman-Kac Formula*. 1996, 245 pp. ISBN 0-7923-3843-X

Y.E. Gliklikh: *Ordinary and Stochastic Differential Geometry as a Tool for Mathematical Physics*. 1996, 206 pp. ISBN 0-7923-4154-6

E.S. Fradkin and M.Ya. Palchik: *Conformal Quantum Field Theory in* D-*dimensions*. 1996, 474 pp. ISBN 0-7923-4158-9

P. Bandyopadhyay: *Geometry, Topology and Quantization*. 1996, xxx pp. ISBN 0-7923-4305-0

Other *Mathematics and Its Applications* titles of interest:

N.E. Hurt: *Quantum Chaos and Mesoscopic Systems.* Mathematical Methods in the Quantum Signatures of Chaos. 1997, 347 pp. ISBN 0-7923-4459-6

C. Cercignani, V.I. Gerasimenko and D. Ya. Petrina: *Many-Particle Dynamics and Kinetic Equations.* 1997, 244 pp. ISBN 0-7923-4696-3

G. Kalmbach: *Quantum Measures and Spaces.* 1998, 453 pp. ISBN 0-7923-5288-2

V. Fomin: *Optimal Filtering.* Volume I: Filtering of Stochastic Processes. 1999, 376 pp. ISBN 0-7923-5286-6

T.S. Vashakmadze: *The Theory of Anisotropic Elastic Plates.* 1999, 256 pp. ISBN 0-7923-5695-0

MIX
Papier aus verantwortungsvollen Quellen
Paper from responsible sources
FSC® C105338

If you have any concerns about our products,
you can contact us on
ProductSafety@springernature.com

In case Publisher is established outside the EU,
the EU authorized representative is:
Springer Nature Customer Service Center GmbH
Europaplatz 3, 69115 Heidelberg, Germany

Printed by Libri Plureos GmbH
in Hamburg, Germany